8/05

THE HUMAN GENOME
SOURCEBOOK

THE HUMAN GENOME SOURCEBOOK

Tara Acharya and Neeraja Sankaran

GREENWOOD PRESS
Westport, Connecticut • London

Library of Congress Cataloging-in-Publication Data

Acharya, Tara.
    The human genome sourcebook / Tara Acharya and Neeraja Sankaran.
        p.   cm.
    Includes bibliographical references and index.
    ISBN 1–57356–529–6 (alk. paper)
    1. Human genome—Popular works.   2. Human genetics—Popular works.   I. Sankaran, Neeraja.   II. Title.
    QH437.A245   2005
    611'.0181663—dc22   2005003396

British Library Cataloguing in Publication Data is available.

Library of Congress Catalog Card Number: 2005003396
ISBN: 1–57356–529–6

First published in 2005

Greenwood Press, 88 Post Road West, Westport, CT 06881
An imprint of Greenwood Publishing Group, Inc.
www.greenwood.com

Printed in the United States of America

The paper used in this book complies with the
Permanent Paper Standard issued by the National
Information Standards Organization (Z39.48–1984).

10 9 8 7 6 5 4 3 2 1

Contents

Preface

Launched in the 1990s in an attempt to decode the "book of life" as it were, the Human Genome Project had a tremendous impact on society, touching virtually every aspect of our daily lives in ways that were probably never imagined or intended by those who began it. Within a few short years, this project took the field of genetics far beyond the realm of scientists and doctors, scientific journals, and conferences, and catapulted it squarely into our daily lives—into newspapers, magazines, television, films, and, of course, the Internet. Today we are all inundated with images of the DNA double helix, and words such as "genome" roll off the newscaster's tongue just as easily as ballgame statistics, weather updates, and stock market reports. There are popular books on nearly every aspect of genetics and genomics, on the ups and downs of the Human Genome Project, on the biotechnology industry it has spawned, on the personal lives of the people involved, and on the social, cultural, and ethical consequences of the advances in genomics.

Wading through this sea of information, fascinating though it is, can be more than a bit overwhelming. One of the biggest dangers of trying to keep up with a field whose frontiers are swiftly advancing is that the foundations are often overlooked in favor of whatever is "hot" and newsworthy. Advances in technology are often discussed without any background on the science or the motives that gave rise to them in the first place. This has certainly been true of genetics and genomics, which seem to hold out promises with one hand and problems with the other. Consequently, readers are frequently plunged directly into the thick of discussions and debates over such issues as "magic bullets," human cloning, and "designer babies" without an adequate knowledge base from which to work.

As a first-stop guide to the human genome, *The Human Genome Sourcebook* tries to provide the information to fill this gap. Designed for the general reader—intelligent and curious, but without an advanced knowledge of genetics—who is interested in going beyond the headlines and buzzwords, this book is about the genes we are made of, how they help us conduct our normal life's activities, and what happens when they are damaged. Our aim is to supply users with the background and knowledge they need to understand the science behind the human genome, the power and the limitations of this knowledge, and its place in today's—and tomorrow's—world.

Organization

This reference volume is organized into five chapters that form roughly two parts. The first part is intended to provide readers with the tools they need to navigate their way through the rest of the book. Chapter 1 provides a general introduction and some historical facts about genetics and the genome. Chapter 2, organized in the form of an expanded glossary, provides readers with the vocabulary to understand the working of the genome. By including information about current debates and discussions, we have tried to convey something of the complexity and currency of the field. This chapter is organized alphabetically by term. Readers might like to refer back to this section if they come across unfamiliar terms in the later chapters, although it may be worthwhile to read through certain terms such as "gene" and "chromosome" before delving into the specifics of genes and diseases. Chapter 3 provides a roadmap of the physical layout of the human genome and provides information about each of the chromosomes. Most of this chapter is presented in tables on the individual chromosomes, which provide snapshots of important known genes on each chromosome and the diseases.

Chapters 4 and 5 make up most of this reference volume. Virtually every facet of our lives, from the invisible molecular reactions that take place inside our cells, to the way we look, and even how we behave, is influenced in some way by our genes. Chapter 4 is a catalog of some of these genes, chosen to represent major functions at every level of activity from the molecular (e.g., DNA replication, protein synthesis) all the way through to the level of the whole organism. For example, a single large entry on the function of "sugar metabolism" looks at the genes for the different enzymes involved in this pathway, while entries on "blood" and "muscle" describe genes for the components of these substances. We chose this type of thematic organization for this chapter rather than gene names (nomenclature) because the labels given to many genes do not necessarily describe their function. In fact, many gene names were chosen quite randomly by the scientists who first discovered or isolated them, and as such there is no single system of nomenclature. Of course, our system is also somewhat arbitrary because biological functions are often interlinked and there is no single intuitive way to consider them. We urge readers to make extensive use of the index and the "*See also*" cross-references when looking up individual entries. (Note: cross-references are identified by chapter number only when the chapter is not the current one.)

The final chapter (Chapter 5) deals with genetic diseases. A common tendency while talking about genetic disorders is to refer to the so-called "disease" genes. This is a mistake because it implies a causal relationship between genes and diseases. Rarely, if ever, do we have genes that encode a disease per se. Genes code for proteins that carry out normal functions in our bodies. Diseases arise when a gene or its corresponding protein is damaged and unable to perform its function properly, or, in some cases, gains certain abilities that make it harmful. In this chapter, we relate specific aberrations in our chromosomes and mutations in our genes to the diseases or malfunctions that result. We have reverted to an alphabetical system of organization for this section. Wherever possible, we use the common rather than the medical name for a disorder. All entries in this section have been cross-referenced to the appropriate entries in the previous chapter so that readers are able to use this reference dynamically and relate diseases to the corresponding loss of function and make connections between form and function that are not immediately obvious.

Acknowledgments

Many people contributed to the successful completion of this project, but there are two in particular that both authors would like to thank for their role. First our editor, Kevin, who displayed just the right mixture of patience and prodding to see us through, and second, Sandy Windelspecht, whose has so beautifully complemented our text with her illustrations and at such short notice. We could not have done this without you both.

Tara would also like to thank all her family members for their support and patience during the writing of this book. Thanks also to Neeraja for her constant enthusiasm and for the opportunity to work together on this book.

Neeraja would like to add a note of appreciation for all those who have lent their moral support and encouragement—Thatha, Patti, Appa, Amma, Mahesh, and Vidya from afar—and closer up, Aditya, Aman, Devesh, and Manager Mike. I dedicate this to Poorna, Bhuvana, and Tejas, three of the best possible reasons for writing this. Of course, none of this would have been possible without the consent and cooperation of my coauthor. Thanks, Tara, for picking up the slack on slow days and gray ones and generally making this a fun ride!

Chapter 1

Introduction

The human genome has come a long way, so to speak. True, it has always been with us—*in* each of us—since we have existed on Earth, but in terms of our knowledge about it, the genome's journey has been quite short, albeit very rapid. For most of human history, the genome had been an unknown presence in our lives. Indeed, the word has been in use only since the 1930s, but the past few decades have made up for this late start. Today, the genome is virtually everywhere—in newspaper headlines and articles, the subject of best-selling books, and most of all, perhaps, on the Internet. Conduct a search on the word "genome" on Google, for instance, and the computer returns nearly 8 million hits in under 1 second!

So just what is the genome and why is everyone talking about it? These are the types of questions that we try to answer with this book. Although we make no claims about providing the final answer, if indeed there is such a thing, we hope to give our readers a feeling for the vastness and complexity of the human genome as we describe some of its genes—genes that play a vital role in determining who we are, how we look, and what we do.

Derived from a combination of the words "*gen*e" (a unit of heredity), and "chromos*ome*" (the physical structure where most genes are located), "genome" is the word we use to describe the entire genetic material of a living organism. Metaphors such as "blueprint," "recipe book," and "book of life" have often been used in describing the genome and convey some sense of its function. This is because the genome quite literally contains all the instructions needed by its owner to live as a unique being.

The word "genome" is typically used in a species-specific manner; we thus talk about the genome of the mouse, fruit fly, rice, or human being in the sense of a composite or representative genome of the organism. Although there is considerable variation among the genomes of individuals of a single species, the gross structure and arrangement of genes in a species is basically the same. For example, all humans have 23 pairs of chromosomes, although individual genetic dif-

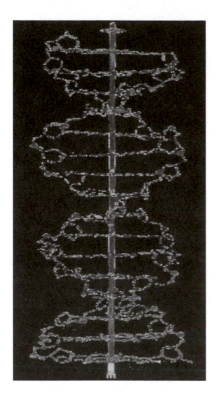

A digitized model of the DNA double helix. [© E. Saul/Custom Medical Stock Photo]

ferences obviously exist among all 6 billion of us. At the level of individuals, we can see many different features so that each one is unique. This uniqueness is reflected in the differences in sequences of genes of the individuals. At the same time, it is possible to form groups on the basis of certain criteria that some members of a population share and others lack. For example, if we choose sex—male and female—as the criterion, we see that there are discernible similarities among the members within each group and an equally obvious difference between members of the two groups. The genomic or genetic bases for this difference are the sex chromosomes—all females have a pair of X chromosomes, whereas males have one X and one Y chromosome. Despite this difference within the species, all humans, whether male or female, form a single group of organisms—known as the *Homo sapiens* species— that differs markedly from any other group (e.g., chimpanzees or gorillas), each type of organism being markedly different from one another and also consisting of unique individuals within the group. Each of these groups differ from one another on one main criterion: a genome of a certain size and gross structure. Take yet another step outward, and the chimpanzees may be grouped together with the humans as being distinct from birds. As we move farther and farther back to include more and more members of the living world, our groups of "similar" and "different" beings grow larger—plants are discernibly different from animals, and both seem markedly different from single-celled bacteria. Such differences are reflected at the level of the genome as well.

Physically, the genome is seldom a single identifiable structure but consists of all the chromosomes as well as any extra genetic material in an organism. With the exception of certain viruses, the chemical nature of this material in all living organisms is deoxyribonucleic acid (DNA). DNA is a complex, twisted ladder of a polymer (a large molecule made up of many smaller molecules—the building blocks) of units known as the nucleotides. The sequence in which these nucleotide building blocks are strung together is what determines the instructions carried by the DNA—sort of like words being strung together to form sentences. There is no fixed size to DNA, and theoretically speaking there is no end to the size of the polymer. In reality, however, DNA molecules are of finite size, and a chromosome is simply a very long DNA molecule that is tightly coiled and wound about itself to form a compact bundle. If stretched out, the DNA in each of our cells would be over 6 feet long, but it is tightly packaged to fit inside a cell that is 1,000 times smaller than the period at the end of this sentence. The number of these chromosome bundles varies from species to species; bacteria and yeasts, for example, have just a single large chromosome, whereas humans possess 23 pairs of chromosomes. The mouse genome has 20 pairs of chromosomes, fruit flies have 4 pairs, and dogs have 39 pairs. In the plant kingdom, the genome of the peanut has 20 pairs of chromosomes, rice has 12 pairs, and sugar cane has 40 pairs.

In addition to the chromosomes, which reside in a special compartment of the cell known as the nucleus, the genomes of most organisms often include some smaller pieces of DNA that lie outside the nuclei. In humans, for instance, the nuclei hold more than 99 percent of the genome, but there is a small yet vital fraction that is found in the mitochondria, the powerhouses or energy generators inside our cells. With the exception of the blood cells, each and every one of the trillion or so cells in our bodies contains all this DNA material.

So, how do DNA and the genome work? Why is it that, despite the fact that almost all living organisms depend on the same chemical molecule—DNA—for passing on life's instructions, living beings are so diverse in their appearance and their abilities? The answer to this lies in the content of the instructions embedded in the genome (i.e., the sequence of DNA building blocks). The genome of each organism consists of units known as the "genes," which contain instructions to make biochemical molecules that carry out life's activities. The products of most genes are proteins. Almost all our cellular functions—respiration, digestion of food, and even the translation of the DNA code into RNA (ribonucleic acid) and proteins—depend on the action of proteins. For example, hemoglobin is a protein that transports and delivers oxygen to different parts of the body, while enzymes such as pepsin and amylase help us digest food. Physiological functions and structures that are not obviously related to proteins also depend on the action of enzymes or other proteins. Our bones are made of calcium, and our eye and hair color comes from certain nonprotein pigments, but the formation of the bones as well as the pigments depends on the action of different proteins, which are encoded by specific genes. Even behavior and other attributes that have no obvious link to proteins are eventually a product of the interactions of different proteins with one another.

The action of all of the gene products together helps determine how an organism looks and behaves. Some genes are common throughout the living world; for example, genes for products that help to preserve the genome in the organism, to access its information, and to pass it on to the next generation. Other genes make products that endow a species, or even an individual, with unique properties. Humans have about 30,000 genes, many of which are still to be identified and understood in terms of their functions. The interesting thing about the human genome is that the genes, despite their great numbers, account for only about 3 percent of our DNA. The rest of the material is "noncoding" DNA, which adds greatly to the complexity of the genome. Consider for instance that the fugu fish has a genome that is only one-sixth the size of ours, but with the same number of genes! We do not yet know exactly what the function of noncoding DNA is, but some scientists guess it probably helps genes determine when and where they should act and make their products.

The mere presence of a gene in an organism is necessary but not sufficient for the appearance of a trait. In order for an organism to exhibit a trait, the gene for that trait must be *expressed*, meaning its information must be translated into a usable form by the organism. The analogy of a book is helpful to understand how the genome works. It does not matter how exhaustively a book covers a subject: Unless it is read, its information is useless. Even though each one of our cells contains the entire genome, only certain genes are expressed in individual cells at any given time. It is this differential expression in different cells, and at different stages of an organism's life, that makes an ear different from a foot or enables a pea plant to develop from a seed. This finely tuned differentiation, achieved remarkably with a single set of genetic instructions, is what makes us complete, complex organisms, with tissues and organs specialized to carry out all the functions of life.

So far we have talked mostly about what the genome does in an individual organism, but this represents only half of its function. We now come to the second half of its function, which is the transmission of information to succeeding generations (i.e., *heredity*). Before the advent of the study of genetics, there were several theories on how traits were passed on from parents to their offspring. For centuries, blood was considered to be the material basis of heredity. A popular theory of heredity in the eighteenth century was preformationism, which was the idea that the reproductive cells in one generation contained fully formed miniature organisms that got passed down to the next generation, and development was simply the growth of these miniatures. In the late eighteenth and early nineteenth centuries, many people believed that each generation of organisms inherited traits that their parents had developed through responding to their environment. The long neck of the giraffe, for example, was thought to be a result of successive generations of animals (in the distant past) having to reach higher and higher for leaves as the leaves nearer to the ground were depleted. Toward the latter part of the nineteenth century, this view—known as the Lamarckian view after Jean-Baptiste Lamarck (1744–1829), the scientist who popularized it—was replaced by Charles Darwin's (1809–1882) theory that evolutionary change results from natural selection acting upon inherited traits. With the advent of genetics in the twentieth century, and the shifting focus to life at the molecular level, Darwin's theory was translated to the language of genes. The genomic DNA in each generation is duplicated and passed down to the progeny. A new property is acquired when the sequence is changed or mutated and then passed on to the progeny. The replication and transmission of the genome ensures the propagation of a species while maintaining the uniqueness of each individual.

The key to the hereditary process lies in the double-helical structure of DNA. As we mentioned earlier, the sequence of DNA bases preserves functional information and needs to be passed down to the next generation. One strand of the helix is perfectly complementary to the other, so that when the two strands separate, the cell's machinery can duplicate each strand of the original DNA through a process called replication. This remarkable ability to self-replicate is what enables heredity in almost all living organisms.

FROM HEREDITY TO THE GENOME PROJECT: A HISTORICAL OVERVIEW

Most genetics texts date the birth of modern genetics to 1866, which is the year that an Austrian monk named Gregor Mendel (1822–1884) published the results of studies that he had conducted on pea plants to determine the way in which certain physical properties such as the color of the peas were passed on from one generation to the next. Mendel's papers were, as far as one can tell, the first recorded systematic look at patterns of heredity in any living organism, but like most other events in human history, the history of genetics is considerably more complicated. Nor can its birth be ascribed to any one point in time. Primarily this is because of the complexity of the subject and the fact that it draws on many different fields. In addition to botany (or, more specifically, plant breeding), such diverse and unconnected disciplines as medicine, entomology (the study of insects), cytology (or cell biology), embryology, microbiology, and immunology have all contributed to the body of current knowledge. Furthermore, even though genetics only gained status as a formal science in the twentieth century, humans have speculated on the nature of heredity—the passing of both physical features and behavioral traits from parent to child—since antiquity. Even without any knowledge of the material basis for heredity, humans have engaged in genetic manipulation—the selective breeding of crops and animals for special properties, for example—for many centuries. In many ways, then, the history of genetics is as old as humanity itself!

Mendel's exalted status in today's historical accounts is also somewhat ironic when we consider the reception of his ideas during his own lifetime. His papers were published in a relatively obscure journal and remained unnoticed for several years. Mendel himself did not live to see the success of his ideas. It was not until 1900 that his ideas received any attention. In that year, by some curious coincidence, three European naturalists, Carl Correns (1864–1933), Hugo de Vries (1848–1935), and Erich von Tschermak (1871–1962), published the findings of experiments that they had conducted—independently of one another—on the nature of heredity. Although their experiments and designs had not been based on Mendel's work, they arrived at much the same conclusions that he had three decades earlier. All three of these scientists gave Mendel his due, albeit posthumous, recognition and reported their findings as evidence confirming his laws. This "rediscovery" of Mendel in 1900 marked the beginning of a boom period for investigations into the nature of heredity.

Meanwhile, medicine had made significant contributions to the theories of inheritance. In the eighteenth century, a French physicist named Pierre de Maupertuis (1698–1759) had noted the tendency of polydactyly—the presence of more than five fingers or toes per hand or foot—to run in families. Certain medical conditions such as hemophilia and color-blindness were seen to be inherited in a very sex-specific manner. In both conditions, boys were affected much more frequently than girls, and physicians noted that these boys came from families in which either the father or the maternal grandfather or uncle had been afflicted. The first person to apply Mendel's laws to human traits was Sir Archibald Garrod (1857–1936), a British physician who in 1908 correlated the disease alkaptonuria with a specific enzyme deficiency that was heritable. He coined the phrase "inborn error of metabolism" to describe this and other inherited disorders such as albinism, pentosuria, and cystinuria.

Today, we automatically associate terms such as "genetics" and "genes" with heredity—but these words did not even exist in the English vocabulary at the beginning of the twentieth century. The use of the term "genetics" as the science of inheritance—or the genesis—of traits was proposed by a British biologist named William Bateson (1861–1926), who was also responsible for giving us the terms (and concepts) allele, heterozygote, and homozygote. His major experimental contribution to the nascent field he named was the discovery, in 1905, of the fact that certain traits were inherited together. (This property, known as genetic linkage, later proved to be an important tool in the physical mapping of genes.) Bateson was also among the first scientists to study heredity in plants rather than animals. Meanwhile, the word "gene"—derived from the Greek word for the concept of "giving birth"—to denote the unit of heredity was coined by a German named Wilhelm Johansenn (1857–1927), about whom little else is known. At the time, the "gene" was still very much an abstract concept—simply a word to represent what Mendel had earlier called "hereditary factors"—with no material basis. Studies on the physical and chemical nature of the gene were rooted in other areas of science and were not incorporated into the Mendelian framework until some decades later.

In addition to being the year of publication of Mendel's papers, 1866 stands out as an important year for the history of genetics because this was when the German biologist and philosopher Ernest Häckel (1834–1919) forwarded the hypothesis that the hereditary information of a cell was transmitted via its nucleus. This idea was corroborated about a decade later, in 1875, by Oscar Hertwig (1849–1922), who showed the importance of the nuclei of cells for cell division. In a series of experiments over the next decade, Walter Flemming (1843–1905) discovered the chromosomes inside cell nuclei and with the help of his colleagues described their behavior during cell division.

The person most responsible for tying together studies of heredity with the chromosomes was probably the American geneticist Thomas Hunt Morgan (1866–1945). Beginning in 1909, Morgan—along with a team of scientists working in what came to be known as the "Fly room" at Columbia University—studied the mechanics of genetics using the tiny fruit fly *Drosophila melanogaster* as a model organism. Much of what we know today about heredity and its mechanisms stems from the work of Morgan. One of the important ideas to emerge from this research was the notion of the linear arrangement of genes in the chromosome.

Morgan's work on the fruit fly yielded a lot of physical information about genes and their role in the inheritance of traits but did not address the issue of the chemical identity of the gene. Because it was obvious that genes contained "instructions" or information, they were assumed to be made of proteins, which at the time were the only molecules believed to have any ability or capacity for storing information. The credit for the discovery of the chemical nature of the gene, or hereditary molecule, goes to Oswald Avery (1877–1955), Colin Macleod (1909–1972), and Maclyn McCarty (1911–   ), medical bacteriologists at the Rockefeller University (then the Rockefeller Institute for Medical Research), who were studying a process known as transformation in pneumonia-causing bacteria. In 1944, they published a paper that effectively stated that the carrier of hereditary information was the chemical DNA, which was first identified by Friedrich Miescher (1844–1895) in 1869 and named as nuclein. At first, they encountered considerable resistance to this idea, but within a decade most of the scientific community had accepted it and became focused on determining the structure of DNA, which they perceived as the key to understanding genetics at the molecular level. In 1953, James Watson (1928–   ) and Francis Crick (1916–2004) used X-ray diffraction photographs of DNA provided by their colleagues Rosalind Franklin (1920–1958) and Maurice Wilkins (1916–2004) to propose the double-helix structure for this molecule.

The decades that followed saw the integration of information from various fields of study to provide a more cohesive picture of genes and the genome. People began to understand the nature of the interactions between different molecules of life (DNA and proteins), and Crick proposed the "Central Dogma" to describe the flow of information in living systems. In 1966, Marshall Nirenberg (1928–   ), Har Gobind Khorana (1922–   ), and Robert Holley (1922–   ) determined the correlation between the arrangement of nucleotide bases on DNA and that of the amino acids on proteins, and thus succeeded in cracking the "genetic code." The 1970s and 1980s saw the development of new technologies that gave scientists access to the genome as never before. Of particular importance in this regard were the techniques developed by Frederick Sanger (1918–   ) and Walter Gilbert (1932–   ) to determine the sequence of nucleotides in DNA and the discovery of enzymes that could cut and patch together pieces of DNA from different sources. The subsequent invention of polymerase chain reaction technology in 1983 gave scientists the ability to generate billions of molecules of DNA from small samples within a few hours, and the automation of sequencing methods sped up access to genetic information by several orders of magnitude. Combined with faster and improved tools for data processing, these technologies armed scientists with the tools that gave them the power to manipulate DNA at the level of individual genes. Along with this came the hope of understanding and curing a host of diseases, and promises for improving the quality of human life in other spheres as well.

In the mid-1980s, spurred by the advances in genetic knowledge, the U.S. De-

Francis Collins and Craig Venter with President Bill Clinton at the White House ceremony on June 26, 2000, announcing the completion of the draft sequence of the human genome. [AP/Wide World Photos]

partment of Energy and the National Institutes of Health set out on an ambitious plan to sequence the entire human genome—all 3.1 billion base pairs of it! Suddenly the genomics revolution was upon us. Riding on the coattails of the information technology revolution (without powerful computers and automated sequencers, the sequencing would probably have taken several decades longer), the Human Genome Project progressed at a gratifyingly rapid pace. When in April 2000 the publicly funded Human Genome Project and the private firm Celera Genomics together announced the completion of the draft sequence of the human genome, Francis Collins, director of NIH's National Center for Human Genome Research, could justifiably boast that it was "ahead of schedule and under budget." This was a truly remarkable milestone in human existence, and the years ahead promise to be even more exciting as scientists embark on the next phase of genomics and gain fascinating new insights about our existence.

The following timeline highlights some of the major conceptual and experimental landmarks in the history of genetics and the genome and gives the reader an idea of the speed with which the field has grown, especially since the 1960s.

Milestones in Genetics and Genomics

1866	Gregor Mendel proposes basic laws of heredity based upon cross-breeding experiments with pea plants.
1868	Friedrich Miescher isolates new material, which he names nuclein (later DNA), from the nuclei of pus cells.
1882	Walter Flemming examines salamander larvae under a microscope and identifies tiny threads in the cell's nucleus as chromosomes.
1900	Carl Correns, Hugo de Vries, and Erich von Tschermak rediscover Mendel's principles, marking the beginning of modern genetics.
1908	Sir Archibald Garrod coins the phrase "inborn error of metabolism" and suggests recessive inheritance of the enzyme deficiency in the disease alkaptonuria.
1905	Nettie Stevens and Edmund Wilson independently describe the behavior of sex chromosomes.
1910	Thomas Hunt Morgan begins studies on fruit flies, finding that some genetically determined traits are sex-linked and also that genes are found on chromosomes.
1918–1926	Herman Muller induces mutations through X-rays, suggesting that the gene constitutes the basis of evolution by reproducing its own internal changes.
1941	George Beadle and Edward Tatum provide experimental proof of the connection between genes and enzymes.
1944	Oswald Avery, Colin Macleod, and Maclyn McCarty show that DNA is the hereditary material in pneumonia-causing bacteria.
1946–1949	Linus Pauling and colleagues trace the cause of sickle-cell anemia to abnormality in hemoglobin (and dub this disorder as the "first molecular disease"). This abnormality is later correlated to a specific mutation in the hemoglobin gene.
1953	The double-helix structure of DNA is proposed by James Watson and Francis Crick.
1958	Arthur Kornberg purifies DNA polymerase I (the enzyme that makes DNA) from *Escherichia coli*.
1960s	Crick uses the expression "central dogma" to describe the flow of information in living systems.
1961–1966	Marshall Nirenberg, Gobind Khorana, and Robert Holley crack the "genetic code" (i.e., the relationship between nucleotide triplet codons and amino acids).
1970	Peter Duesberg and Peter Vogt discover the first oncogene (cancer-inducing gene) in a virus.
1972	Hugh McDevitt identifies genes that control immune responses to foreign substances, suggesting that certain genes might cause susceptibility to some diseases.

1973	Bruce Ames develops a test to identify chemicals that damage DNA, and the Ames test becomes widely used to identify carcinogenic substances.
1975–1977	DNA-sequencing techniques are developed independently and using different approaches by Walter Gilbert and Frederick Sanger.
1976	J. Michael Bishop and Harold Varmus identify oncogenes on animal chromosomes and show that changes in their expression result in cancer.
1979	John Baxter reports the cloning of the gene for the human growth hormone.
1980	The U.S. Supreme Court rules that life forms can be patented.
1982	Genentech's human insulin is the first product of recombinant DNA technology to reach the market.
1985	Kary Mullis and others at Cetus Corporation invent polymerase chain reaction (PCR) technology, which allows the generation of billions of DNA copies in hours.
1987	The National Institutes of Health sets up GenBank, a national computerized data bank, to store nucleic acid sequences.
1989	The gene coding for cystic fibrosis transmembrane conductance regulator protein (CFTR) on chromosome 7 is discovered. When mutated, this gene causes cystic fibrosis.
1990	The Human Genome Project is launched to map the entire human genome.
1992	The first physical maps for chromosome 21 and chromosome Y are published.
1993	Genes for hereditary colon cancer, Huntington's disease, hyperactivity, Lou Gehrig's disease, Alzheimer's disease, and diabetes are discovered.
1995	Physical maps of chromosomes 3, 11, 12, 16, 19, and 22 are published.
1997	The first successful cloning of a mammal—"Dolly" the sheep—is achieved.
2000	The Human Genome Project and Celera Genomics announce completion of the draft sequence of the human genome.

THE GENOME AND THE FUTURE: SOCIAL AND ETHICAL CONSIDERATIONS

Like any new discovery or technology, the rapidly advancing field of genomics poses a double-edged sword for humanity. On the one hand, the discovery of different genes holds promise and hope for curing diseases, improving crop yields, and improving the quality of life in other spheres. On the other hand, the rapid progress of the science has been a cause for some concern—both reviving old fears and presenting new ones. Although some of these issues might yet be the stuff of science fiction and others are premature in light of the current state of knowledge, there are an equal number of legitimate concerns that should certainly be considered and discussed both by the scientists engaged in genomic sciences and by the wider public. It is true that with knowledge comes power, but power goes hand in glove with responsibility, which is a corollary that we often lose sight of but must be careful to remember, especially in the age of genomics.

One of the greatest dangers posed by the new genomic science is the possibility of what some call "genetic determinism," namely the idea that our genes determine who we are. This is a point that needs stressing because while genes play a big role in our identity, they are by no means a solo act, or even the main players in life's scenarios. Our environment and the nature of the interaction between our genes and environment also play equally important roles in determining the final outcome of the expression of a given gene. Often, however, this fact gets lost in the furor that follows in the wake of an information explosion, and many concerns about the abuse of privacy and the misuse of knowledge are quite valid.

Many fears about the Human Genome Project come from a concern about how genetic information—which goes to the very heart of our existence and could expose our fundamental vulnerabilities—can be used. For many, the specter of a genomic "McCarthyism" is all too real. For a society in which issues of health care and health insurance are pressing concerns, the possibility that insurance companies could use information regarding the genetic predisposition of individuals to specific diseases to deny insurance coverage is a matter of great concern. Similarly, at a time when unemployment is on the rise, there are fears that prospective employers could discriminate against certain job applicants on the basis of their genetic profiles. Activist groups both in the United States and abroad have raised concerns over the possibility of the misuse of genomics and genetic technology in dealing with developing countries. For instance, it has been suggested that multinational companies could run clinical trials on ethnic groups to understand the interactions between drug response and genetic makeup without providing adequate compensation to the participants or without giving these groups a share in any potential profits emerging from the research. Are such scenarios real possibilities or just false alarms? That they are not so far-fetched is evident when one considers discriminatory practices such as ethnic or racial profiling. For many people, memories of the infamous Tuskegee experiment—treatments were withheld from African American syphilis patients in order to study the course of the disease in this minority—serve as powerful reminders of the abuse of power in the name of science. Does the Human Genome Project then merely transfer issues of bigotry and discrimination to the invisible level of our genes?

Gene therapy is a notable example of the Janus face of genetic technology. When the first clinical trials began in 1990, there was great hope in its potential as a new type of miracle cure for a variety of genetic diseases, but since then little progress has been recorded. In 1999, the approach came under attack with the unexpected death of 18-year-old Jesse Gelsinger, who was taking part in a gene therapy trial for a rare genetic disease characterized by a deficiency in the enzyme ornithine transcarboxylase (OTC). Gelsinger's death is believed to have been caused by a severe immune response to the adenovirus that was used as the vehicle to transfer the normal (therapeutic) *OTC* gene. Another incident took place in 2002 when a child treated in a French gene therapy trial for X-linked severe combined immunodeficiency disease developed a leukemia-like condition. The U.S. Food and Drug Administration (FDA) reacted to these unfavorable results almost immediately by placing a halt on all gene therapy trials using retroviral vectors in blood stem cells. In light of these events, it should be no surprise that the FDA has not yet approved any human gene therapy product for medical use.

Reproductive technologies are another area of heated debate about the potential dangers of genomics. Many people now have the option (if they can afford it) of using preimplantation genetic diagnosis to gauge their risks for passing on a genetic disease to their offspring. Although well-intentioned, this technology does involve the destruction of embryos, and the selection of others with certain genetic profiles—in other words, "designer babies." In 2004, a British couple gave birth to a baby boy who had been genetically selected to be a perfect match for his brother who needed a bone marrow transplant. Although many hailed this as a tremendous advance in science, others denounced it as a dangerous first step toward designing babies for spare parts. What is perhaps even more disturbing about these decisions is that the selection processes are based on the knowledge of a very small fraction of the genome! As we mentioned earlier, less than 3 percent of our total genome is coding sequence, while the rest, which we have (rather arrogantly) dubbed "junk DNA," is material about which we know little, if anything. That we can and sometimes do make decisions concerning our children based on so little information is a sobering thought indeed!

The possibility of human cloning, which received a lot of media attention after scientists in South Korea reported their success in cloning human embryos in 2004, brings with it the fear of a new wave of eugenics. Embryonic stem cell research is yet another controversial topic, with some in favor of it for its therapeutic potential and others opposed to it because of the necessary death of human embryos. All of these issues have triggered tremendous debate among the many stakeholders: individuals, families, scientific researchers, governments, insurance companies, pharmaceutical companies, law enforcement, and employers.

It is important, as the science races ahead, for the social, ethical, and legal policies to keep up to maximize the benefits of genomics and minimize its risks. Cognizant of the need to address such a diverse variety of issues and concerns in a timely manner, the Human Genome Project, when launched in 1990, established a branch called the Ethical, Legal, and Social Implications (ELSI) Research Program. With a budget of more than $10 million (U.S.) per year in federal funds, ELSI is the largest supporter of bioethics research in the world. Many other national governments, academic centers, and international organizations, such as the World Health Organization, have also set up programs to explore the potential risks and benefits of genomics and to formulate appropriate policies to maximize its benefits and minimize its risks.

Despite the potential pitfalls, there is no denying that the field of genomics has had an enormous practical impact on our lives and will continue to do so over the coming years. It has already completely transformed the way scientists study biology. Researchers now have vast amounts of genetic data and have the use of many new technologies with which to study the information. Today, identifying genetic mutations that cause disease is much simpler than it used to be and takes much less time as well—with publicly available genome databases, powerful software programs to comb the data, DNA-sequencing machines, and DNA microarray technologies that can track the expression of tens of thousands of genes in one shot. It will probably take at least a few years before this wealth of information becomes routinely translated into health benefits—in the guise of new medical tools to prevent, diagnose, and treat noncommunicable and communicable diseases—and improve people's lives worldwide. But we are well on our way on the exciting journey toward this goal.

Chapter 2

Terms and Concepts in Genetics

ALLELE

Any gene is capable of assuming multiple forms due to possible variations in its DNA sequence. These alternative forms of the same gene are called its alleles.

The difference between two alleles of a gene can be as small as a single nucleotide pair or extend to large tracts of sequence. When we speak of "mutant" genes, we are actually referring to an allele whose sequence is so different from the normal gene that its function is altered as well. Such defective mutants aside, each gene has several alleles that are essentially normal but whose sequences show minor differences from one another. The most common of these allelic variants (i.e., the version of the gene that is distributed most widely in the population) is called the wild-type allele. Because human chromosomes are present in pairs, each person can have two alleles of each gene at any given site (known as the locus) of the chromosome. One of the alleles resides on the chromosome inherited from the mother, and the other is located at the corresponding position or locus on the chromosome from the father. If both alleles in a given locus are identical in an individual, he or she is said to be *homozygous* for that allele; if they are different, the individual is *heterozygous*. Although an individual can possess a maximum of two alleles at any given locus, most genes have more than just two alleles distributed across the human population. It is this allelic variation that helps population geneticists track diseases or historical migratory patterns in various populations.

See also: Chromosome, Gene, Heterozygote, Homozygote, Locus.

ANTICIPATION

Anticipation is a phenomenon observed in genetic diseases that occur over several generations in a family whereby the symptoms appear earlier (at a younger age) and with greater severity in each successive generation. It is believed to be related to genetic instabilities that arise due to the presence of mutations in the form of repeated nucleotide units in the affected gene. The most dramatic example of anticipation has been observed in myotonic dystrophy and, to a lesser degree, in Huntington's disease and Friedreich's ataxia. These diseases are caused by an expansion of a trinucleotide in the affected gene, and the increasing severity of symptoms over generations is correlated with the size of this repeated sequence. For example, where normal individuals have 5 to 35 copies of the GCT (or CTG) unit in their myotonin gene, an affected person may have anywhere from 50 up to even 1,000 repeats. The number of these repeats increases over successive generations and with it the severity of disease.

See also: Genetic Disease/Disorder; Friedreich's Ataxia, Huntington's Disease (HD), Muscular Dystrophy—Myotonic Dystrophy (Ch. 5).

ANTISENSE

A gene, or "gene sequence," consists of the specific sequence of nucleotides that determines the sequence of amino acids of the protein encoded by that gene. Because a DNA molecule is a double-stranded molecule with the two strands running in opposite directions, a gene actually resides in only one of the two strands. This strand contains what is called the "sense" sequence or the actual coding sequence for the protein. The same sequence—albeit with uracils (Us) instead of thymines (Ts)—is also present in the messenger RNA (mRNA) molecule that is formed during gene expression. The complementary sequence to the gene/mRNA is called the *antisense* sequence. It is the sequence present in the strand of the DNA duplex that does not encode the gene. Single-stranded pieces of RNA that are complementary to the mRNA molecule are known as antisense RNA molecules.

What is the function of antisense molecules? Interestingly, it is the antisense (noncoding) DNA strand that serves as the actual physical template from which mRNA molecules are made during the transcription of a gene. (See the entry on transcription.) The biological function of antisense RNAs is not fully understood, but scientists have found ways in which to apply antisense RNA molecules in biotechnology and as possible agents for gene therapy. An antisense RNA molecule can bind (through complementary base-pairing) to mRNA. This action prevents translation of the mRNA for two reasons: first, by making the mRNA inaccessible to the ribosomes; and, second, because the duplex RNAs are very quickly degraded by enzymes in the cell. Thus, antisense RNA molecules may be used as inhibitors of specific genes and proteins.

A common example of antisense RNA in biotechnology is the "Flavr Savr" tomato, which has been genetically modified to have a longer shelf life than other tomatoes. Normal tomatoes contain enzymes that would degrade their flesh and cause them to spoil if left too long in storage or while being shipped. The Flavr Savr varieties contain a gene for a specific antisense RNA that blocks the mRNA for these enzymes. The resulting tomatoes produce only about one-tenth as much of the enzyme as the unmodified fruit, which gives the farmer a longer window of opportunity for getting unspoiled tomatoes from the field to the customers.

Antisense RNA molecules are also being investigated as possible agents for human therapy. Suppose a person has a disease that is caused by a mutation in a specific gene encoding a protein. By introducing antisense RNA molecules that are complementary to the mutant messenger RNA (mRNA), it may be possible to stop the translation of the mRNA into mutant protein, thereby preventing the disease symptoms from manifesting. Investigations are under way to examine the potential of an antisense RNA that is complementary to the proto-oncogene BCL-2 as a therapy for certain B-cell lymphomas and leukemias.

See also: Deoxyribonucleic Acid (DNA), Gene, Gene Therapy, Ribonucleic Acid (RNA)—Messenger RNA (mRNA), Transcription.

APOSTOSIS

Consider a living organism containing thousands, even millions, of cells that are constantly undergoing division and (hence) multiplication. There must be some mechanisms for some of these cells to die because otherwise the organism would continue to grow out of all conceivable proportion! To preserve its integrity, an organism must devise a mechanism whereby certain cells are killed and others kept alive, and the overall functioning continues undisrupted. This is achieved by a programmed system of cellular death or suicide that is called *apoptosis*.

The machinery of apoptosis is as intrinsic to the cell as any of its normal "living" functions. A cell may be induced to commit suicide either because of a cessation of the signals it normally receives (through cytokines or growth factors) in order for it to continue functioning or due to the receipt of certain external messages (e.g., oxidants) that tell it to shut down. This results in the activation of certain enzymatic pathways in the cell causing in turn the degradation of various parts. The cells shrink, their mitochondria break down and release cytochrome c into the cytoplasm, and enzymes catalyze the fragmentation of nuclear DNA. This last activity is considered a landmark of apoptosis. Eventually, certain key phospholipids are exposed on the external membrane, which leads to the phagocytosis of the cells. Because apoptosis is a normal event, the phagocytic cells also release certain substances to inhibit inflammation.

Apoptosis is especially important during the developmental phases of an organism. The proper formation of fingers and toes in a fetus or the connections between neurons in the brain require the apoptosis of specific cells. An example of an important instance of apoptosis in an adult is the sloughing off of the inner lining of the uterus at the start of menstruation. Organisms also employ apoptosis as a means of getting rid of virus-infected cells or cells that have become cancerous or show other forms of DNA damage. Researchers Sydney Brenner, Robert Horvitz, and John Sulston, whose pioneering work in the area of apoptosis has enabled us to better understand this vital process, won the 2002 Nobel Prize in Physiology or Medicine for their achievements.

See also: Cell, Cell Cycle, Organism.

BACTERIOPHAGE

This is the term given to the group of viruses that infect bacteria. The bacteriophages, or phages, are among the simplest known viruses, consisting of a single piece of nucleic acid—either DNA or RNA—that is encased in a protein coat. The infecting phage typically attaches itself to a host bacterial cell and injects its genome into the cell, leaving the protein coat outside the host. Because of their relatively small genome size and simplicity of structure and sequence, phages have been widely used as cloning vectors by molecular biologists.

There are two main classes of bacteriophages, distinguished mainly by their life cycles. Virulent or lytic phages, so named because they eventually cause the lysis of their host cell, remain physically separate from the bacterial chromosome throughout their infective phase. An example is the T4 phage of the bacterium *E. coli*. The second class of bacteriophages are called temperate or lysogenic phages. These phages establish a stable relationship—known as lysogeny—with their host by integrating a copy of their genome into the bacterial chromosome. Typically, a lysogenic phage confers its host with immunity against infection by other related phages. A famous and widely used temperate bacteriophage is phage lambda.

See also: Biotechnology, Virus.

BIOINFORMATICS

Bioinformatics applies computer science to the study and interpretation of the vast quantities of complex life sciences data that are available to biologists today. It draws from several fields, including the biological sciences, mathematics, statistics, and informatics.

The NIH Biomedical Information Science and Technology Initiative Consortium has defined bioinformatics as follows: "the research, development, or application of computational tools and approaches for expanding the use of biological, medical, behavioral or health data including those to acquire, store, organize, archive, analyze or visualize such data."

Bioinformatics has several useful applications, but some of the more common ones include the collection and organization of biological data in large, searchable databases, such as protein or DNA sequence libraries, and the analysis of large-scale data for an understanding of biological networks—for example, of protein expression in tissues—as opposed to conventional biological research, which involves the study of individual systems.

See also: Biotechnology, Genomics, Pharmacogenomics/Genetics.

BIOTECHNOLOGY

Biotechnology refers to techniques that use living organisms (or parts of organisms) to make products, change their qualities, such as by improving the yields of plants (crops), and develop microorganisms for specific uses. In one sense, humans have practiced biotechnology for centuries—traditional animal and plant breeding, such as the cultivation of rice or the domestication of the dog, are examples of early biotechnology practices. Beer brewing and wine and cheese making are also uses of biotechnology, namely the use of microbes. Modern biotechnology includes techniques such as recombinant DNA technologies, cloning, genetically modifying animals and plants, gene therapy, developing transgenic animals, and many more.

See also: Chimera, Cloning, Gene Therapy, Transgenic.

CARRIER

In genetics, a carrier is described as a person who carries a mutated or disease-causing gene without showing any outward symptom of that genetic defect. The reason for the carrier state is heterozygosity; only one of the two chromosomes in the carrier has a copy of the mutated gene, while the other chromosome has a normal gene. Obviously, the carrier state is only achievable with autosomal recessive conditions, since dominant conditions will express themselves even in heterozygous individuals. Carriers have the ability to pass on the defective genes to some of their offspring. In Figure 2.1 the genotype of a nonsymptomatic carrier is represented as "Aa," where "A" represents the normal gene and "a" its bad copy. As shown, there are three possible outcomes among the offspring genotypes: "Aa" offspring, who are carriers like the parents; "aa" offspring, who, being homozygous for "a," will have the disease; and homozygous "AA" offspring, who are neither symptomatic nor carriers of the disease.

See also: Genotype, Inheritance, Phenotype.

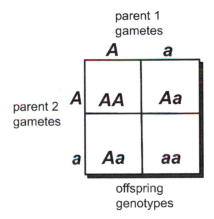

Carrier genotype: Aa

Figure 2.1. Genetic carriers. [Ricochet Productions]

cDNA

The shorthand notation for complementary DNA, cDNA is the term given to DNA molecules that are copied from messenger RNA (mRNA).

In the normal progression of life's information from DNA to RNA to protein, there is usually the removal of some amount of information from the original genomic DNA. For example, introns—which are present in most of our genes—are removed (or spliced) and the exons stitched together to form a single mRNA sequence that is then translated into the appropriate protein. A cDNA molecule is synthesized using this processed mRNA sequence as the template and so contains only the information that is directly translated into the peptides.

cDNA technology was developed after the discovery of the retroviruses and the enzyme reverse transcriptase. The ability to synthesize cDNA molecules in vitro has been invaluable to the field of biotechnology. The BAC and YAC systems, used so extensively in the Human Genome Project, are good examples of the application of this technology. Also, because cDNA genes lack introns, they can be expressed in prokaryotic cells regardless of their original source. This gives us a very simple and convenient system for synthesizing and purifying eukaryotic proteins.

See also: Deoxyribonucleic Acid (DNA), Ribonucleic Acid (RNA)—Messenger RNA (mRNA), Reverse Transcriptase.

CELL

Whereas the gene is the fundamental unit of heredity and information in a living organism, the cell is the fundamental unit of life itself. It is the smallest and simplest structure that is capable of sustaining life in an independent manner. The cell is the smallest entity that has the ability to exist independently by utilizing energy from its surroundings and has the capacity to produce another being identical or nearly identical to itself, which in turn is capable of sustaining and reproducing itself.

All true living things are made of cells. The number of cells in an organism ranges from one (e.g., organisms such as bacteria and yeasts) to several billion (as in the case of humans, animals, and plants). The fundamental components of any cell include a set of instructions—usually in the form of DNA—for it to follow in order to "live," the means to read and execute these instructions, typically in the form of proteins, and a membrane to encase these components within a defined boundary. Whereas the cell in a unicellular organism must perform all of the functions of its life, different cells of a complex, multicellular organism are usually specialized to perform different functions. The specialization of functions is controlled by the organism's DNA.

The internal structure of a cell depends on the type of living organism to which it belongs. Single-celled prokaryotes, such as bacteria, have a single compartment that contains all of the components suspended freely in the intracellular matrix. In contrast, eukaryotic cells (in both uni- and multicellular organisms) are divided into two main parts: the *nucleus*, which is a membrane-bound compartment containing the organism's genetic material in the form of chromosomes, and the *cytoplasm*, which is the remaining intracellular matrix.

The cytoplasm contains a number of different organelles for performing various cellular functions. Distributed throughout the cytoplasm of the eukaryotic cell is an interconnected network of tubules, vesicles, and sacs known as the *endoplasmic reticulum* (ER). The ER serves many specialized functions in the cells, including protein synthesis, sequestration of calcium, production of steroids, storage and production of glycogen, insertion of membrane proteins, detoxification of harmful substances, and transport of various materials within the cell. The ER may be rough or smooth depending on the presence of *ribosomes* on its surface. Another system of tubules and sacs, usually found close to the nucleus, is the *Golgi apparatus*, which controls the glycosylation of proteins as well as trafficking of different types of proteins within the cell. There are also some other proteins, such as lysosomal enzymes, that may need to be sequestered from the remaining constituents because of their potentially destructive effects. Other organelles of interest in a cell include the mitochondria (see separate entry) and chloroplasts—which are the sites for photosynthesis in plant cells but are absent from animal cells.

See also: Eukaryote, Gene, Mitochondria, Nucleus, Prokaryote.

CELL CYCLE

Most cells are in a constant state of alternating between two main phases, a phase where they are actively dividing into two new progeny cells and a nondividing state when they are growing and metabolizing and also preparing for the next round of division. The series of events that a cell goes through, from the time of its "birth" as a new cell until the end, when it divides to form two new cells, is called the cell cycle.

In unicellular organisms, the cell cycle is the same as the organism's own life cycle. Obviously, this is not true for multicellular organisms, where cell division is necessary throughout the life of the organism to provide new cells for various parts during development (e.g., in the embryo), and later to maintain the structure and function of various tissues and organs.

The basic form of the cell cycle is divided into two phases. The interval between two successive divisions is known as the *interphase*. This phase is, in turn, divided into several stages, each marked by the enhanced activity of one type in the cell. In the first phase, the cell's energy is channeled toward various metabolic activities, including its own growth and its normal function in the organism (e.g., melanin production by the skin cells, or enzyme production in the liver). This gap is followed by a period of active DNA replication in the nucleus. Then the cell enters a phase where its volume and cytoplasmic content is roughly doubled, making it ready to enter the cell-division phase.

Most somatic cells undergo the division process known as *mitosis* to form two new nearly identical progeny, each of which in turn enters interphase. However, there are also many exceptions to this general pattern. In the human body, for example, there are cells, such as the nerve cells, that once formed do not divide in the adult animal. These cells never undergo DNA replication and assume a steady state instead. Still other cells—particularly those in stages of embryonic development but also, for example, cells in the endometrium of the adult female—undergo *apoptosis*, or programmed death, in order to ensure the proper formation and functioning of the organism. Finally, the sex cells, or gametes, represent a special case where the cells in interphase undergo *meiosis* instead of mitosis to form haploid cells, which do not undergo further division. The cycle of these cells is completed, in effect, during fertilization, when a male and female gamete fuse together to form a cell (zygote) in which the original chromosome complement is restored. Of course, only one out of several thousands (even millions in the case of sperm cells) of these cells actually completes its cycle in fertilization—the rest are discarded. The processes of cell division and death (apoptosis) are described in independent entries.

See also: Apoptosis, Meiosis, Mitosis.

CHIMERA

In classical Greek mythology, the chimera was a monster that had the head (or many heads) of a lion, the body of a goat, and the tail of a dragon. Inspired perhaps by the multiple sources of its various parts, modern-day biologists have used the term as a moniker for any biological entity that contains parts from more than one source. A cDNA made from two different mRNAs, or a cloned piece of DNA containing material from nonadjacent segments, is a chimera. The bacterial artificial chromosomes (BACs) and yeast artificial chromosomes (YACs) used with such facility in sequencing the human genome are also examples of chimeras. In mouse genetics, chimeras have been made by mixing embryonic stem cells from one mouse with a genetically distinct blastocyst (i.e., an embryo of another mouse). Such chimeras have been used in research to recover targeted mutations produced in the stem cells.

See also: cDNA, Clone, Transgenic.

CHROMOSOME

Each living organism contains nucleic acid—DNA in most cases, RNA in some— that holds the instructions it requires for living. This nucleic acid or genetic material must be packaged in the cell in a manner such that it is stored with minimal damage or alteration while simultaneously making its information available for use during the organism's lifetime. In addition, the information must also be passed on to the next generation.

The physical structures that house this genetic material are called the *chromosomes*. If we think of a cell or organism as containing a library of instruction books that it requires for carrying out its various activities, then the chromosomes may be viewed as the individual books, each with specific chapters, that make up this library. The number and organization of the chromosomes present in an organism are characteristic for a given species. A single-celled yeast has just one chromosome, whereas fruit flies (*Drosophila*) have 4, an onion 16, and human beings 46. This number is called an organism's karyotype. Although the karyotype may often indicate the degree of complexity of an organism (i.e., the more complex an organism, the more chromosomes it has), this is not always the case. A dog, for example, has 78 chromosomes to our 46 but has about the same, or lesser, degree of complexity.

To better understand the concept of chromosomes, let us take a brief look at the story of their discovery. In the late nineteenth century, scientists studying the process of fertilization under the microscope noticed that the nuclei of eggs and sperm underwent a transformation during the process. Rodlike structures of different shapes and sizes appeared in the nuclei, and the number, shapes, and sizes of these structures were similar in both eggs and sperm. These same structures were also observed later during the processes of cell division—mitosis and meiosis. Scientists found that these rods were easier to see when the cells were stained with certain dyes, and so called them "chromosomes," which means "colored bodies."

Virtually all living organisms, with the exception of viruses, have chromosomes, although the organization, distribution, number, and size vary vastly across the living kingdom. The simplest chromosome organization is found in single-celled organisms such as bacteria that typically contain just one single chromosome, which contains only DNA molecule and lies naked or unsequestered in the cytoplasm of the cell. Such organisms are called the *prokaryotes*. The prokaryotic chromosome is usually a large, circular DNA molecule with no physically discernible beginning or end. The organization of genes in a bacterial chromosome is quite simple; each gene is a discrete segment of DNA with a specific beginning and end, no overlaps, and no interruptions.

The chromosomes of higher-order organisms, from the single-celled yeast up to the human, are contained within a special membrane-bound compartment inside the cell. This compartment is called the *nucleus*, and the organisms are labeled as *eukaryotes* (for "true" nucleus). Chromosomes of eukaryotes are considerably more complex than those of the prokaryotes. In many organisms, they may be found in association with special proteins, called histones, that function as the cores around which the DNA material is wound up like thread on a spool. In addition, the organization of individual genes within eukaryotic chromosomes is also more complex. The structure and organization of the human chromosomes is described in more detail in Chapter 3.

Chromosomes of eukaryotes may be present either as single sets or in pairs. This property depends on the phase of the life cycle of an organism. When the cells of an organism contain a single set of chromosomes they are in a *haploid* state, whereas cells with chromosome pairs are said to be *diploid*. Many simpler organisms, such as fungi, live out most of their life cycle in the haploid mode, whereas most animals we know, including ourselves, exist in a diploid state, with only the reproductive cells (sperm and eggs) in the haploid phase. The plant kingdom exhibits even more complexity with respect to ploidy, with many plants containing several sets of chromosomes in a single cell or nucleus. The transformation from one phase to another requires certain special events to occur in the course of cell division and multiplication: Diploid cells undergo meiosis to form haploid cells, and the fertilization of haploid gametes produces diploid cells.

The main advantage offered by diploidy or higher ploidies is the possibility for greater genetic variability. This variability is increased even more during the process of meiosis, when the pairs of chromosomes separate out and are segregated into different gametes. Because the chromosomes segregate independently of one another, the variation factor is immense: In humans, for instance, a single diploid cell may distribute its chromosomes in one of a total of 8,388,608, or 2^{23}, possible combinations. Diploidy is also the underlying reason for the appearance of dominant and recessive traits and for Mendelian inheritance. Organisms with higher ploidy will not follow simple Mendelian inheritance patterns because of the multiple ways in which their chromosomes can segregate during cell division.

See also: Diploid, Gene, Haploid, Heredity, Meiosis, Mitosis, Nucleus, Segregation; and also Chapter 3 (for more details about the human chromosomes in particular).

CLINICAL HETEROGENEITY

Clinical heterogeneity refers to the property of certain diseases which, although classified in the same group or treated under a single label, actually vary widely in their prognosis and in their response to treatment. Often this heterogeneity may be traced to a multiplicity of causes. An example is cancer. Although the basic problem of cancer is the same—namely, the uncontrolled growth and multiplication of cells—it is extremely heterogeneous in virtually all other respects. It may be triggered by any number of unrelated causes, including radiation, chemicals, and viruses. It is true that all of these agents induce mutations, but more often than not these mutations occur in different genes. In addition to the multiplicity of causes, cancer is also heterogeneous with respect to location. It can manifest in many different cell types of different tissues and organ systems, which in turn causes the response to therapy to be quite unpredictable and treatment very challenging.

CLONE

A clone is a genetically identical copy of any biological entity, from a piece of DNA to a complete organism. Basically, a clone is obtained by the faithful duplication of one cell or organism's complete complement of DNA. A clone of a cell, for example, may be derived through mitosis (or simple cell division) from a single parent cell. The occurrence of small mutations during mitosis, however, prevents the formation of clones in every cycle of mitosis. Perhaps the best known example of naturally occurring clones are identical twins. Such twins come about as the result of splitting a single zygote, for which reason they are also known as *homozygotic* twins. (Twins arising from the simultaneous fertilization of two independent eggs by two different sperms are called fraternal or heterozygotic twins and are not clones of each other.) In microbiology, the organisms in a single isolated colony are most likely clones because they arose from a single bacterial cell. An mRNA clone is an identical copy of the mRNA that corresponds with a particular gene. There are various procedures to clone different biological entities, some of which (e.g., PCR) have been used extensively in the laboratory to create large quantities for experiments.

CLONING

Few subjects in biology raised as much of a hubbub in the twentieth century as cloning. The idea of producing an exact duplicate of oneself has long fueled human imagination, as manifested, for example, in the glut of films concerning the subject—from 1978's *The Boys from Brazil* to the 2002 *Star Wars* offering titled *Attack of the Clones*. The report in 1997 of the creation of "Dolly," a sheep cloned from the cells of adult sheep by Ian Wilmut and colleagues at the Roslin Institute at Glasgow, Scotland, seemed to bring this idea from the realm of fantasy to reality. The cloning of higher mammals had not been possible until then. If a sheep could be cloned successfully, how much longer before science developed the first cloned human being? And what were the ethical and legal implications of cloning humans? A host of such questions greeted the news from Scotland.

Although Dolly brought the subject of cloning to high visibility in the popular press, the process of cloning itself is not new. For centuries, plant breeders have used cloning techniques to develop stocks and special seeds that are propagated generation after generation without alterations. Laboratory scientists who grow and maintain "pure" bacterial cultures are cloning these bacteria. In fact, the term cloning refers to more than just duplicating an organism. Biologists use the term to describe any process that leads to the duplication of any biological material. Therefore, even the use of a bacterial or yeast vector to create several copies of a particular sequence of DNA is a form of cloning. As for humans, when concerns regarding possible human cloning were raised in discussions that followed the announcement of Dolly, more than one scientist pointed out that human clones have always existed—in the form of identical twins! Scientific definitions notwithstanding, the concerns over human cloning as a possible means of immortalizing a human being or as a means of creating "organ farms" for ex-

Dolly (1997–2003) was the first mammal to be genetically cloned. [AP/Wide World Photos]

isting individuals feature strongly in the minds of different people. Consequently there are many debates under way as to the ethics of cloning research.

The simplest form of cloning is simply the exact replication of any DNA sequence; for example, through PCR. The purpose of this type of cloning is usually to make large quantities of material to facilitate its study. Another type of cloning technique is called blastomere separation, or twinning, which involves the splitting of a developing embryo to produce two identical embryos.

The technique that created Dolly involves the removal of the nucleus of an unfertilized egg cell followed by the transfer of the nuclear material (chromosomes) of an adult somatic cell into the egg and stimulating its division. Because the injected nucleus is diploid, the egg will now function as a zygote, which means its division will give rise to an organism with a DNA composition identical to that of the organism from which the nucleus was derived—in other words, a clone! Although this type of cloning is simple in principle, the failure rate is very high. Aborted attempts to clone primates suggest that cloning higher animals such as humans is going to be challenging, notwithstanding ethical and legal restrictions. Also, there is the possibility that the nuclear material used in cloning has undergone somatic mutations, which means that the so-called clone is not really a clone of the original animal at all! Dolly, for example, developed arthritis. This is unusual for a young sheep, but scientists cannot be certain whether this happened because Dolly was a clone of an adult or because of unrelated reasons. It remains unclear whether genetic changes in the cells used to obtain nuclei will lead to adverse effects on the health of the cloned animals.

Meanwhile, some examples of the uses of cloning include the large-scale reproduction of high-yielding varieties of farm animals and the production of rabbits whose genes have been modified to produce biological drugs and vaccinations for treating human diseases. The production of immunologically "safe" organs for transplants is another possibility that continues to attract various researchers. Whether these benefits outweigh the fears of misuse and the various ethical objections to cloning still remains to be put to the test.

See also: Clone, Deoxyribonucleic Acid (DNA), Fertilization.

COMPLEX TRAIT

A phenotype or trait that is controlled by more than just one gene is known as a complex, or polygenic, trait. Examples include height in human beings and the color of grain in wheat. The effect of the genes is often cumulative. In addition, there are often significant environmental effects—height, for instance, is strongly influenced by factors such as diet and general health. The differential action of the polygenes in different individuals accounts for a range of possible phenotypes in a population rather than just two extreme manifestations. This is why we see a continuous variation—typically with a bell curve distribution—of traits such as height among a population.

Polygenic or complex genetic diseases are those that are caused by interactions between two or more mutant genes. Many common adult diseases, including diabetes mellitus, hypertension, and coronary heart disease, as well as psychiatric disorders such as schizophrenia and manic depression, are polygenic in nature. Congenital defects such as cleft lips and cleft palates are also the result of defects in multiple genes.

Complex traits and diseases do not follow a simple Mendelian pattern of inheritance, and the latter pose an enormous challenge to medical researchers who are trying to decipher the causes of these diseases.

See also: Gene, Genotype, Phenotype.

CONSENSUS SEQUENCE

A consensus sequence is a nucleotide sequence—usually quite short—that is most often found in a specific site on a DNA molecule, such as a promoter or a splice site. Such sites, although not necessarily identical, are very similar, even in vastly distant species or organisms. They are represented in terms of a "consensus" of the most frequently appearing bases at each of the positions. They play an important role in regulating gene expression in some manner. An example of a consensus sequence is TATAAAA (also called the TATA box), which appears about 25 to 30 base pairs upstream of the transcription initiation site of a gene and is the site for the binding of RNA polymerase.

See also: Conserved Sequence, Transcription.

CONSERVED SEQUENCE

A conserved sequence is a sequence of bases in a DNA molecule, or an amino acid sequence in a protein, that has remained essentially unchanged throughout evolution. The conserved area might be a very short stretch of the DNA molecule, such as a promoter, transcription factor binding site, or intron–exon junction, or may comprise an entire gene (and thus an entire protein). Quite often, the conservation of a sequence mirrors a conservation of function through the course of evolution. For instance, vital processes such as protein synthesis have remained essentially unchanged throughout evolution. Consequently, the genes or proteins for these functions have also been conserved. Similarly, the 180 base-pair homeobox encoding a 60 amino acid protein called the homeodomain is a transcription binding site present in virtually all known eukaryotes.

See also: Consensus Sequence, Homeodomain.

DEOXYRIBONUCLEIC ACID (DNA)

DNA, or deoxyribonucleic acid, is the molecule of heredity in almost all living organisms. It is the chemical form in which most organisms, barring a few viruses, store their genetic information.

DNA is a polymer, which is defined as a long molecule made of repeating subunits or building blocks. The subunits that make up DNA are called nucleotides, and therefore DNA is a *poly*nucleotide molecule. Each nucleotide unit consists of a 5-carbon sugar (deoxyribose), a nitrogen-containing organic compound, called a base, attached to the sugar, and a phosphate group that forms the link between one nucleotide and the next. The difference in the units lies in the structure of the bases. There are four bases—adenine, thymine, guanine, and cytosine—and the nucleotides in a DNA molecule are represented by the single-letter abbreviations A, T, G, and C (see Figure 2.2).

DNA is the molecule of life! It contains genes that encode all the instructions that the cell needs to perform all of its functions. If a cell needs to make some quantity of a protein, the appropriate gene is transcribed into mRNA, which is then translated into protein.

The genetic information stored in any piece of DNA (and hence a gene) depends on the specific order in which the bases are arranged in the molecule. The possible variety in sequence has an exponential relationship with the number of units present: Thus a piece of DNA four units in length can be arranged in one of 256 ways (4^4), while a five-unit piece may contain one of a possible 1,054 (4^5) sequences.

Strictly speaking, DNA is not a single polynucleotide but a pair of polynucleotide strands twisted around each other. One can imagine the two strands as a spiral staircase with the sugar–phosphate backbone making up the railings and the bases as the steps of the staircase. This spiral staircase structure is known, rather famously, as a "*double helix*." This structure is possible because of specific rules governing the types of bonds (hydrogen bonds) that are possible between the bases and that hold the two strands of DNA together: A always pairs with T, and C with G, across the two strands.

The specific nature of base pairing makes the sequence of the two strands de-

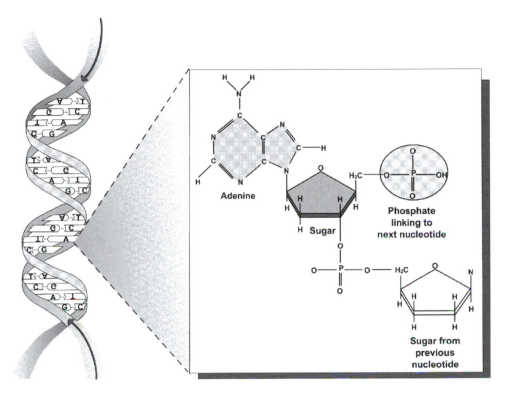

Figure 2.2. DNA structure. [Ricochet Productions]

pendent on one another, and they are said to be *complementary*. This complementarity allows a DNA molecule to contain different genes on its two strands; for example, a strand that reads as GATTC in one direction would have as its complement a piece with a sequence that reads as GAATC, which "spells" an entirely different word in gene-speak. The complementarity of the bases is also essential for one of the most remarkable features of DNA—its miraculous ability to replicate itself.

This double-helical structure is the most common form in which DNA is present in various living things. Each of our cells, for instance, contains millions of base pairs worth of DNA, neatly coiled and packaged into chromosomes that are organized into various genes. Nature chose this form as the most stable storage unit for information. However, certain viruses contain their genetic information in the form of single-stranded DNA. Occasionally, during the course of gene expression or replication, DNA also assumes certain triple- and quadruple-stranded forms.

See also: Chromosome, Gene, Gene Expression, Genetic Code, Ribonucleic Acid (RNA); DNA Synthesis/Replication and Repair (Ch. 4).

DEVELOPMENT

Development refers to the complex series of biological steps that take a single fertilized zygote to a fully functional multicellular organism with specialized tissues and organs. Every cell in our bodies is virtually identical in terms of genetic information because all of our cells arose from the same original fertilized egg. However, as we know, all the cells are not the same (e.g., a cell from your skin is very different from a nerve cell), and once these cells are differentiated in the embryo, they cannot revert to their original, undifferentiated state. In most animals, there are three main development stages: establishment of the dorsal (back) and ventral (front) axes of the body; formation of the main body parts, such as the limbs; and, finally, the specialization of tissues and formation of major organs. The study of development in various life forms, particularly higher animals, forms a separate discipline known as developmental biology.

See also: Gene Expression.

DIPLOID

When the cells of a living organism contain a pair of its complete chromosomal complement, they are said to be diploid. The somatic cells in most "higher" multicellular forms of life, including humans, are diploid. The diploid chromosome number for humans is 46.

See also: Chromosome.

DNA DIAGNOSTICS

DNA diagnostics is the practice of testing a person's DNA (either by looking at the sequences of specific genes or loci or for specific markers) in order to diagnose various medical or physiological conditions. For example, a person's genetic profile for the protein beta-globin can establish whether or not he or she has beta-thalassemia.

See also: Genetic Screening; Thalassemia (Ch. 5).

DNA REPLICATION

DNA replication is the process by which a DNA molecule generates a second molecule identical to itself. DNA replication typically takes place when a cell is ready to divide and form progeny cells, both in single-celled and multicellular organisms. The formation of many copies of a viral genome while it is inside its host cell is also the result of DNA replication. The ability of DNA to replicate it-

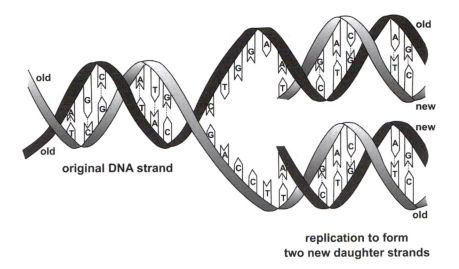

old

old

new

new

old

old

original DNA strand

**replication to form
two new daughter strands**

Figure 2.3. DNA replication. [Ricochet Productions]

self is essential in order for it to perform one of its two key functions, namely the transfer of information (or the instructions for living) from one generation to the next. Gene expression (discussed separately) sees to the other function—executing the functions that enable an organism to live.

The most important feature of DNA replication is that it is *semiconservative*. This means that half (i.e., one strand) of the parent molecule is physically conserved within each progeny molecule. This type of duplication is possible because of the complementary double-helical structure of DNA. Each strand of the helix functions as a template on which a complementary second strand is built (conforming to the rules of base pairing), giving rise to two daughter double helices with identical base sequences. Each progeny molecule thus contains one strand from the parent and one newly synthesized strand (see Figure 2.3).

There are many steps in DNA replication, and they are catalyzed by different enzymes. Because DNA is present in a very tightly wound form, the first step is the unwinding of the molecule to make it accessible to the various replicating enzymes. This unwinding—which occurs in small segments of DNA at a time—is catalyzed by enzymes called *helicases*. After the molecules are unwound, the two strands are separated so that each may be used as a template. The synthesis of complementary strands against the template is carried out by an enzyme called *DNA polymerase*. The raw materials required by the enzyme include a short primer to get started and the four nucleotides. The primers—which are short (RNA) oligonucleotide sequences complementary to the DNA at various locations—are synthesized by an enzyme called *primase*. They anneal to their complementary sequence on the DNA and provide the anchor or scaffolding on which the new chain is built.

DNA polymerase proceeds to build the DNA chain one base (nucleotide) at a time by catalyzing the formation of a phosphodiester bond between the 5' end of the growing chain and the 3' phosphate group in a nucleotide. Because the two strands in a helix are antiparallel, they are synthesized at different rates. The template strand running in the 3' to 5' direction requires just a single primer and is synthesized more or less continuously from one end to the other. It is called the leading strand. The other strand, called the lagging strand, has to be synthesized in short, discontinuous strands (working backward from the point of separation— see Figure 4.3 for details). A second type of DNA polymerase comes in to replace the RNA nucleotides with DNA where the primers were. This is followed by the action of an enzyme called *DNA ligase* that stitches all the fragments together to form a continuous molecule.

DNA replication in bacteria is a pretty straightforward process because it starts at a single fixed point called the replication origin. DNA replication in eukaryotes can start at many points (a relief because the average human chromosome contains 150 million nucleotides) and usually takes about 1 hour for completion.

See also: Cell Cycle, Deoxyribonucleic Acid (DNA), Gene Expression.

DNA SEQUENCING

Laboratory methods that are used to determine the exact sequence of nucleotides (or bases) of a given piece of DNA are called DNA sequencing methods. The techniques for determining DNA sequence at this finely tuned level were first developed in the mid- to late 1970s, and provided geneticists and molecular biologists with a level of precision and exactness as never before. The inventors of the basic sequencing techniques, Frederick Sanger and Walter Gilbert, shared the 1980 Nobel Prize in chemistry for this fundamental contribution.

The Sanger method is a biochemical or enzymatic approach for determining DNA sequence. Here, the double-stranded DNA to be sequenced is first converted to single-stranded molecules. Either of these strands is then used as the template in an enzyme-directed synthesis reaction in which it is incubated along with DNA polymerase, an appropriate oligonucleotide primer, the normal deoxynucleotide (dNTP) building blocks of DNA (dATP, dGTP, dCTP, and dTTP) and small quantities of molecules known as the dideoxynucleotides (ddNTPs). The ddNTPs are virtually identical to their dNTP counterparts, but with one crucial difference: they lack an –OH group at the carbon 5 (C5) position of the sugar molecule. This difference is not recognized by the DNA polymerase, which can substitute a ddNTP for a normal dNTP during chain elongation. But once a ddNTP is incorporated into a DNA strand, further synthesis stops because the lack of the –OH group at C5 means that there is no place to add the next base (refer to the structure of the DNA molecule in Figure 2.2). Because ddNTPs are incorporated randomly, DNA synthesis in their presence generates a mixture of fragments of all possible sizes, each with a specific ddNTP at the 5-prime end. When the sequencing reaction is

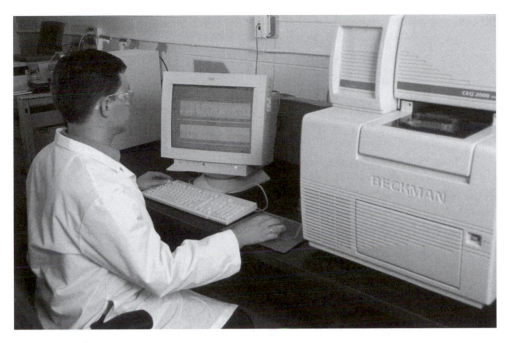

Technician with automated DNA-sequencing equipment. [© T. Bannor/Custom Medical Stock Photo]

complete, these fragments can be separated according to size (from the longest to shortest) by gel electrophoresis. By tagging each of the four ddNTPs with a different fluorescent dye that fluoresces a different color when illuminated by a laser beam, it is possible to determine the base at each position of the sequence from the colors of polynucleotide fragments that are generated during the reaction.

The sequencing technique developed in Gilbert's laboratory uses chemical reagents to carry out base-specific cleavage of DNA, instead of enzyme-directed synthesis. Here, the DNA to be sequenced is radiolabeled at one end. Next, chemical reagents are added that specifically alter one of the bases in the DNA strand. The altered base is then removed from the sugar phosphate backbone of DNA, and the DNA strand is treated with the chemical piperidine, which cleaves it at the position of the missing base. Each reagent produces a set of different lengths of end-labeled DNA. These fragments can then be separated on a gel and analyzed to obtain the base sequence of the original molecule.

When first developed these sequencing methods were laborious, long-drawn-out processes involving large electrophoretic gels and the visual interpretation of the results. As recently as the 1980s most laboratories were still using manual techniques, and sequencing even a single gene was a long project, lasting over several weeks or months. With advances in automating various techniques, scientists have been able to reduce both the time as well as the amounts of material needed for sequencing by several orders of magnitude—for instance, down to a few hours. The acceleration of the Human Genome Project is a resounding testimonial to the progress made in sequencing techniques.

See also: Deoxyribonucleotide Acid (DNA), Gene, Oligonucleotide, Polymerase, Primer; DNA Synthesis/Replication and Repair (Ch. 4).

ENVIRONMENT

In a genetic context, the environment refers to the sum total of the conditions surrounding an organism (and hence its genes). Not only does this include broader geographic, climatic, and biological factors but also subtler influences, including even psychological factors (at least in the case of humans), as well as conditions local to one part or organ system in the body. Thus, anything from the microenvironment of the developing fetus in the amniotic sac to the amount of sunlight absorbed by our skin on a day at the beach, or even the food we eat, constitutes an environmental influence on our being. In short, the environment includes just about every aspect of an organism's life other than its genes.

The debate about whether it is genetics or environment that has a greater effect on our phenotypes or various traits is a heated one, both within and beyond the scientific community, and is far from being resolved. Certainly, our environment has a profound effect on our phenotype, as can be seen from the fact that one gets a tan upon exposure to the sun. So, even though their genetic makeup is virtually identical, monozygotic twins would appear different (e.g., with respect to their skin coloring) if one lived in Alaska and the other in Barbados! On the other hand, if these twins had, say, a genetic propensity for melanoma, then even the person exposed to less sunlight would be more likely to develop the cancer than, say, a third, unrelated individual in Barbados who does not carry that genetic predisposition.

Thus, it would be fair to say that in general, both genes and environment play an important role in specifying any trait or predisposition. In addition, many scientists say that a third component, namely the *interaction* between genes and the environment, also plays an important role in determining a trait.

See also: Gene, Genetic Disease/Disorder, Predisposition.

ENZYME

An enzyme is a biological or biochemical catalyst. This means that it enables or facilitates a biochemical reaction that would occur very slowly in the absence of the enzyme. The enzyme itself appears to have undergone no net chemical change at the end of the reaction.

All living organisms, from the simplest bacteria to complex beings such as plants and animals, depend on various enzymes to catalyze vital chemical reactions involved in processes such as respiration, photosynthesis, digestion, and replication. Among the many digestive enzymes in the human gut, for example, are enzymes to break down sugars, fats, and proteins. Inside the cell, activities such as protein synthesis and DNA replication are mediated by a variety of other enzymes. The action of an enzyme is specific—which means that a given enzyme catalyzes a specific type of reaction. For instance, different sugars, such as lactose, amylose (a breakdown product of starch), and maltose, are attacked by different enzymes, named lactase, amylase, and maltase, respectively.

Chemically, most enzymes are proteins, and, in fact, until the discovery of catalytic RNA molecules in the 1980s, it was thought that *only* proteins could function as biological catalysts. Although the vast majority of enzymes known to us are indeed proteins, we now know that some of the fundamental reactions in living systems, such as protein synthesis, are catalyzed by RNA enzymes (ribozymes).

See also: Protein, Ribozyme.

EUGENICS

The term eugenics has many negative connotations in today's world, associated as it is with ideas of genocide and ethnic cleansing. Stripped to its bare essentials, however, it translates as "good breeding." The word was first coined in 1883 by the British biometrician Sir Francis Galton, who proposed that the human race could be improved in much the same manner as for plants (i.e., by selective breeding). As conceived by Galton, eugenics had two faces: positive eugenics, which promoted the breeding of people with favorable traits; and negative eugenics, which discouraged breeding among those with undesirable ones.

Obviously, the idea of "human improvement" was (and remains) fraught with problems, particularly because there could be no consensus as to which characteristics should be selected and which should be bred out of the human race. Further complicating the issue was the fact that human genetics and inheritance was, and indeed still is, incompletely understood. Galton, for instance, believed strongly in the heritability of personality traits such as intelligence and leadership, whereas many, especially today, would dispute that idea.

Regardless of these intrinsic problems, Galton's ideas of eugenics were embraced quite widely and gained much popularity in many parts of the world. However, even those people who agreed with the soundness of the concept in theory—notable examples include such prominent members of the British intellectual community as Julian Huxley and J.B.S. Haldane—felt that it could not be put into practice because human genetics was simply too complex to play with. Notwithstanding the diversity of opinions on the heritability as well as desirability of various traits, the authority to choose the traits has more often than not been accorded to the rich and powerful regardless of their knowledge base, which has naturally led to discriminatory outcomes. People have used the banner of eugenics to promote various agendas, such as racial and socioeconomic segregation, on the grounds that such traits (i.e., race, class, and poverty) were also genetic. Involuntary sterilization in mental hospitals and genocide are some examples of the extreme consequences of putting eugenic ideals into practice.

With the coming of age of the Human Genome Project and techniques for prenatal genetic testing and genetic manipulation, we are certainly entering a new era of possibilities vis-à-vis eugenics. In this new era, the ethical concerns voiced in earlier times over potential abuses of eugenics are perhaps more valid than ever.

See also: Inheritance.

EUKARYOTE

The term "eukaryote" is derived from the Greek and means "true kernel." It refers to organisms whose cells contain a true nucleus or a specialized membrane-bound compartment that houses cellular DNA. Eukaryotes are distinct from bacteria and Archaea in this respect, as the latter do not have a nuclear membrane surrounding the genetic material. Eukaryotic cells also contain other membrane-bound structures, such as the Golgi apparatus, mitochondria, and chloroplasts, for performing specific functions, whereas the prokaryotes lack these organelles. In addition, the two categories of organisms have distinct types of ribosomes.

See also: Cell, Nucleus, Prokaryote.

EVOLUTION

Arguably the single most revolutionary concept in biology, the subject of evolution continues—more that a century after it was first put forward—to spark debate among its proponents and dissidents. Witness, for instance, the 1998 decision by a school board in the state of Kansas to allow different schools to choose whether or not to teach evolutionary theory in their classrooms. Even Galileo's views about the universe, condemned as heresies in his own time, did not arouse the same level of controversy as evolutionary theory.

Perhaps the most widespread notion about the theory of evolution can be summed up in the pithy statement that "man descended from apes." But, in fact, neither Darwin's original idea nor today's evolutionary theory make any such claims. Rather, what the theory says is that all life on Earth in the very distant past—several eons ago—probably had a common origin. All extant life forms—that is, living creatures today—are thus related to one another in varying degrees through common ancestral forms. So while neither apes nor monkeys gave rise to humans, it is likely that they shared a more recent common ancestor than any of them did with dogs, for example, and even more distantly with a pine tree.

The mechanism for the evolution of species is by the accumulation of changes in the gene pools of living populations of species over time (i.e., several generations). Virtually every genome undergoes some changes or mutations as it passes down from one generation to the next. However, a few changes, whether random or not, seldom have an impact on the identity of a species. It takes several generations of the stable accumulation of mutations to effect sufficient differences in a genome for the emergence of a new species. Changes may take the form of sequence changes or the addition and deletion of genes and sequences to the genome. In general, it is believed that mutations that result in a survival advantage for organisms that possess them are more likely to spread and persist than mutations that do not result in a survival advantage or result in a survival disadvantage. It should also be noted that evolution proceeds at a much slower rate as genome size increases, because the relative change in a genome rather than gross differences has a greater impact; that is, a 3-kilobase (3,000 bases) change in a piece of genomic DNA of 3 million base pairs has a much greater impact than a similar change in genomic DNA that is 3 billion base pairs long.

See also: Heredity, Mutation.

FERTILIZATION

Fertilization is the process by which the nuclei of two gametes (sex cells) fuse with one another to form a cell known as the zygote. In humans, these cells are the sperm and egg cell (ovum). Fertilization is a hallmark of sexual reproduction in living organisms and marks the change from the haploid to the diploid phase in an organism's life cycle.

See also: Cell Cycle, Diploid, Haploid, Nucleus, Zygote.

FINGERPRINTING

In molecular biology, fingerprinting refers to certain techniques for identifying individuals on the basis of their macromolecular (DNA or protein) components. The term "fingerprinting" likely derives from the long-established detective technique of using actual fingerprints—namely the patterns of ridges and whorls on the tips of fingers—as markers or signatures for identifying criminals. First developed in 1985, DNA fingerprinting relies on the principle that, like fingerprints, the sequence of base pairs in our DNA is unique for every individual (except, of course, identical twins). Coincidentally enough, one of the early applications of DNA fingerprinting was also in the field of forensics. It has also been used successfully in paternity identification, with a higher degree of accuracy than any previously used method.

The basic technique of DNA fingerprinting involves the treatment of the DNA with restriction enzymes, followed by the analysis of the enzymatic digest by electrophoresis. Each restriction enzyme attacks at specific points (restriction sites) on the DNA molecule where specific patterns of nitrogen bases occur. Everyone has these points in their DNA but often not at the same places because of naturally occurring variation. Because the restriction sites occur in different places in different individuals, the length of the fragments differ from person to person. In very rare instances, two people might have the identical pattern of fragment lengths produced by one restriction enzyme. Therefore, in DNA fingerprinting, samples are tested with more than one enzyme. The probability that two or more people will produce identical patterns when their DNA is cut by two or more restriction enzymes is so small as to be practically nil.

After electrophoresis, the DNA fragments are immobilized on nylon filters (see the entry on Southern blotting) and hybridized with radioactive probes derived specifically from human DNA. This yields a pattern or "fingerprint" that is specific for a given individual. By comparing the pattern of the accused criminal with the pattern of DNA retrieved from the crime scene, the identity of the perpetrator may be determined. Paternity determination is somewhat more complicated, as it involves the analysis of inheritance probabilities of various bands rather than a straightforward comparison of band patterns.

See also: Intergenic DNA, Restriction Fragment Length Polymorphism (RFLP), Southern Blotting, Variation.

GAMETE

Gametes are the mature reproductive cells in living organisms. Human gametes include the sperm in males and egg cells (ova) in females. Gametes are produced in the body by a special process of cell division called meiosis, which results in the halving of the number of chromosomes (i.e., gametes are haploid). Thus, for instance, a human sperm or egg cell contains only 23 chromosomes—22 autosomes and a sex chromosome. Only a sperm may carry the Y chromosome. Only one from each pair of the chromosomes of a diploid cell is represented in a gamete. When the two gametes—egg and sperm—are fertilized to form a zygote, the full number of 46 chromosomes is restored. The formation of gametes is necessary in order to produce progeny with chromosomes from both parents.

See also: Cell Cycle, Chromosome, Meiosis.

GENE

The simplest definition of a gene—the one to be found in most textbooks and genetics-related Web sites—is one that describes the gene as "the fundamental physical and functional unit of *heredity*" in a living organism. However, the simplicity of this statement belies a host of complexities both in the concept and the history of the word. The definition is barely adequate, not only because it presupposes a conception of heredity on the reader's part but also because it obscures almost as much information as it conveys.

Even if we were to address the first objection by providing a definition of heredity (see the entry in this chapter), we are left with the question of what is inherited or passed on from one generation to the next. Is it a physical characteristic, such as height or eye color, a quality such as intelligence, or a type of behavior? The basic definition implies that the gene itself is passed on—note the "physical" component in the definition—but tells us nothing about what the gene consists of or what it looks like. Also, the definition conveys no sense of how the gene acts as a carrier of heredity.

With these inadequacies of the basic definition in mind, it might make more sense to expand the definition of the gene to that of a physical entity that specifies an observable trait in a living organism and carries this information over to the progeny of that organism. A gene encompasses functions of heredity as well as *information.*

In order to understand how a gene performs these dual functions, it is useful to turn from a conceptual or functional definition to a physical one. Within this framework, a gene may be defined as a region of DNA within a chromosome, which when expressed or transcribed leads to the production of RNA (ribonucleic acid). The DNA within our cells contains the information for everything that occurs within each cell—every action, every substance made, every event, every response—everything! The same thing holds true for single-cell organisms such as bacterium—the DNA within a bacterial cell has *all* of the information necessary for the life of the cell. There are some important differences, however, between the genes of simpler organisms, such as the bacteria, and those of multicellular ones, such as plants or animals. In bacteria, a single gene is a discrete, continuous piece of DNA. This picture is often confounded in so-called higher organisms, where a gene is not necessarily present as a single piece of DNA but instead has several pieces interspersed in other DNA sequences that must be patched together in order to become functional. And, strictly speaking, it is not the DNA itself that undergoes these changes most of the time but rather another molecule, called messenger RNA, which functions as the working copy of the instructions during the execution of the functions encoded in the gene.

The mere presence of a gene in an organism is necessary but not sufficient for the appearance of a trait. In order for an organism to exhibit a trait, the gene for that trait must be *expressed*, or its information be translated into a usable form by the organism. Consider a book: No matter how exhaustively it covers a subject, its contents are meaningless unless it is read. The expression of a gene is akin to the reading of individual chapters or paragraphs of a book. Furthermore, it is even incorrect to think of a one-to-one relationship between a gene and a trait because a gene does not automatically endow its carrier with a trait. Rather, each gene carries—in a special script or code—the information or recipe for a single protein that is produced during gene expression. Each of these proteins is then capable of performing a specific function. Unless a particular property is the direct outcome of the action of a single protein—for example, the breakdown of milk sugar to simpler sugars, which is the result of the action of a single enzyme called lactase—we cannot relate a single trait to a single gene. The idea of a single gene for "blue eyes" is incorrect because eye color is not determined by a protein. The gene (or genes) encodes one (or more) enzymes that work in the cell to produce a pigment that imparts color to the iris. The pigment itself is not encoded into a gene.

While we may continue to use, for practical purposes, the working definition of the gene as a fundamental structural and functional unit of information and heredity, we see that in reality the concept is considerably more elastic.

See also: Deoxyribonucleic Acid (DNA), Gene Expression, Heredity, Protein, Ribonucleic Acid (RNA).

GENE EXPRESSION

Gene expression refers to the process through which the information in a gene, in the form of a DNA sequence, is made accessible to (i.e., transformed to a usable form in) a cell. The end result of gene expression is the formation of a gene product, which may be in the form of either RNA or protein. The majority of known gene products are proteins—which account for a large proportion of the structural and functional components of a living organism. However, certain important cellular components, such as transfer RNA molecules and ribosomal components, are RNA molecules encoded by specific genes.

The pathway of gene expression depends on the nature of the gene product. The first step is *transcription*, through which information in DNA is transformed into RNA. In cases where the end product is a protein, there is a second step, known as *translation*, during which genetic information transcribed and processed into RNA molecules is further processed to form proteins, or more specifically peptide sequences. These processes are discussed separately.

Gene expression is the process that relates an organism's phenotype to its genotype, and no trait manifests itself unless the gene or genes governing it are properly expressed. The relationship between gene expression and phenotype is quite straightforward in haploid organisms. In diploid organisms, which contain two copies of each gene, the picture is somewhat more complicated, particularly when the organisms are heterozygous (i.e., with two different gene copies in each of their chromosomes). In these instances, the phenotype corresponds to the dominant allele or gene. There are genes for certain traits (e.g., the color of certain flowers) that are not strictly dominant or recessive; the expression of such combinations of genes gives rise to phenotypes that reflect a combination of the gene products! Phenomena such as genomic imprinting and the clinical heterogeneity of diseases are often caused by improper or imbalanced gene expression.

Gene expression in a cell is an exquisitely controlled process that is influenced by a host of intrinsic and extrinsic factors. Although some genes are expressed in all cells, others are specific to certain tissues or to the stage of the organism's development. Still others require special environmental conditions or triggers in order to be expressed properly. The interplay of the various factors that control gene expression is far from being completely understood.

See also: Deoxyribonucleic Acid (DNA), Gene, Protein, Ribonucleic Acid (RNA), Transcription, Translation.

GENE FAMILY

Many functional protein-encoding genes in our genome (and most likely in the genomes of other species as well) occur in small clusters of closely related sequences known as gene families. Whereas satellite DNA is present in clusters of hundreds of thousands of sequence repetitions, a gene family (with the exception of those for ribosomal RNAs) comprises only a few versions of the gene.

A gene family is the result of gene duplication over the course of evolution. Very few genes in a given cluster are identical, although there may be a few exact repeats. Usually, the duplication was followed by modifications so that the structure and function of the genes of one gene family are somewhat, but not completely, altered. The resulting gene products also display a relatedness in function. On occasion, the mutations are so many as to inactivate gene activity—such copies of the gene become the pseudogenes.

The discovery of gene families grew out of the concept of isoenzymes (enzymes of similar activity but slightly differing specificities), which was first introduced in 1959 by Clement Markert and Freddy Moller. An example of a well-characterized gene family in the human genome is the alpha-globin cluster on chromosome 16. It consists of seven genes: two genes for alpha-globin itself, a gene each for two other globin proteins (designated as theta- and zeta-globins), and three pseudogenes. A second, unlinked cluster of related genes, called the beta-globin family, consists of six genes. The apolipoproteins that are involved in cholesterol transport are also encoded by a family of genes organized as two unlinked clusters on chromosomes 11 and 19.

See also: Pseudogene.

GENE FREQUENCY

Also known as "allelic frequency," this is a quantitative term used by population geneticists to express the relative occurrence of a particular allele of a gene within a population. It is generally expressed in terms of a percentage. Mathematically, the frequency of an allele within a population is equal to the number of alleles of a specific type within the population divided by the total number of possible alleles that may be found at the locus where it is found. If 200 A alleles and 400 a alleles are found within a given population, then the frequency of A alleles is 200 /

(200 + 400) = ⅓ = 0.33. Gene frequencies are useful to track phenotypes or diseases in populations and in general to describe genetic variation in the gene pool.
See also: Allele, Gene, Population Genetics.

GENE THERAPY

Gene therapy is the term applied to methods of preventing or treating a disease by genetic manipulation, namely by altering the DNA sequence of a patient.

The idea behind gene therapy is either to restore a gene function lost by mutation or to balance the negative effects of a mutation by providing the normal gene so that it can be expressed in the diseased individual (usually in a target organ or tissue). For example, if a particular disease is caused by a mutation that inactivates a gene, then simply introducing copies of the "normal" gene into the target organ or tissue might help solve the problem. On the other hand, if the mutation alters the gene product to cause harmful effects, it may be necessary to employ a dual approach—first to inactivate the harmful gene (e.g., by antisense therapy) and then to supplement this treatment with copies of the normal gene to restore its regular function.

There are two main approaches to gene therapy that depend on the nature of the cells targeted: somatic cell therapy and germ-line therapy. Somatic cell therapy refers to the modification of the DNA in the somatic cells of an individual with a particular disease so that the defective (mutant) cells no longer cause the disease. It is a possible approach for diseases where the problem is directly attributable to a defect in a particular gene. Biochemical diseases arising from mutations to a single gene (e.g., cystic fibrosis) are potential candidate diseases for this sort of gene therapy. Somatic cell alteration is a form of genetic remedy that does not affect gametes but simply changes a diseased person's gene by manipulating the somatic cells that cause the disease. Quite often, this kind of gene therapy involves altering the genetic makeup of only those cells where the disease is manifested—consequently it may be considered a form of "genetic surgery."

The second type of genetic therapy involves permanent changes to a person's genome by germ cell alteration. This procedure, known as germ-line gene therapy (GLGT), results in the modification of heritable characteristics. These changes can be passed on to progeny. As one may well imagine, the idea of being able to tamper with the genes of an unborn child—even for therapeutic purposes—raises many ethical, social, professional, and personal issues that must be dealt with before such therapy can be employed.

Because it does not tamper with the genetic makeup of future generations and involves only the affected individual, there are perhaps fewer ethical objections to somatic cell gene therapy, but, even so, its progress has also not been smooth because of various technical problems. One of the biggest challenges has been the development of a system to deliver the genes to the target tissue or organ. An ideal delivery system should be specific to its target, be able to express the gene at an appropriate level for a prescribed time period, and, above all, should pose no risk to the patient. Scientists have used three major approaches to deliver genes: viral vectors, nonviral vectors, and physical gene transfer. Viral vectors are currently the most effective means for efficient gene transfer but also the most problematic because of the diseases they may bring. In ex vivo gene therapy, cells with defective genes are removed from a patient for transfection with wild-type copies of the affected gene. The treated cells are then returned to the patient. In vivo therapy involves direct administration of a gene or packaged gene to the patient.

Gene therapy suffered a major setback in 1999 when 18-year-old Jesse Gelsinger died while undergoing gene therapy for a rare genetic disorder of the liver. Although the cause of death was unclear, the prime suspect was the adenovirus used to deliver the therapy, and the FDA stepped in to end the trial. Following this, a major debate on the risks of gene therapy swept the nation. Today, despite the promises it holds, medical scientists are proceeding very cautiously with the wholesale application of gene therapy. It is still under intensive investigation in various laboratories throughout the world and is by no means a standard method of disease treatment.

See also: Antisense, Genetic Disease/Disorder, Vector.

GENETIC CODE

The genetic code describes the relationship between information stored by living organisms in their DNA (or genes) and the information present in the protein molecules of these organisms. DNA molecules store their information in the form of linear sequences or arrangements of their four nucleotide building blocks (designated as A, T, G, and C). These sequences serve as codes for the sequence of the amino acids of the corresponding proteins. The relationship between the two types of sequences is not simple, however, because proteins in living things are made of a possible 20 amino acids, in contrast to the four possible building blocks of the nucleic acids. To deal with this disparity, nature devised a special relationship between the two species of molecules: Each amino acid is encoded by a triplet of nucleotides known as a *codon*. For example, the triplet TTT codes for the amino acid lysine, and CAT codes for methionine.

Let us now see why a codon is made of three bases and not more or less. Suppose each base (i.e., A, T, G, and C) could code for an amino acid. Then, we would have a situation where proteins could be made of merely four amino acids, which as we know is not the case. What about a two-letter code? This still leaves us short four amino acids because there are only 16 ways to arrange the four bases (4×4) in pairs. With a triplet codon, we have 64 possible arrangements of bases ($4 \times 4 \times 4$), which is certainly enough to accommodate 20 amino acids.

In fact, a triplet code gives us many more codons than amino acids, which has led to a redundancy in the genetic code. All of the amino acids, with the excep-

Table 2.1. The genetic code: DNA triplet codons and the amino acids they encode

2nd base → 1st base ↓	T		C		A		G		3rd base ↓
T	TTT	Phe (F)	TCT	Ser (S)	TAT	Tyr (Y)	TGT	Cys (C)	**T**
	TTC	Phe	TCC	Ser	TAC	Tyr	TGC	Cys	**C**
	TTA	Leu (L)	TCA	Ser	TAA	Stop	TGA	Stop	**A**
	TTG	Leu	TCG	Ser	TAG	Stop	TGG	Trp (W)	**G**
C	CTT	Leu (L)	CCT	Pro (P)	CAT	His (H)	CGT	Arg (R)	**T**
	CTC	Leu	CCC	Pro	CAC	His	CGC	Arg	**C**
	CTA	Leu	CCA	Pro	CAA	Gln (Q)	CGA	Arg	**A**
	CTG	Leu	CCG	Pro	CAG	Gln	CGG	Arg	**G**
A	ATT	Ile (I)	ACT	Thr (T)	AAT	Asn (N)	AGT	Ser (S)	**T**
	ATC	Ile	ACC	Thr	AAC	Asn	AGC	Ser	**C**
	ATA	Ile	ACA	Thr	AAA	Lys (K)	AGA	Arg (R)	**A**
	ATG*	Met (M)	ACG	Thr	AAG	Lys	AGG	Arg	**G**
G	GTT	Val (V)	GCT	Ala (A)	GAT	Asp (D)	GGT	Gly (G)	**T**
	GTC	Val	GCC	Ala	GAC	Asp	GGC	Gly	**C**
	GTA	Val	GCA	Ala	GAA	Glu (E)	GGA	Gly	**A**
	GTG	Val	GCG	Ala	GAG	Glu	GGG	Gly	**G**

*The ATG codon also functions as the start codon for translation. So methionine is always the first amino acid in a newly synthesized peptide chain, although it is often removed in the post-translational modification stage.

tion of methionine and tryptophan, are encoded by two to five different codons (see Table 2.1). Three of the codons—TAA, TAG, and TGA—do not code for amino acids at all but constitute a special group known as the stop codons. They serve as punctuation marks, signaling the end of a gene (and hence a protein) within a strand of DNA.

It may interest readers to know that the triplet code was in fact predicted by several scientists (including the co-discoverer of the double helix, Francis Crick) some years before its existence in nature was determined. The credit for actually "cracking" the code and determining the specific codon–amino acid correspondences goes to biochemists Marshall Nirenberg, Robert W. Holley, and Har Gobind Khorana, who shared the 1968 Nobel Prize in Physiology or Medicine for this achievement.

See also: Deoxyribonucleic Acid (DNA), Gene, Protein.

GENETIC DISEASE/DISORDER

A genetic disease or disorder is one that is known to be inherited (i.e., passed along from one generation to another) via the genes. It may be caused by the presence of a defective gene (i.e., one that contains a mutation) that causes the loss of the gene's normal function, by the development of a new, harmful gene product, or both. Common examples include cystic fibrosis and sickle-cell anemia.

Genetic diseases may be inherited through one or several genes. Those caused by the interaction of several genes or mutations are called polygenic or complex diseases. Examples include hypertension and diabetes mellitus. It has recently come to light that even seemingly simple, single-gene diseases may arise from more than one type of mutation. In other words, mutations of unrelated origin may result in the same or a similar mutant phenotype. The heterogeneity of simple diseases falls into two main categories: *Interlocus* heterogeneity refers to situations where the same disease or set of symptoms results from mutations in one of many genes. An example is albinism, which is manifested by the body's inability to make the pigment melanin. Because melanin synthesis is a multistep process, with each step catalyzed by a different enzyme, a mutation affecting any one of these steps will have the same end result. Phenylketonuria (PKU) is another example. The second type of heterogeneity, known as *intralocus* heterogeneity, refers to cases where a disease may be caused by a variety of mutations within the same gene.

Although some diseases are *believed* to be almost purely genetic (e.g., Huntington's disease and cystic fibrosis), many others, especially polygenic diseases such as diabetes and cardiovascular disorders, depend heavily on environmental factors as well. Research in the field of disease genetics now points toward diseases lying on a continuum of genetic and environmental factors—ranging from simple monogenic Mendelian diseases with no discernible contribution from the environment to complex multigenic diseases with a strong environmental component.

See also: Association, Inheritance, Linkage.

GENETIC ENGINEERING

Genetic engineering is the process of modifying the DNA of an organism in the laboratory using various molecular biological techniques. The basic principle of genetic engineering is to identify the segment of DNA in an organism's nuclear DNA, remove it using special DNA-cutting enzymes (called restriction enzymes), and replace it with a piece of DNA from another source. The use of artificially constructed plasmids in order to express specific genes in a bacterium is one example of genetic engineering. Gene therapy is an example of genetic engineering in which corrected or normal genes are introduced into target cells of a person with a disease in order to compensate for a malfunctioning or mutant gene.

See also: Biotechnology, Gene Therapy, Restriction Enzyme.

GENETIC EQUILIBRIUM

An important concept in population genetics, genetic equilibrium refers to the state of a population in which the relative frequencies of different alleles of a gene remain constant over successive generations of that population. Also known as the Hardy–Weinberg equilibrium (after the two men who independently derived the mathematical equation in 1908), it shows how certain recessive alleles (e.g., the allele for sickle hemoglobin—see the entry for sickle cell anemia in Chapter 5) may persist in a population and why they are never altogether replaced by dominant alleles for the same trait. Specifically, if an infinitely large, random mating population is free from outside evolutionary forces such as mutation, natural selection, and migration, then the gene frequencies will not change over time. The Hardy–Weinberg equation states that for a single gene trait with two alleles, A and a, with allele frequencies of $p(A)$ and $q(a)$, respectively, where $p + q = 1$, the frequencies of the three possible genotypes are given by the equation $p(AA) + 2pq (Aa) + q(aa) = 1$. From this equation we can see that the frequency of the recessive allele a may become very low but will never go away completely. It would be helpful to retain the recessive allele in the event of a sudden change in environmental conditions that make it advantageous for the population.

The Hardy–Weinberg equilibrium has useful applications in medical genetics, making it possible to gauge or predict the frequency at which a certain disease gene/allele may be carried in a population. For example, cystic fibrosis—a recessive disease traced to the defect in a single gene—occurs in approximately 1 out of 2,000 people of Northern/Central European descent. According to the equation, this means that 4 to 5 percent of the individuals in the population are carriers for this disease, an important factor to keep in mind when offering genetic counseling to families at risk for this disease.

See also: Gene Frequency.

GENETIC IMPRINTING

Also known as parental imprinting, genetic imprinting is a phenomenon whereby the expression of a trait varies according to the parental source of its gene. In other words, the expression of the trait depends on whether its gene was inherited from the mother or the father. Although not yet completely understood, imprinting is thought to occur during gamete formation, when a gene (and sometimes an entire chromosome) is marked for either lesser or greater expression in one parent, and passed down to the next generation. The biochemical basis for this marking is not known. Imprinting is not a case of the simple dominance or recessiveness of a gene and is also distinct from mutation because it can be reversed from generation to generation.

Whereas most (99 percent) genes are not imprinted, there are about 100 genes that are imprinted (i.e., either the mother's or the father's copy is permanently silenced). An example of imprinting is manifested in the age of onset of such neurodegenerative disorders as Huntington's disease and myotonic dystrophy. In the former, early onset occurs when the disease gene is inherited from the father, whereas the latter appears to be associated with early onset when the mutant gene comes from the mother.

Two rare genetic diseases are associated with genetic imprinting in humans. In both cases, the disease appears to map to an identical deleted segment of chromosome 15. When the masculine half of this gene is missing, the maternal copy controls development and the result is a child suffering from Prader–Willi syndrome. The child will be mentally retarded, placid, tend to overeat, and later have little interest in sex. When the female half of the gene is missing, the result is Angelman syndrome—a male gene trying but failing to control cortical development so the child is retarded, has poor control of movement, and cannot speak. From this disease pattern, we deduce that the specific region of chromosome 15 is imprinted differently in male and female gametes and that both maternal and paternal regions are required for normal development (see Figure 2.4).

Father: AA' (A' is a recessive mutation)
Mother: A$_i$A' (A$_i$ is the imprinted gene)

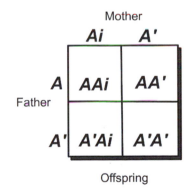

Figure 2.4. Inheritance of imprinted genes. [Ricochet Productions]

GENETIC SCREENING

The testing of a population (or a subset of a population) for the presence of a specific gene, allele, or mutation is known as genetic screening. Genetic screening may be conducted for any number of reasons, for instance to assess a population's susceptibility to a particular genetic disease, to gauge the impact exposure to a possible mutagen (e.g., a chemical or radiation), or to predict the likelihood of a child of two parents inheriting certain disorders. Depending on the circumstances, screening may be conducted prenatally, on newborns, or on adults.

Prenatal screening is a diagnostic measure undertaken only on fetuses known to be at risk for certain genetic diseases (e.g., the fetuses of women over 35 who may be at an increased risk for trisomy 21 or Down syndrome). Nowadays, there are two widely available methods for obtaining genetic material without putting either the fetus or the mother at risk—*amniocentesis*, in which fetal cells are obtained by drawing out the amniotic fluid from a pregnant mother in her sixteenth to twentieth week of pregnancy, and *chorionic villus sampling* (CVS), in which fetal cells are obtained by removing pieces of the chorionic villus, which is a membrane of fetal origin outside the fetus. Chorionic villus sampling may be performed between the eighth and twelfth weeks of pregnancy.

Genetic tests on postnatal individuals are conducted on DNA material derived from their white blood cells. *Newborn* (or neonatal) *screening* is the practice of conducting genetic tests on newborn babies for known genetic disorders. It is now a common practice in many developed countries in the world. The idea behind conducting such tests is to detect such conditions early so that any possible interventional measures may be taken *before* the disease has a chance to develop fully. Examples include congenital hypothyroidism, galactosemia, and sickle-cell anemias.

The first neonatal test for a genetic disease was not a genetic test at all but relied on a prominent phenotypic characteristic, namely the presence of a chemical called phenylalanine in the blood, for the detection of phenylketonuria (PKU). The disease threatens the newborn child with mental retardation and premature death, but fortunately a simple therapy exists (restricting phenylalanine from the diet) if it is detected in time. Today, all newborn babies are tested for PKU.

Perhaps the most common reason for testing adults is to detect heterozygotes. This is useful for prospective parents who may be at risk for having a child affected with some serious autosomal recessive disease, such as hemophilia. In addition, screening may also prove useful in assessing the effects of inadvertent exposure to potential mutagens.

More recently, screening has also been used to determine the risk of developing certain genetic disorders such as Huntington's disease. This type of testing is known as predictive genetic testing. It is most useful for diseases or conditions that are caused by disorders in a single gene, but the concept is being extended to complex multigene diseases, such as cardiovascular disease, diabetes, and various cancers.

Although genetic screening can be useful, it also has the potential for serious ethical misuse. For example, insurance companies could use this information to deny coverage to women who are at high risk for breast cancer. Similarly, employers with access to a person's genetic profile could use the information as grounds for firing him or her. Not only genetic, but also general psychiatric, counseling is strongly recommended for people before going through a diagnostic genetic test.

See also: Gene Frequency, Genetic Disease/Disorder, Heterozygote.

GENOME

As we defined it at the outset of this book, the genome is an organism's complete set of DNA. In addition to the main chromosomal DNA—which in eukaryotic organisms is present within the nucleus—the genome will also include DNA present as plasmids (in prokaryotes) or in organelles such as the mitochondria and chloroplasts.

Genomes vary widely in size: That of a typical bacterium is about 600,000 base pairs long, whereas eukaryotic genomes often run to billions of base pairs. The bulk of the genomic DNA is physically packaged into chromosomes, and the larger the genome of an organism, the more chromosomes it will contain. A general rule is that greater genomic size corresponds with higher complexity, but this is not always so. For example, the lungfish has 140 billion base pairs compared with a human's 3.3 billion! This does not necessarily mean that the lungfish has more genes than a human. In fact, scientists believe that lungfish have 15,000 genes, whereas humans are estimated to have 30,000. Some lilies and grain plants (such as wheat, rice, and corn) have even larger genomes.

In addition to size, the genomes of prokaryotes and eukaryotes also differ in the manner in which their information is organized. In bacteria, for instance, the genome is composed entirely of genes that encode proteins. In higher organisms, however, this is not the case. For example, less than 5 percent of the human genome encodes proteins, while the rest of the DNA makes up long noncoding regions found both between and within genes.

See also: Chromosome, Deoxyribonucleic Acid (DNA), Gene.

GENOMICS

A relatively new term, genomics refers to the study of genomes of organisms. It is a broad field, encompassing many areas of study, including, but not restricted to, genetics, statistics, informatics, biochemistry, and medicine. The main distinction between genomics and the more conventional methods in molecular biology and genetics is the scale: Instead of dealing with a single isolated gene at a time, genomics takes into consideration several different genes and intergenic DNA simultaneously in an attempt to tease apart their biological relationships. The genomics revolution has made it possible for scientists to begin to understand the processes that make up a complex organism. This is an exciting step forward not only in the study of diseases and their potential treatments but also in the study of life's processes.

See also: Bioinformatics, Biotechnology, Pharmacogenomics/Genetics.

GENOTYPE

The genotype of an organism is the complete genetic constitution of an individual. This means all the genes present in its genome, regardless of whether these genes are expressed or not. The genotype represents the potential of an individual to exhibit various traits, whether or not these traits are exhibited or seen. Recessive traits may often be carried through generations without ever being expressed, but they are nevertheless a part of the genotype of the organism.

See also: Gene, Gene Expression, Genome, Heterozygote, Phenotype.

GERM LINE

The germ line of any animal consists of its gametes—sperm and ova in the case of humans. These haploid cells contain half the number of chromosomes as cells from the rest of the body (somatic cells). When fertilization occurs, nuclear DNA from the sperm enters the nucleus of the ovum, restoring diploidy. Mutations that occur in these cells are known as germ-line mutations and are passed down to the next generation when the mutated gamete participates in fertilization.

See also: Meiosis, Somatic Cell.

HAPLOID

The haploid state of existence of living organisms is the state in which their cell(s) contain just a single set of the total chromosomes. Many simpler organisms, such as fungi and algae, live out the majority of their life cycle in the haploid phase. More advanced organisms, such as green plants and animals, have a mostly diploid life cycle and only produce haploid gametes. Human gametes have 23 chromosomes.

See also: Chromosome, Diploid.

HAPLOTYPE

A combination of closely linked alleles that are found in a single chromosome tend to be inherited together. A specific set of such alleles in an individual is known as his or her haplotype. Every individual's haplotype is unique because of various factors, including somatic mutations, recombination, and crossovers that might occur during cell division, as well as the virtually nonexistent possibility that two people might randomly inherit identical sets of alleles.

Interestingly, alleles do not have to be physically close on the chromosome in order to be inherited together, although physical proximity does increase their chances of linked inheritance. Pharmacogenomics, a new field that addresses individual responses to pharmaceutical drugs, uses haplotypes to help define how a person may react to a drug.

See also: Allele, Pharmacogenomics/Genetics.

HEREDITY

Heredity is the transmission of traits from one generation to another. In general, when we speak of heredity, we are referring to the phenotype or the physically discernible traits that are passed from parent to progeny. For instance, parents with blue eyes will typically have children with blue eyes also. The underlying basis for heredity is the gene. Any trait may be passed on as long as it is on a gene, regardless of whether the organism is prokaryotic or eukaryotic, single-celled or multicellular, haploid or diploid.

See also: Gene.

HETEROZYGOTE

A heterozygote is an individual who has two different alleles of the same gene at a given locus in a pair of chromosomes. Humans, and indeed most diploid organisms, are heterozygous for a vast number of their genes.

In any given population, there is usually one allele that occurs more frequently than the rest and appears to be the most normal with respect to function. This allele is designated as the "wild-type" and the others as variants or mutant alleles. The most common type of heterozygous genotype is the presence of one wild-type and one variant allele. When someone has two different mutant alleles in one locus, she or he is said to be a compound heterozygote. In instances where there are multiple copies of the gene, an individual may have two mutant alleles of the gene at different loci, each paired with a normal or wild-type copy. In that case, he or she is a double heterozygote.

Heterozygosity is the reason different individuals with genes in common may often exhibit different phenotypes (or traits) and is also the basis for Mendelian inheritance. To understand this better, let us take the classic example—used by Mendel in his famous experiments—of the color of the seeds (peas) in pea plants. A pea plant with yellow seeds often has one "yellow" allele and one "green" allele, with one allele inherited from each of its two parents. The fact that you only see the yellow color is because yellow happens to be dominant over green, which is said to be recessive in this case. Of course, it goes without saying that plants homozygous for yellow will also have yellow seeds. It should be noted, however, that not all alleles exhibit this dominant/recessive relationship and that many alleles are codominant (i.e., expressed equally). See Figure 2.5.

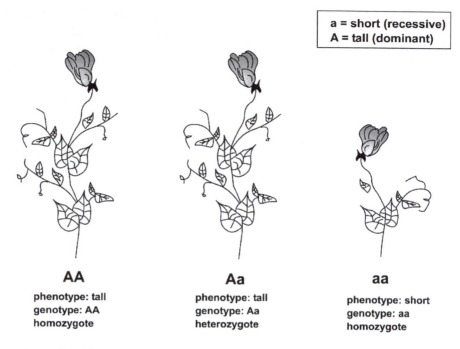

a = short (recessive)
A = tall (dominant)

AA

phenotype: tall
genotype: AA
homozygote

Aa

phenotype: tall
genotype: Aa
heterozygote

aa

phenotype: short
genotype: aa
homozygote

Figure 2.5. Heterozygote versus homozygote. [Ricochet Productions]

Heterozygosity can be advantageous compared with homozygosity because it offers greater possibilities for genetic variability. An interesting example is that of the sickle-cell anemia mutation on the hemoglobin gene. Individuals homozygous for this mutation are of course at a disadvantage because they will suffer from the sickle-cell anemia. However, this mutation has also been associated with a higher resistance to malaria. A person who is heterozygous for this allele will therefore have an advantage over those homozygous for the normal hemoglobin gene.

See also: Allele, Homozygote, Inheritance, Mutation.

HOMEODOMAIN

The homeodomain is a sequence of DNA 180 base pairs in length that encodes a 60 amino acid peptide chain that functions as a DNA-binding transcription factor. It is a key regulator of embryonic development, and evidence suggests that it may also contribute to evolution. First discovered in 1983 independently by M. Scott and Amy Weiner at Indiana University and W. Gehring's group at the University of Basel in Switzerland, this conserved sequence has since been found almost universally among the eukaryotes.

See also: Conserved Sequence.

HOMOLOG

A homolog is a structural or functional counterpart of a gene or protein in one organism that is also found in another organism. Different proteins in humans and the fruit fly *Drosophila*, for example, perform identical functions of signal transduction. Scientists estimate that 99 percent of our genes have animal homologs, which leaves only about 300–600 genes that seem to be unique to humans.

See also: Gene.

HOMOZYGOTE

A homozygote has the same alleles of the same gene on a pair of chromosomes. Recessive alleles of a gene must be homozygous in order for their phenotype to manifest itself in an organism. For example, we know that a guinea pig with white fur is homozygous for fur color because all other coat colors happen to be dominant over white (see Figure 2.5).

See also: Allele, Heterozygote, Inheritance.

HOUSEKEEPING GENES

Because it would be too wasteful for a cell to maintain constant levels of all of its proteins, many proteins are translated only when they are needed. Even so, all of the cells of an organism need to maintain a basic level of activity in order to stay alive. These activities entail certain routine metabolic functions such as the production of energy and consequently require the presence of a certain baseline concentration of the enzymes that perform these functions. The level of these enzymes is maintained by the constant expression of the corresponding genes in the genome of the organism, known as the housekeeping genes. For example, all cells maintain a basic level of enzymes involved in the Krebs cycle and in the protein machinery involved in mRNA translation.

See also: Cell, Gene, Metabolism, Ribonucleic Acid (RNA)—Messenger RNA (mRNA).

HYBRID

There are many different definitions for the term "hybrid," but essentially it refers to the product of the fusion or pairing of two distinctly different entities. For example, complementary DNA and RNA strands can pair together to form a DNA–RNA hybrid helix. The mating of individuals that belong to genetically different populations or species can form a hybrid, such as the mythical griffin (half lion, half bird) and the real mule (half horse, half donkey). A chimera is another example of a hybrid.

See also: Chimera, Southern Blotting.

INBREEDING

The breeding or crossing of two organisms that are related to each other is called inbreeding. Although this practice is often adopted in laboratories and agriculture in attempts to produce genetically "pure" strains of plants or animals, inbreeding is not, for the most part, desirable in the real world. It reduces genetic variability, which is a disadvantage for the survival of a species, and has often been associated with a higher propensity for genetic disorders.

The underlying reason for the higher incidence of genetic diseases in the offspring of related parents has to do with probability. In general, when two parents are heterozygous for a gene with a disease-causing mutation, there is a 25 percent chance that the offspring will be homozygous for that mutation and will suffer from the disease. However, the more closely the parents are related, the more likely they are to have inherited the same mutations from *their* parents. This increases the chances for their offspring, in turn, to inherit two mutant alleles (i.e., to be homozygous for the disease-causing allele). Of course, in reality it is not this simple because most diseases are caused by several genes acting together, but the basic idea is the same.

See also: Heterozygote, Homozygote, Inheritance.

INDEPENDENT ASSORTMENT

Independent assortment refers to the independent behavior of chromosomes with respect to one another during segregation in meiosis. It is this segregation that accounts for the observation that the inheritances of many different traits from one generation to the next are unrelated to each other. Although this observation is in direct contrast with the phenomenon of the linkage of certain other phenotypes or traits, the same basic phenomenon governs both outcomes. Traits specified by genes on a single chromosome tend to show linkage, whereas those on different chromosomes will assort independently.

The rule of independent assortment was first proposed in the nineteenth century by the Austrian monk Gregor Mendel following extensive experiments performed with pea plants. He analyzed the pattern of inheritance of seven pairs of physical traits. He started by crossbreeding a plant that was pure-breeding (and hence, in modern terms, homozygous) for round (RR), yellow (YY) seeds with one that was breeding true for wrinkled (rr), green (yy) seeds. Mendel found that all of the offspring of this first-generation mating were round and yellow, indicating that these traits were dominant. Quite evidently, all the progeny in this generation were dihybrids; that is, heterozygous for each pair of alleles (RrYy). Mendel then crossed these dihybrids with each other. In the next generation (F2), he observed all possible combinations of seed color and texture: round and yellow; round and green; wrinkled and yellow; and wrinkled and green. He repeated these experiments many times and observed that the F2 generation always produced the same ratios of offspring: nine-sixteenths of the offspring were round and yellow, three-sixteenths were round and green, three-sixteenths were wrinkled and yellow, and one-sixteenth was wrinkled and green. This meant that the inheritances of the characteristics were independent—which was what Mendel proposed.

Although independent assortment was a brilliant discovery, we now know that the rule is only true for genes that are on different chromosomes (i.e., not linked). Genetic linkage was the reason why when Bateson and Punnett tried to carry out Mendel's experiments on another set of traits of pea plants, they did not observe the 9:3:3:1 ratio. Using fruit flies, Thomas Morgan later showed that sometimes genes can be inherited in blocks, preventing independent assortment.

See also: Linkage, Segregation.

INHERITANCE

Inheritance describes the manner in which the genes and hence various physical traits (phenotypes) of a living being are passed along from one generation to the next. Although any gene or trait has the potential to be passed along, the manner in which various traits are inherited depends on several factors, especially the genetic constitution of the parents and, more importantly, of the specific gametes participating in a fertilization.

The basic laws of heredity (and the hint of genes) were discovered by the Austrian monk Gregor Mendel (1822–1884). His work remained obscure and received little attention until it was resurrected in the early 1900s, most notably by Thomas Hunt Morgan. Mendel's experiments with pea plants in his monastery garden laid the foundation for the science of genetics. His approach was to make hundreds of crosses of pea plants with specific observable traits by means of artificial pollination. He kept careful records of the plants that were crossed and of the offspring, and generalized these findings in the form of three main laws of heredity, namely the association of traits in pairs (e.g., tall/short and smooth/wrinkled), the occurrence of dominant and recessive traits, and the independent segregation of different traits during reproduction. With these laws, Mendel found he could predict the nature and frequency of different traits from one generation to the next.

When rediscovered at the turn of the twentieth century, Mendel's laws were found to be applicable to humans but with multiple exceptions. Sex-linked traits and linked genes defy his rules. Other exceptions have also been found, including sporadic cases, nonpenetrance, variable expressivity, and preferential parental transmission.

See also: Gene, Heredity, Independent Assortment.

INTERGENIC DNA

Intergenic DNA is the term given to the long stretches of DNA that are found between known gene sequences. It is the basis for the observation that the genome size of a eukaryotic organism does not necessarily correlate with the number of genes it contains or its complexity in terms of possible functions. For example, the size of a newt's genome is six times that of a human's! Much of this size is due to the vast amount of intergenic DNA—specifically satellite DNA—in the newt genome. Even the human genome can be said to be vastly intergenic, as only about 2 percent of the entire DNA content is known to code for various functions. Aside from a few promoter functions, the function of most intergenic DNA is unknown. "Noncoding sequences" is another label often given to these sequences (see Figure 2.6).

See also: Deoxyribonucleic Acid (DNA), Gene, Genome.

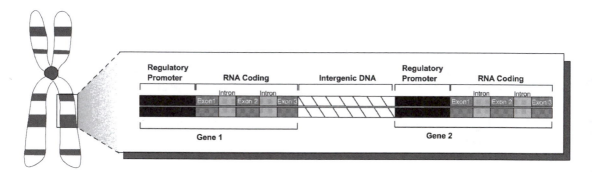

Figure 2.6. Chromosome with genes and intergenic DNA. [Ricochet Productions]

KARYOTYPE

The karyotype of an individual refers to the representation of his or her complete chromosomal constitution (i.e., the total number of autosomes and sex chromosomes, as well as any abnormalities in the number or morphology of the chromosomes). The term is also used for a photomicrograph of chromosomes that have been stained and arranged in order. The preparation of a karyotype (called karyotyping) is often used by cytogeneticists to examine the number, size, and

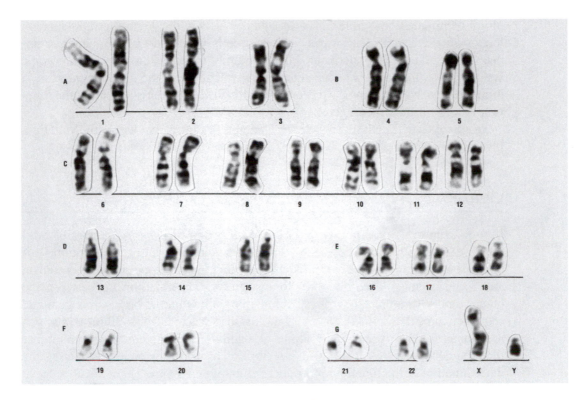

Human karyotype. [© SIU BioMed/Custom Medical Stock Photo]

shape of the chromosomes and correlate any abnormalities with phenotypic disorders.

See also: Chromosome.

KNOCKOUT MUTANT

A knockout mutant is a genetically engineered organism (most often, but not exclusively, a laboratory mouse) in which a specific gene has been removed or otherwise inactivated and thus "knocked out." Typically, the target gene for a knockout is one whose role in a disease is known or suspected but whose normal function is unknown. The idea is to compare the functions of the knockout mutant with a normal organism (i.e., one that is genetically identical to the knockout in all other respects). Any abnormalities observed in the knockout may be assigned to the loss of the target gene, and the normal functions can be deduced from the information.

Transgenic and knockout mutant organisms are important for understanding gene function as well as for developing therapies for genetic diseases. The cloned gene of interest is chosen from a genetic library. Through standard molecular biology techniques, its sequence is disrupted with another gene that contains selection markers that make it easy to select cells where the gene of interest has been knocked out. The knockout cells are then introduced into mouse embryonic stem cells so that they develop into a mouse that lacks the gene of interest. A number of such organisms have been created, allowing scientists to study the function of several interesting genes.

See also: Genetic Disease/Disorder, Genetic Engineering, Genotype, Mutation, Phenotype, Transgenic.

LIBRARY

In molecular biology, a library describes a collection of DNA sequences or fragments that have been inserted into a particular cloning vector. A "genomic" library, for example, is a library made by chopping up the genome of an organism and inserting the pieces into the cloning vector. The relationship between the clones can be established by physical mapping, allowing the organism's genomic sequence to be determined. Another kind of library is a "cDNA" library, in which the cloned fragments are cDNA copies of a genome or segment of genome rather than the genomic DNA itself. A cDNA library has useful applications such as the determination of the function or products of unknown genes. There are also "gene libraries" for specific organisms or strains of organisms. Such a library represents a collection of gene sequences of a particular strain, including some of the unique genomic sequences of the strain, that have been cloned in a host cell. All publicly available DNA sequences are documented by and stored in NCBI's GenBank. In December 2004, the data bank contained approximately 44,576,000,000 bases in 40,600,000 sequence records.

See also: cDNA, Clone, Vector.

LINKAGE

Linkage is the tendency of certain genes or alleles to be inherited together. Generally speaking, two genes tend to be linked if their loci are close together on a chromosome. However, there are exceptions to this. Sometimes, there can be a recombination "hotspot" between two adjacent genes, which may result in two genes that are relatively far apart on the chromosome being linked. Even two genes on two different chromosomes may be linked by such recombination events!

LINKAGE DISEQUILIBRIUM

The technical definition of linkage disequilibrium is the nonrandom association of two alleles in a population. In other words, a specific allele of a gene appears in association with an allele in a nearby locus at a higher frequency than would be expected purely by chance. As an example, suppose there are two loci on a chromosome, one bearing the gene A and the other bearing the gene B. Gene A has only one variant, but gene B occurs in two forms, B and b (i.e., B has two alleles). One person can have gene A and allele B (AB), and another person can have gene A and allele b (Ab). Normally we would expect to see the two combinations at a more or less equal frequency in the population (i.e., there is a 50 percent chance that any person is either AB or Ab). If, however, we find that the combination AB is found in 90 percent of people, we may conclude that gene A is somehow linked with allele B and that they are in a state of linkage disequilibrium.

What is the reason for linkage disequilibrium? There are a number of potential causes, including recombination, genetic drift, and a high mutation rate. For example, a high rate of recombination between loci A and B decreases their linkage disequilibrium because random association increases. The property is very useful in genetic studies of population diversity.

See also: Allele, Linkage.

LINKAGE MAP

A linkage map is a map of a chromosome that shows the relative locations of various genetic features, such as specific genes for known traits or distinctive DNA sequences (e.g., trinucleotide repeats), on the basis of how frequently these features are linked (i.e., inherited together). Although a linkage map may often be close to the physical map of a chromosome, it is not identical because of phenomena such as recombination.

See also: Linkage, Mapping.

LOCUS

In genetics, locus (plural: loci) is the term given to a specific location on a chromosome. When we speak of a gene at a particular locus, it is as if we were marking its place by chapter and page number (even line number in the page) in a long book. It is a useful concept in medical and population genetics as an address locator of genes for a given trait or disease. It should be noted, however, that the idea of a genetic locus is only meaningful when we are dealing with individuals within the same species because only these organisms have a comparable genetic makeup. That is, they all have the same number of chromosomes with specific genes at roughly the same loci. It would be pointless to compare the genes of two unrelated species (e.g., an ape and a human being) in terms of loci because of the differences in both the number and organization of their chromosomes.

See also: Chromosome, Mapping, Synteny.

MAPPING

In genetics, as in geography, mapping refers to the pictorial or schematic representation of the subject matter—in this case genes or chromosomes rather than land—in terms of their positions relative to one another. A genome map describes the order of genes or other markers and the spacing between them on each chromosome. There are different types of chromosome or genome maps constructed at different levels of resolution.

At the coarsest resolution are *genetic linkage maps*, which assign relative chromosomal locations to genetic landmarks—which could either be genes for known traits or distinctive sequences of DNA—on the basis of how frequently they are inherited together (i.e., linked). Genes and markers that are physically close to one another on the chromosome are said to be tightly linked because they are less likely to be separated by recombination than are gene markers that are located far apart. However, their physical proximity cannot be taken for granted because of the possible occurrence of recombination hotspots. Recombination frequency provides an estimate of how close two markers are to each other on a chromosome.

A *physical map* is more akin to a geographical map in that it shows the actual physical relationship between various genetic landmarks—again, genes or distinctive sequence tracts—in a piece of DNA. Physical maps can vary greatly in resolution—for instance, the lowest-resolution map of the human genome shows the banding pattern on each of the different chromosomes, whereas the highest-resolution map depicts their complete nucleotide sequence.

On the genetic map, distances between markers are measured in terms of centimorgans (cM), named after the American geneticist Thomas Hunt Morgan. Two markers are said to be 1 cM apart if they are separated by recombination 1 percent of the time. A genetic distance of 1 cM is roughly equal to a physical distance of 1 million base pairs (1 Mb).

Genetic linkage maps are used to assign a gene to a relatively small area on a chromosome. The next step is to learn the gene's precise location using physical maps. To construct a physical map, a chromosome is first broken into smaller pieces of DNA. The pieces are copied (or cloned) in the laboratory to obtain millions of identical copies of specific DNA segments. The clones are lined up to reflect the order that existed on the original chromosome. Information about the location and known genetic content of these unique and ordered DNA fragments (called contigs) is stored in a computer, and clones of the ordered pieces themselves are stored in laboratory freezers. When genetic linkage maps indicate that a gene lies in a particular region, scientists can sequence the stored clone to identify the order of every chemical base in the gene.

The value of the genetic map is that an inherited disease can be located on the map by following the inheritance of a DNA marker present in affected individuals (but absent in unaffected individuals), even though the molecular basis of the disease may not yet be understood nor the responsible gene identified. Genetic maps have been used to find the exact chromosomal location of several important disease genes, including those for cystic fibrosis, sickle-cell disease, Tay–Sachs disease, fragile X syndrome, and myotonic dystrophy.

See also: Library, Linkage, Linkage Map.

MARKER

A marker is a gene or other identifiable portion of DNA whose pattern of inheritance can be followed through generations. For example, the blood groups A, B, AB, and O are markers because their inheritance can be followed from one generation to the next. There are many markers of various types (single-site mutations, repeated sequences, etc.) for various observable human phenotypes. Markers may be used for genetic mapping in a manner similar to milestones along a road.

See also: Mapping.

MEIOSIS

Meiosis is a special type of cell division that gives rise to gametes or sex cells (sperm and ova in humans) in organisms that reproduce sexually. The main feature that distinguishes this type of cell division from regular cell division (mitosis) is the fact that the progeny cells arising from this event are haploid. This means that they consist of only one set of chromosomes—22 autosomes and one sex chromosome. Meiosis is actually a two-stage cell division with chromosome reduction occurring during the first division. The separation of different chromosome pairs during this division is completely random and is independent of where they originally came from (i.e., the mother or father). Meiosis thus serves as a way to introduce variations—at least in terms of possible gene combinations—into the next generation. The cycle of meiosis is completed in a sense when the gamete participates in a fertilization to produce a diploid zygote.

See also: Cell Cycle, Diploid, Fertilization, Haploid, Independent Assortment, Mitosis, Zygote.

MITOCHONDRIA

Biology textbooks from grade school onward describe the mitochondrion (pl.: mitochondria) as the "powerhouses" or "energy generators" of the cell. Present in the cytoplasm of all eukaryotic cells, the mitochondria are the organelles where the cell produces the energy it needs to perform its various activities. In other words, they are the sites of cellular respiration. A cell may contain anywhere from 2 to 100 mitochondria.

Ranging in size from 0.3–1.0 μm by 5–10 μm, mitochondria may be seen only under an electron microscope, where they appear in various shapes from small rod-

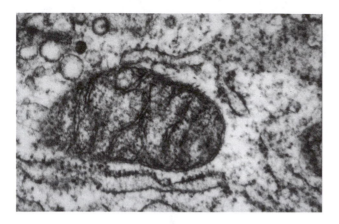

Mitochondrion (under transmission electron microscope). [© Roseman/Custom Medical Stock Photo]

like packages to large branched structures. A mitochondrion consists of four parts: outer and inner membranes, an intermembrane space, and the internal matrix. The outer membrane is studded with transport proteins, which help large molecules enter the mitochondrion. The inner membrane is folded into many projections, called cristae, which form the actual site for respiration. The fluid matrix within the sac contains various proteins and enzymes that catalyze the various respiratory and other reactions.

From the perspective of this book, mitochondria are interesting because of their genetic autonomy from the rest of the cell. They contain their own DNA molecules, which function and replicate independently of the nucleus. This mitochondrial DNA (mtDNA) is present in the form of circular chromosomes (plasmids), 16.5 kb in length, that number between 5 and 10 per organelle. The chromosome contains the genes for various protein enzymes used by the mitochondria as well as the genes for ribosomal components and tRNA molecules.

It should be noted that the rate of mutation of mitochondrial DNA is about 20 times more than that of the cytoplasm because of its location (i.e., in an environment containing many unstable high-energy molecules) and relative lack of protection by the DNA repair enzymes that are present in the nucleus. Diseases associated with mutations or defects of the mitochondrial genes are known as mitochondrial diseases. One of the best-known examples is Leber's hereditary optic neuropathy.

Although they are to be found in all eukaryotes, the mitochondria (along with the chloroplasts, which are the site of photosynthesis in plant cells) seem to resemble prokaryotes more closely with respect to their genetic organization. In addition to being independent of the nucleus, these organelles have haploid chromosomes, and their ribosomes are like prokaryotic (70S) rather than eukaryotic (80S) ribosomes. Indeed, some evolutionary theorists believe that these organelles were once free-living prokaryotic organisms (bacteria) that became incorporated into eukaryotic cells quite early in evolution and now exist in a symbiotic relationship with the latter.

See also: Cell, Eukaryote, Prokaryote.

MITOSIS

Mitosis is the process of normal cell division whereby a single cell duplicates its contents and then splits to form two nearly identical progeny cells. The main site of mitosis is the cell's nucleus, where the entire chromosomal contents must be duplicated so that each progeny cell gets all the genetic information stored in the parent cell. Mitosis actually occurs in a series of five distinct steps: prophase, prometaphase, metaphase, anaphase, and telophase. It is an essential function in adult multicellular organisms and maintains the structure and functions of various organs and tissues during the lifetime of the organism. It is also the mechanism for asexual reproduction by single-celled organisms.

See also: Cell Cycle.

MUTAGEN

A mutagen is any agent, such as a chemical, ultraviolet (UV) or other radiation, or even a virus, that causes a change in genetic material (i.e., a mutation). This change can either be corrected when the cell divides or it can be passed on to the next generation. The mutation may or may not be harmful. A *carcinogen* is a mutagen that specifically induces mutations that lead to the development of cancer (e.g., coal tar and nicotine are among some well-established chemical carcinogens, and UV rays have been associated with melanomas or skin cancer). A *teratogen* is a mutagen that causes mutations that are particularly harmful to a fetus, or unborn child. The rubella (German measles) virus, for instance, is a notorious teratogen. Pregnant mothers exposed to this virus, especially during the first trimester of pregnancy, give birth to children with numerous serious birth defects, including mental retardation and blindness.

See also: Mutation, Reversion.

MUTATION

A mutation is any change in the base sequence of a DNA molecule. It may be regarded as a failure of a cell or organism to store its genetic information faithfully. Mutations may or may not have any perceptible effect (i.e., change in the phenotype of the organism). Whether or not a mutation causes a visible change depends on many factors, including, in some instances, environmental conditions.

Mutations vary greatly in form and scale from a single base (called point mutations) to the level of whole chromosomes, and they may be caused by many different factors acting alone or in concert. At the molecular level, a mutation can take the form of the substitution of one or more bases for others or the physical insertion or deletion of nucleotides in a DNA strand.

Point mutations, or single-base substitutions, are most often random events and usually caused by mistakes made by DNA polymerase during the process of replication when a nucleotide is mispaired with one that is not its complement. Because this is a random event, the change may be either in a gene, a regulatory sequence, or in other noncoding regions. Depending on the location of this change, the mutation may appear silent (i.e., have no effect on the phenotype) or manifest as a loss of gene function or a defective function.

Silent mutations within a gene are possible because of the redundancy of the genetic code. Thus, for example, if the third base of the triplet GTT were replaced by any of the three other nucleotides, it would still code for the amino acid valine. Consequently, there would be no change in the structure of the protein (i.e., the phenotype remains unchanged). The change of the codon TAC (for tyrosine) to TAA, on the other hand, would cause a very drastic effect because the latter is a stop codon. Such a mutation is called a *nonsense* mutation, and its translation usually results in the synthesis of a truncated protein product that is usually functionless.

A mutant allele that produces such a nonfunctional protein is called a *null allele*. Null alleles may also be produced in the case of *missense mutations*, which occur when the base substitution results in the change in the amino acid specified by the mutant triplet (e.g., TTA to TTC would result in the substitution of a phenylalanine with leucine). Often, however, missense mutations produce a protein with at least some of the activity or with altered activity. An example is the replacement of a specific histidine molecule (in position 28 of the protein) with glycine in alpha-globin, which is a part of the hemoglobin molecule. The result of this mutation is a drastic change in the ability of hemoglobin to bind oxygen, which in turn leads to anemia. Typically, a point mutation that results in a change in an amino acid that is critical in binding a substrate within an enzyme's active site may well shut down the enzyme's activity. But if the mutation is in a noncritical part of the protein and results in a minor change in the local structure of the protein, it may not have a drastic effect on the activity.

Point mutations in noncoding regions of the genome may lead to changes in restriction sites (also seen within genes) as well as other special sequences such as promoters and intron–exon junctions. Changes in the latter would interfere with transcription and translation and thus produce different types of diseases.

Mutations in a gene that involve the insertion or deletion of portions of the DNA are seldom silent. The most common outcome of a small-scale event (e.g., involving one or two nucleotides) is known as a *frameshift mutation* because it disrupts the location of the rest of the gene within the open reading frame. Consider, for example, the following stretch of sequence within a (hypothetical) gene:

****ATTGGCCCAAATATAGGGCATCCC**** (asterisks indicate the bases flanking the stretch of sequence used for this example)

Assuming it to be in frame, its codons and corresponding amino acids would read as follows:

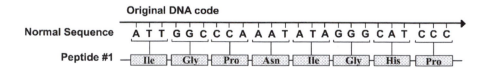

Now suppose that a mutation caused the deletion of the tenth or eleventh A. The sequence of codons and hence amino acids would be altered as follows:

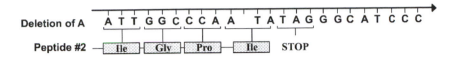

Because TAG is a stop codon, the translation of the gene with this deletion would result in a truncated protein product!

The insertion of an A (labeled as Ai) at the same position would also lead to a frameshift, albeit with a different outcome. Instead of the premature truncation that occurred with a deletion, the insertion of an A leads to an altered peptide sequence beyond the proline residue at the third position in this sequence.

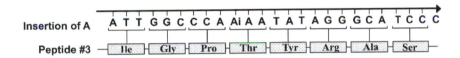

Large-scale insertions or deletions in DNA may lead to the expansion or removal of entire sequences, which may manifest itself in any of the types of outcomes described earlier with reference to substitutions. The insertion of large tracts of repeat sequences, for instance, is often associated with genetic disorders such as Huntington's disease, and Duchenne muscular dystrophy is often the result of a large deletion in the X chromosome. A large majority of alpha-thalassemia cases are also caused by large-scale deletions in the alpha-globin genes on chromosome 16.

The various mechanisms involved in generating deletions and insertions are not yet completely understood. Excisions of sequences might also be caused by high-energy radiation. One well-documented source of insertion and deletion is the unequal crossover of sequences during recombination events that take place during cell division (mitosis or meiosis). This inequality frequently occurs when there is a misalignment of large repeated sequences. Still another mechanism for causing insertions and deletions are the transposable elements, which can cause entire genes to "jump" from one location to another. Both unequal crossover and transposition pair the events of deletion and insertion (i.e., one locus loses bases while another locus gains them).

Any of the mutations described may be broadly classified as spontaneous and induced mutations depending on their source. *Spontaneous mutations* arise randomly in nature, usually due to simple errors made by the DNA polymerase enzymes during DNA replication cycles. Any changes in the environment of the cell or organism that favor increased chemical reactivity create conditions conducive to the occurrence of these mutations. In contrast with spontaneous events, the exposure of the DNA to a mutagen (e.g., a chemical, radiation, or virus) results in what is known as an *induced mutation*. Depending on the nature of the mutagen, induced mutations may appear at random locations in a DNA molecule (e.g., chemicals or radiation) or may target specific sequences (transposons).

When dealing with eukaryotes, mutations may also be classified according to whether they occurred in the somatic or germ cells of the organism. A *somatic mutation* (e.g., in a skin cell exposed to ultraviolet light) typically affects the cells in the immediate vicinity of the mutagen exposure and, unless gametes are involved, are not passed on to the next generation of the organism. But if a gamete (sperm or egg cell) is exposed to a mutagen and then participates in fertilization, then any mutations introduced into its genes (i.e., the *germ-line mutations*) will be seen in the offspring and may also be passed along in successive generations. In diploid organisms, the visibility of these mutations also depends on the nature of the mutation (dominant or recessive) as well as its location (autosomal or X-linked).

Finally, we must ask the question of why mutations exist in nature. At first glance, they may seem like completely undesirable events that are responsible for a variety of diseases in different organisms. However, mutations are of fundamental importance in the living world because they are the ultimate source of variation as well as speciation. In other words, mutations may be viewed as the basis for evolution. Without the ability to mutate, organisms would lose their ability to adapt to sudden changes in the environment. Of course, not all mutations are favorable, but the most harmful types of mutations do not survive for many generations (i.e., they are removed by the process of negative selection). Positive selection ensures the propagation of favorable mutations through the generations.

See also: Gene, Genetic Code, Genetic Disease/Disorder, Genotype, Phenotype, Reading Frame, Translocation.

NUCLEUS

The nucleus is the membrane-bound compartment in eukaryotic cells that gives these organisms their name (*karyon* means nut or nucleus in Greek). The nucleus houses the bulk of the cellular DNA packaged in the form of chromosomes. It also contains an organelle called the *nucleolus*, which is the site of ribosome biosynthesis. The nucleolus is usually found in association with the nucleolar organizer region (NOR), which is the region of the chromosomes containing the ribosomal RNA (rRNA) genes.

The nuclei of most cells may be distinguished from the rest of the cell (the cytoplasm) by staining with specific dyes. Every cell of a eukaryotic organism possesses a nucleus—the red blood corpuscles (RBCs) being the notable exception. Most cells in most organisms contain just a single nucleus, although there are some exceptions. The cardiac muscle cells of the human heart, for instance, may be considered to have many nuclei, as the tissue is more or less continuous, without defined membranes around individual cells. Fungal mycelia are another example of cells with multiple nuclei. Certain types of viruses (e.g., the respiratory syncytial virus) induce the fusion of their host cells so that they form multinucleate syncytia.

The number of chromosomes in a nucleus is characteristic for the organism and is known as its karyotype. Depending on the stage of an organism's life cycle, the nucleus may be haploid or diploid (i.e., contain a single set or a pair of chromosomes, respectively).

The nucleus functions as the storage compartment for the information an organism needs for living, which it passes down to the next generation. Genes are transcribed within the nucleus, but the RNA transcripts are quickly transported into the cytoplasm so that protein synthesis takes place outside the nucleus. During cell division, the nucleus is the first part of the cell to show perceptible changes, first by duplicating its DNA and then by splitting. The fusing of two nuclei is the key event in the fertilization of two haploid cells to form a diploid zygote.

See also: Cell, Cell Cycle, Chromosome, Fertilization, Zygote.

OLIGONUCLEOTIDE

An oligonucleotide is a short segment of DNA or RNA ranging from a few to a few hundred nucleotides in length. Oligonucleotides may be generated within a cell when a long piece of DNA or RNA is degraded into smaller fragments by enzymes or even mechanically. However, oligonucleotides of specific sequences—to be used as primers in sequencing or PCR reactions, for instance— are synthesized in automated machines.

See also: DNA Sequencing, Polymerase Chain Reaction (PCR), Primer.

ONCOGENE

An oncogene is a gene whose activity leads to the uncontrolled proliferation (i.e., growth and division) of a eukaryotic cell. In other words, it is a gene capable of inducing tumors. Why does a cell contain such a gene in spite of its obvious harmful effects? This is because the oncogene itself is actually an altered form of a normal gene called a proto-oncogene. Generally, a proto-oncogene encodes a protein involved in signal transduction and so plays a role in the cell cycle. The expression and activity of these genes are under very tight control in a normal cell. But when the oncogenes are mutated or improperly regulated, they disrupt the normal functioning of the cell and cause it to lose such properties as specialization and contact inhibition, which leads to unchecked growth and eventually the formation of tumors.

See also: Signal Transduction.

ORGANISM

In its broadest sense, an organism may be described as an entity containing a network of biochemical processes that function in a concerted fashion to maintain life in that entity. And how is "life" defined? Although there is no single answer, for the purposes of this book, we may think of life in terms of two abilities: first, a capacity to sustain oneself through the process of converting various forms of energy; and second, to create another entity similar to itself that is capable of doing the same.

An organism may contain all of these abilities in a single cell or may possess many cells that function together to perform the necessary functions. All living beings are organisms.

See also: Cell, Deoxyribonucleic Acid (DNA), Protein.

PATENT

Most of us are probably aware of what a patent is in the context of an invention. According to the United States Patent and Trademark Office (USPTO), a patent is the official grant of property rights to the inventor. What USPTO grants is not the right to make, use, offer for sale, sell, or import but the right to exclude others from making, using, offering for sale, selling, or importing the invention. With such a definition, it may come as a surprise to see mention of patents in a book about the human genome. And yet, because of the quantities of data that have emerged from the biotechnology industry and genomics in particular, there has been an overwhelming push to protect newly uncovered genetic information with patents. For example, biotechnology companies move to patent any genes that have been discovered to play a role in particular diseases so that they can have exclusive rights over diagnostics and therapies that involve such genes.

Not surprisingly, because of the nature of the information, this idea of patenting genes or genetic information has generated a great deal of controversy. Those opposed to the idea of patenting genes take the position that because genes are already in existence no one can really "invent" a gene, and so genetic information cannot be protected by patents. Those who support the idea of patents do so on the grounds that the "discovery" of a gene is tantamount to its invention. They also counter the argument that a gene cannot be patented with their position that the patented object is actually not the gene but a cDNA copy of the gene, which because it is constructed in the laboratory is technically an invention!

Although affording protection of intellectual property rights has the advantage of encouraging innovation in research, there is substantial concern about the potential restriction on research and development and the consequent impact of patents on public health. For example, the company Myriad Genetics was granted a European patent on the BRCA1 gene sequence, which allows it to prevent any external institution from testing for BRCA1 gene mutations. One of the arguments against allowing this is the allegation that Myriad's test does not cover all the potential mutations in the gene and therefore may fail to diagnose cases correctly.

The controversy is further exacerbated by the fact that many companies as well as academic researchers move to patent genes for which no functional information is yet available, thereby restricting opportunities for doing research on those genes. To an extent, the USPTO has been lenient with the demonstration of a gene's uses, given that much of the research still remains to be done. The broad nature of these patents also affords less than optimal protection but is better than no protection at all, as far as the patent holders are concerned. The very idea of patenting a gene, however, remains unpalatable to many, whose opinion is that genes are representative of life and nobody should hold exclusive rights over them or the information contained therein.

See also: Gene.

PEDIGREE

In human genetics, a pedigree (derived from the French expression "pied de grue" for crane's foot) is a diagrammatic representation of a family history of the transmission of various traits or diseases over generations. Pedigrees can be quite useful to geneticists as a tool in determining the nature of an inherited disease (e.g., whether it is dominant or recessive or whether the disease gene is autosomal, X-linked, or mitochondrial). In addition, it may be a tool for predicting possible inheritance in successive generations of a family.

See also: Inheritance.

PENETRANCE

Penetrance is a term used by population geneticists as a way of measuring the relative effect of a certain mutation on the phenotype of the individual. In other words, it is a measure of the effectiveness of a mutant gene in manifesting itself. In mathematical terms, it is the ratio of the number of people afflicted with a given genetic disease to the number of people who actually possess a copy of the disease-causing mutation or allele.

A disease-causing mutation is said to demonstrate *incomplete penetrance* when an individual with the gene has a less than 100 percent chance of developing the disease. An example of such a gene is BRCA1, whose carriers have only a 50–80 percent chance of developing breast cancer. In addition, many autosomal dominant conditions may also show less than full penetrance. *Nonpenetrance* is the complete lack of the mutant phenotype despite the presence of a mutant gene (allele) in the individual's genotype.

There are many possible factors influencing penetrance. The penetrance of some genes is age- or sex-dependent. Sometimes, mutations in other genes are required for the gene of interest to show penetrance. The environment can also play a role in incomplete penetrance by acting as a trigger for disease development in some cases and suppressing it in others.

See also: Genotype, Phenotype.

PHARMACOGENOMICS/GENETICS

As many of us know from personal experience, several pharmaceutical drugs currently on the market are effective for some people but can cause side effects, some serious, in others. There are also certain people in whom the drug may have no effect at all! Pharmacogenomics, a product of the fusion of the terms *pharma*cology and *genomics*, aims to tailor drugs to the genetic makeup of patients so they will have close to 100 percent response with no side effects. Close to 100,000 people every year in the United States alone die due to medical errors, and a substantial percentage of these deaths are caused by adverse drug reactions. The ultimate potential for pharmacogenomics is to develop therapies or even preventive approaches based on the genetic risk factors of people. More immediately, pharmacogenomics can be used to study the response of carefully selected groups of patients to drugs that are already on the market. These anticipated benefits are expected to play a role in eventually reducing pharmaceutical research and development costs and in improving health care.

See also: Bioinformatics, Genomics, Proteomics.

PHENOTYPE

Phenotype refers to the observable result of the expression of a gene in an individual's genome. Any given phenotype is the outcome of an interaction between the expression of a gene and its environment. A possible analogy to consider is a child's bedtime story. Although the text is the same, the story will take on many different forms, depending on who is reading it aloud and who is listening. Even the same person is unlikely to repeat his or her performance exactly the second time—the inflections may change for instance or he or she might skip over certain parts. Another, more direct example to consider is the case of identical twins. They are born with the same genotype, but although they resemble each other greatly, we can usually tell them apart because they also have many divergent traits. In the language of genetics, they exhibit different phenotypes despite having the same genotype.

The converse of this situation—cases where the genotypes for two outwardly similar organisms (i.e., with the same phenotype) are different—occurs frequently, especially when the trait has dominant (A) and recessive (a) genes. Only those individuals who have an aa genotype (i.e., are homozygous for the recessive gene) will display the phenotype for the recessive trait, whereas individuals with all other possible genotypes will have the dominant phenotype.

The distinction between genotype and phenotype is an important one in genetics because it demonstrates the difference between what can be and what is, or between potential and actuality. A good illustration of the genotype–phenotype difference is the expression of adaptive enzymes produced by bacteria for fermentation of various sugars. These enzymes are defined by the ability of their genes to be turned on or off according to an organism's need for energy. By default, most bacteria utilize glucose as their carbon (and hence energy) source, as it is abundant in the biosphere. But, in the absence of glucose, they will turn to other sources, such as lactose (the sugar in milk) or maltose, provided their genome possesses the genes for the appropriate enzymes. Only an organism deprived of glucose and grown in a lactose-containing medium would exhibit the lactose-fermenting phenotype. A clone that has been grown in glucose-rich conditions will not produce the lactose-degrading enzymes. Diploid organisms such as humans have an even greater gap between their genotype and phenotype because of the phenomenon of dominant and recessive traits. Thus, the gene for a recessive trait may be part of the genotype of several generations and never manifest into a phenotype.

A final point worth noting is that genotype and phenotype bear an unequal relationship to one another. Although an organism never expresses all of its genes at once (and thus does not exhibit the phenotype for all of its genes), any given phenotype can only be expressed if the appropriate gene is present in the genotype of the organism.

See also: Gene, Gene Expression, Genome, Genotype.

PLASMID

Plasmid is the term given to small pieces of extra chromosomal DNA that replicate autonomously (i.e., independently of the cell's chromosome[s]). They may be found either in prokaryotic or eukaryotic organisms. Most often circular in shape (some linear plasmids have been isolated from certain bacteria), plasmids carry only a few genes, which typically encode for nonessential functions in the organism. An important exception is found in the mitochondria of eukaryotes. Although it is about 1,000 times smaller than the smallest human autosome, the mitochondrial plasmid encodes some respiratory chain proteins that are vital for the proper functioning of the cell. The best-known example of plasmid function in the prokaryotes is as a carrier of genes for traits such as antibiotic resistance. Because of their small size, plasmids may be manipulated more easily than large chromosomes. They have been widely used in molecular biology (e.g., in creating clones of genes) and in the expression of targeted genes into proteins.

See also: Chromosome, Deoxyribonucleic Acid (DNA), Gene.

PLEIOTROPY

Earlier we described the concept of a complex trait as a single phenotype determined by multiple genes. The phenomenon of pleiotropy is the converse; here a single gene is responsible for multiple phenotypes or traits. Quite often, this is just the result of the gene being expressed in different organs or tissues. A good example is the cytokine or growth factor IL-6 that acts on many different types of cells in the body. Depending on where it is produced, IL-6 can induce the differentiation of B cells to become antibody-producing cells, the differentiation of myeloid leukemic cell lines into macrophages, the development of osteoclasts, and acute-phase protein synthesis in hepatocytes. In addition, IL-6 inhibits the growth of myeloid leukemic cell lines and certain carcinoma cell lines.

Mutations, like genes, can also exhibit pleiotropy (i.e., cause many simultaneous seemingly unrelated effects). Marfan syndrome, for example, is due to mutations in the gene for fibrillin. However, the symptoms are quite variable and include various skeletal defects, problems with the eyes, and cardiovascular abnormalities. No patient exhibits all of the same abnormalities or to the same degree.

See also: Gene, Gene Expression; Marfan Syndrome (Ch. 5).

POLYMERASE

The polymerases are members of a class of enzymes that catalyze the synthesis of nucleic acids—DNA and RNA—from their nucleotide building blocks. In addition to the function of chain elongation (by adding nucleotides to the free 3' end of a growing chain), these enzymes also catalyze functions such as DNA and RNA repair.

Most polymerases use DNA as templates for duplication (i.e., they are DNA-dependent). The first polymerase to be isolated—in Arthur Kornberg's laboratory at Stanford University—was DNA polymerase I (from *E. coli*). Reverse transcriptase, which is RNA-dependent, displays the unique ability to make DNA from RNA templates. The accurate reproduction of all living matter depends on the proper functioning of the polymerase enzymes.

See also: Enzyme, Reverse Transcriptase, Transcription; DNA Synthesis/Replication and Repair (Ch. 4).

POLYMERASE CHAIN REACTION (PCR)

The polymerase chain reaction is a laboratory technique to amplify a specific DNA sequence millions of times within a few hours by mimicking enzyme-mediated reactions used by living, dividing cells.

DNA duplication (see entry on DNA synthesis/replication and repair in Chapter 4) in a cell requires the presence of various enzymes, as well as other raw materials (nucleotides and primers), in order to begin and proceed properly. A very important requirement, which has been exploited for the purposes of PCR, is the primer. This is a short and specific sequence of nucleotides that must be able to base pair with a portion of the template so that the DNA polymerase can build the rest of the new DNA strand in a stepwise fashion. The sequence of a primer determines the initiation point of DNA synthesis.

A PCR reaction mixture contains all of the necessary components for DNA duplication: the template DNA, the four nucleotides, primers, and DNA polymerase. (The first and most widely used polymerase, called Taq, was isolated from the high-temperature-dwelling bacterium *Thermus aquaticus*). The primers are chosen such that their complementary sequences flank the DNA segment of interest (i.e., the segment to be amplified) on either side. So when the reaction begins, DNA synthesis is initiated off both strands and in both directions!

PCR is a cyclical technique, with each cycle consisting of three steps. The first is the separation of the two DNA chains in the double helix, which is achieved by heating. The vial is then cooled so that the primers can bind or "anneal" to their complements on the DNA strands. In the final step of the reaction, the temperature is raised again so that Taq polymerase can add nucleotides systematically to the primer and eventually make a strand of DNA that is complementary to the original template. This completes one PCR cycle, at the end of which each original DNA piece has been duplicated. When this cycle is repeated, there are now four templates for synthesis rather than two, and this number increases exponentially with each cycle. The advantage of using the enzyme Taq polymerase is that the cycle can be repeated several times without inactivating the enzyme. At the end of 30 cycles there are one million copies of the original template! These copies of DNA may be used for a variety of purposes, including sequencing, cloning into vectors, etc.

Invented in 1983, PCR very quickly revolutionized molecular biology with its far-reaching applications; for example, in forensics. In 1993, the inventor of PCR, Kary Mullis, received the Nobel Prize in Chemistry for developing the technique.

See also: Cell Cycle, Polymerase; DNA Synthesis/Replication and Repair (Ch. 4).

POLYMORPHISM

In simplest terms, a genetic polymorphism (derived from the Greek words *poly* for "many" and *morph* for "form") is a difference or variation in DNA sequence at a particular locus. For example (although this is an oversimplification), the "blue" and "brown" alleles for eye color are actually caused by polymorphisms in a single gene. Another example are the blood groups, which were the first polymorphisms to be scientifically documented.

By the broad definition just given, even a single base difference between two alleles of a gene constitutes a polymorphism. For the sake of convenience, however, geneticists have defined polymorphism in more stringent terms. Variants of a gene are not considered to be polymorphisms of that gene unless the rarest variant of the group is present in more than 2 percent of the population containing the gene.

Polymorphisms are not restricted to gene coding sequences but rather are distributed randomly throughout any given sequence. Indeed, a comparison of a noncoding autosomal region from 10 different samples will reveal, on average, 1 polymorphism in every 200 to 400 base pairs. Polymorphisms in genes are perhaps easier to detect in terms of phenotypic differences between two individuals, but random polymorphisms have found broader applications—for example, in identification techniques such as RFLPs and DNA fingerprinting.

See also: Allele, Fingerprinting, Mutation, Restriction Fragment Length Polymorphism (RFLP).

POPULATION GENETICS

The branch of genetics that studies genes at the level of entire populations rather than individuals is called population genetics. A population in this case is defined as a group of individuals of the same species (e.g., humans or wolves) who do or can interbreed among each other. Rather than looking at the inheritance of a single gene or trait, or the effect of mutation on the phenotype of one person, population geneticists study things such as the distribution of genes in a population and the various factors that influence the frequency of genes (or their alleles) and genotypes over successive generations in that population. Because changes in gene frequencies are at the heart of evolution and speciation, population and evolutionary genetics are often studied together.

See also: Evolution, Gene Frequency.

PREDISPOSITION

A person is said to have a predisposition for a condition when he or she has the necessary genetic makeup, namely a specific combination of alleles or mutations, that makes it possible for that condition to develop. A genetic predisposition does not, however, automatically mean that the individual has the condition, only that there is a higher risk of developing it during the person's lifetime. Typically, certain external triggers or specific environmental conditions are required to interact with the genes in order for the condition to manifest. The specific genes, alleles, or mutations are thus necessary but not sufficient causes of the disease. A person with *BRCA 1* or *BRCA 2*, for instance, has a predisposition for breast cancer. Other conditions associated with predisposition include complex genetic diseases such as hypertension and diabetes.

See also: Complex Trait, Environment.

PRIMER

A primer is a short string of nucleotides (i.e., an oligonucleotide) that is used to begin or "prime" a DNA synthesis reaction. It is an absolute requirement for DNA synthesis both in vivo and in vitro (e.g., a PCR reaction). Virtually any small piece of DNA can act as a primer, although the most efficient ones are those that are complementary to short sequences in the template strand.

See also: Oligonucleotide, Polymerase Chain Reaction (PCR); DNA Synthesis/Replication and Repair (Ch. 4).

PROBE

In the context of molecular biology and genetics, a probe is a molecule such as a piece of DNA, RNA (oligonucleotide), or protein that is used to identify or pick out target molecules in a mixture. Examples of target molecules include genes, gene families, and various proteins or other gene products. A successful probe must have two important properties. First, it should be marked or labeled in some way to allow for its tracking or detection in an assay. Radioactive and fluorescent dyes are the most widely used labeling agents. Second, a probe must have a structural relationship with the material one seeks to identify. DNA and RNA probes, for instance, will be complementary to their target sequences—in Southern and Northern blots for example—whereas protein probes are typically antibodies that have the ability to bind specifically to the target protein (antigen) being sought in a mixture, as in a Western blot.

See also: Fingerprinting, Oligonucleotide, Southern Blotting.

PROKARYOTE

In the simplest possible terms, prokaryotes are organisms that lack a nucleus in their cells. Prokaryotes are most often unicellular, and their DNA material lies free within the cell and is not sequestered within a membrane. These organisms also typically lack other membrane-bound organelles such as mitochondria. They do, however, possess ribosomes, albeit somewhat smaller than those of the eukaryotes. Because they have no nucleus, prokaryotes tend to divide by simple fission and do not undergo mitosis or meiosis. Examples include the bacteria, blue-green algae, and archaea.

See also: Eukaryote.

PROMOTER

The promoter is a site in a DNA molecule to which RNA polymerase can bind and begin transcription of a gene into mRNA. Promoters have specific sequences of nucleotides that are recognized by specific transcription factors, which are proteins that help RNA polymerase bind to the DNA and begin transcription. Different promoters vary in their ability to initiate transcription. Strong promoters are recognized by RNA polymerase more frequently and have sequences close to the optimal sequence of DNA. Weak promoters are not recognized as frequently due to their disparity from the consensus.

See also: Consensus Sequence, Transcription.

PROTEIN

If the nucleic acids are the carriers of the information of life, then proteins are the executive force that translates these instructions into action and makes life possible. If it were not for proteins, living things would have no way to access the information carried in the genes! Virtually all life processes other than heredity and a small (albeit essential) type of catalysis are carried out by proteins. They form the basis of many structures (e.g., the keratin in hair and nails), catalyze a variety of biochemical reactions, such as digestion (e.g., pepsin in the stomach helps to break down proteins, and insulin, produced in the pancreas, controls blood sugar levels). Proteins also transport important molecules from one part of the body to another (e.g., hemoglobin transports oxygen from the lungs, where the gas is absorbed, to different parts of the body). Proteins are also responsible for structural, catalytic, and transport functions at the cellular level. Muscle cells, for instance, are made up of proteins such as actin and myosin that enable the cells to transform energy. Virtually all cells contain proteins embedded in their membranes to allow for the selective passage of materials into and out of the cells. Fundamental cell-to-cell communication processes as well as the duplication of various cellular components are also mediated by proteins.

Because of their greater diversity of function, proteins need to be considerably more variable and complex in chemical structure than the nucleic acids. Proteins, like DNA and RNA, are polymers but are made up of combinations of 20 building blocks (instead of four). The building blocks of proteins are called the *amino acids*. Consecutive amino acids in a protein molecule are held together by a special link known as the peptide bond between a carboxyl (–COOH) group of one anino acid and an amino group ($-NH_3$) of the next.

The greater number of building blocks alone increases the possibilities for differences in protein sequence (and hence information) over that of nucleic acids by several orders of magnitude. A trinucleotide can have one of 64 possible sequences (4^3) but a trimer unit of a protein can be arranged in one of 8,000 (20^3) ways! But the amino acid sequence of a protein chain (a peptide), known as its primary structure, is only the first source of its variation. The 20 amino acids constitute a vastly more heterogeneous group of chemicals than the four nucleotides. They may be acidic, basic, or neutral, hydrophilic or hydrophobic, large or small, and can interact with one another and with their microenvironment in many possible ways. So the structure and properties of a protein depend not only on the sequence of amino acids in a peptide but also on the chemical nature of the constituent amino acids and how these amino acids might interact with each other *within* the protein chain. These interactions form the basis of yet another level of complexity in a protein molecule, namely the formation of different three-dimensional configurations (or shapes) known as the secondary and tertiary structures. These structures can, in turn, have a dramatic impact on the functioning of the molecule. Common secondary forms include the alpha helix and beta sheet. The catalytic properties of most enzymes reside in special sites on the molecule that are created by their secondary and tertiary structures.

The formation or synthesis of proteins is under the control of genes (see the entries on transcription and translation). This means that, as far as we know, proteins are not formed de novo (i.e., without first being specified by a gene). Thus far, scientists have not observed any examples in nature where a new protein is formed by the random combinations of amino acids with one another, although they have succeeded in the chemical synthesis of new peptides. In nature, a new protein is formed when a gene for an existing protein is mutated.

See also: Deoxyribonucleic Acid (DNA), Enzyme, Gene, Gene Expression, Genetic Code, Translation.

PROTEOMICS

Proteomics is a field of study that has arisen quite naturally from genomics, which is the study of living organisms at the level of the entire genome. Proteomics extends this branch of investigation to the large-scale study of the structures, functions, and interactions of all the proteins expressed by the genome. In other words, genomics is the study of the instructions and proteomics is the study of how they are carried out. Not surprisingly, then, proteomics is a much more complicated science than genomics. One reason for this is that proteins are much more complex than genes in structure and function. Furthermore, the genome of an organism is largely constant and unchanging during its entire life except for a few spontaneous mutations. On the other hand, the "proteome"—the full complement of proteins in an organism—is constantly in flux, changing from moment to moment as an organism moves, eats, or rests. A single organism can have totally different proteins in different parts of its body, at different times of the day and during different stages of development.

Scientists estimate that humans have about 200,000 proteins—much more than the estimated 25,000 genes that code for them. The proteome probably gets its diversity from alternative splicing of these genes as well as from post-translational modifications. Among the biggest challenges in the field of proteomics is the relative paucity of technologies and equipment for conducting large-scale experiments. An example of a proteomics technique that has been developed is the isotope-coded affinity tag (ICAT) developed at the University of Washington, Seattle, by Rudolph Aebersold and coworkers. ICAT enables a comparison of the proteomes of two samples (for example, one with disease and one normal). It uses isotopically labeled reagents and tandem mass spectrometry (MS) to identify, quantitate, and compare expressed proteins from the two samples. This makes it possible to detect abnormal quantities of proteins in the diseased sample. A very young science, proteomics is a field whose full potential is only just beginning to be realized.

See also: Genomics.

PSEUDOGENE

The genomes of many eukaryotic organisms consist of large amounts of DNA that are not encoded into expressed genes. Less than 5 percent of human chromosomal DNA, for instance, codes for various genes, whereas scientists have not yet discovered any discernible function for the remaining 95 percent or so found between and within various genes. Some of this DNA even seems to have the same structure as in other genes that are expressed but that do not code for any known protein (or RNA) products themselves. Geneticists think that these sequences—which they call *pseudogenes*—were once functional genes that lost their protein-coding abilities during the course of evolution due to the accumulation of mutations.

See also: Gene, Gene Expression, Genome, Mutation, Protein.

READING FRAME

Because the genetic code is in the form of a triplet code—with each amino acid encoded by a nucleotide triplet—there are three possible ways in which any DNA sequence may be read or translated into a peptide. For example, the sequence ****ATTGGCCCAAATATAGGGCATCCC**** (asterisks denote nucleotides flanking the sequence shown) may be read as any of the following three series of codons, which, as shown, code for three entirely different series of amino acids:

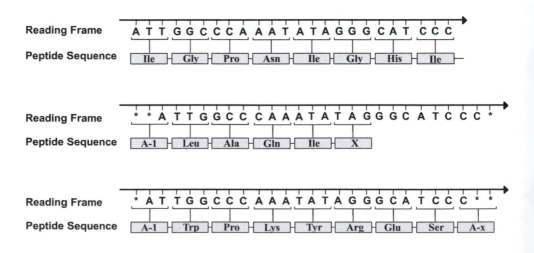

Each of these alternatives represents a different "reading frame."

The presence of more than one reading frame allows for different genes in a DNA molecule to overlap one another, thereby increasing the amount of information a single molecule can store. However, in order for a gene to be translated into a protein, it must be in what is known as an *open reading frame* (or ORF), with a start codon (almost always ATG) and a stop codon in synchrony with one another. This makes the analysis of the DNA strand and the identification of unknown genes very complex. Bioinformatics has given us sophisticated computer programs to analyze DNA sequences for open reading frames and putative proteins.

See also: Bioinformatics, Genetic Code.

RECEPTOR

Receptors are special molecules on the cell surface that act like sensitive antennae, receiving and relaying information about extracellular space to the inside of the cell so that the cell can react appropriately. Typically, the receptors are proteins or glycoproteins (i.e., proteins in association with sugar molecules) that span the lipid bilayer of the cell membrane and have exposed domains at either surface. The outer part of the receptor recognizes and becomes attached to chemical signals—such as ions, hormones, antigens, and cytokines—known as ligands. When the external portion of a receptor binds with its ligand, its internal domain also undergoes changes, and this sets off a chain of events that eventually tell the cell how to react to the stimulus. Different receptors are specific for different ligands and set off different (although often interconnected) pathways within the cell.

See also: Signal Transduction.

RECOMBINATION

Recombination refers to the formation of new combinations of linked genes from one generation to the next. It occurs as a result of the physical breakage and rejoining of DNA molecules (chromosomes) at different loci during meiosis, which is the special process of cell division where a diploid cell produces haploid progeny cells (i.e., gametes, which in humans are the sperm and eggs).

Recombination is one of the underlying reasons why all offspring of the same parents are not identical. During meiosis, chromosomes become paired and intertwined. The intertwining is called "crossing over." When this happens, different parts of the chromosomes often get exchanged (i.e., recombination occurs). When, in the final phase of meiosis, the chromosomes separate, they are no longer identical to the parent chromosomes because each chromosome has picked up new genetic material from its partner. This swapping occurs at every meiosis, so a person's gametes are constantly generating new combinations of his or her genetic material! Furthermore, because recombination almost never occurs in an identical manner in two cells, each meiotic division will yield somewhat different egg or sperm cells (i.e., these cells are not clones of one another). Subsequent fertilization between a sperm and an egg gives rise to more combinations, resulting in siblings that are quite different.

Consider the following hypothetical example of the possible progeny of two heterozygotic parents. Let us assume four different possible alleles of the gene—namely A1, a1, A2, and a2—two in each parent. Ordinary meiosis (without recombination) already gives rise to the possibility of four different combinations of diploid progeny. After crossing over and meiosis, the number of possible gametes in each parent increases to four, resulting in a possible 16 genotypes for the next generation!

Parent 1:

Original genotype: A1a1
Gametes after meiosis (without recombination): A1, a1
Gametes after meiosis with recombination: A1', a1'

Parent 2:

Original genotype: A2a2
Gametes after meiosis (no recombination): A2, a2
Gametes after meiosis with recombination: A2', a2'

Offspring (tabulated):

	Sperm			
	A1	a1	A1'	a1'
A2	A1A2	a1A2	A1'A2	a1'A2
a2	A1	a1a2	A1'a2	a1'a2
A2'	A2A2'	a1A2'	A1'A2'	a1'A2'
a2'	A1	a1a2'	A1'a2'	a1'a2'

(left margin label: Egg)

The type of recombination event described is the most common example of this phenomenon to occur in nature. However, recombination can also occur between bits of DNA from physically separate units or chromosomes. In some cases, this is due to special sequences called transposons (see the entry on transposable elements) that are capable of moving around the genome. A less frequent type of recombination occurs when broken pieces of one chromosome rejoin with a piece of another broken chromosome, giving rise to new or abnormal chromosomes. Such events are known as *translocations* and are often associated with diseases (see Chapter 3, on chromosomes). The Philadelphia chromosome, formed by the translocation of pieces of chromosomes 9 and 22, is an example of recombination associated with the development of certain types of blood cell cancers.

See also: Cell Cycle, Inheritance, Translocation; Cancer—Chronic Myelogenous Leukemia (Ch. 5).

RELATIVE RISK

Relative risk is a term used by population geneticists and epidemiologists to correlate possible causes or agents of a disease with its actual incidence. In mathematical terms, it is the ratio of the outcome in a group exposed to a suspected agent of the outcome to the risk in an unexposed group. For example, in a study to gauge the effects of smoking on lung cancer, the relative risk is calculated as [a/(a+b)]/[c/(c+d)], where a and b are the number of smokers with and without lung cancer, and c and d represent the number of nonsmokers in that population with and without the cancer. The higher the ratio, the more likely it is for the disease to be linked to the factor.

RESTRICTION ENZYME

Bacteria in the environment are under constant attack by bacteriophages, viruses that infect bacteria and eventually "eat" them. To protect themselves and remain viable, the bacteria have evolved mechanisms to get rid of these viruses. Because the main threat from a phage is its nucleic acid portion, bacteria must have a means of getting rid of this material. They do so with specific enzymes, called the restriction endonucleases, that recognize and cut up the DNA of an infecting phage.

The term "restriction" was applied to these endonucleases because they restrict foreign DNA from infecting the cell. There are many restriction endonucleases found in different bacterial species. Each of these enzymes recognizes a specific DNA sequence and cuts it precisely in one place. For example, the enzyme EcoRI cuts the sequence GAATTC between the G and the A (see Figure 2.7). These target sequences—called restriction sites—are short enough to occur randomly at least a few times in a piece of DNA several hundred base pairs long, so that most phages would be susceptible to attack by at least some restriction enzymes.

The same properties of a restriction site that make it a good target for attack in a phage—namely short length, easy recognition, and random occurrence in any DNA segment—make it a good target in the bacterial DNA as well. So how do bacteria protect their own genes from the restriction enzymes? They manage this by chemically modifying their DNA with methyl groups, adding them onto the adenine or cytosine bases. The methyl groups are bulky enough to block the action of the restriction enzymes without interfering with normal cellular processes such as DNA replication and transcription.

Restriction sites, including the EcoRI site mentioned earlier, are often, although not always, in the form of short palindromes. This means that they will be present on both strands of a DNA sequence. Cleavage with the appropriate restriction endonuclease results in the production of DNA pieces with overhanging or "sticky" ends, which are capable of rejoining with their complementary ends. Two pieces of DNA generated by a single endonuclease can join with each other even if they were not contiguous sequences in the original sequence. Indeed, even frag-

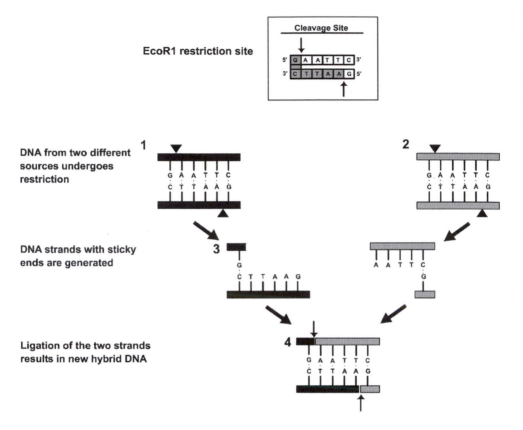

EcoR1 restriction site

Figure 2.7. Restriction site and formation of hybrid DNA. [Ricochet Productions]

ments of DNA from completely different sources—such as a bacterium and a human—may be joined to one another if they are first treated with the same restriction enzyme!

Since their discovery, the restriction enzymes have been used widely in research and biotechnology. Many techniques, including recombinant methods, Southern blotting, cloning, DNA fingerprinting, and RFLPs, rely on fragments generated by different restriction enzymes.

See also: Bacteriophage, Chimera, Restriction Fragment Length Polymorphism (RFLP), Southern Blotting.

RESTRICTION FRAGMENT LENGTH POLYMORPHISM (RFLP)

As we have indicated earlier, any long piece of DNA has a number of restriction sites distributed more or less randomly throughout its length. This means that when treated with a restriction enzyme, it will be cut at those sites to yield a number of smaller fragments of different sizes. So when the genome of a specific organism is treated with a specific enzyme and subjected to gel electrophoresis (which is a technique for separating different pieces of DNA according to size), it generates a highly specific pattern or fingerprint.

Because genomic DNA is very long, the occurrence of polymorphisms—particularly single-base changes—at random intervals along its length is also quite high. These polymorphisms may, on occasion, create or destroy a restriction site (e.g., a change from the sequence GAATTC to GCATTC will render it resistant to cleavage by the enzyme EcoRI). The occurrence of such polymorphisms in a piece of DNA alters the length and number of fragments generated when it is treated with a restriction enzyme. This, in turn, alters the fingerprint of the DNA somewhat. These variations in DNA are called restriction fragment length polymorphisms (abbreviated as RFLPs). RFLPs provide convenient markers for linkage analysis and are excellent resources for DNA identification (see the entry on fingerprinting).

See also: Fingerprinting, Linkage, Polymorphism, Restriction Enzyme.

REVERSE TRANSCRIPTASE

Reverse transcriptase is the name given to a family of enzymes that catalyze the synthesis of DNA from an RNA template. The enzyme actually performs three different catalytic functions: RNA-directed DNA polymerization, RNase H activity, and DNA-directed DNA polymerization. First, a DNA strand is formed using a piece of RNA as the template. Then RNase H activity degrades the template, leaving the DNA strand to act as the template for the synthesis of its complementary second strand by the DNA polymerase. An example of reverse transcriptase is the enzyme responsible for replicating the HIV genome.

The discovery and isolation of reverse transcriptase in the early 1980s by Howard Temin and David Baltimore caused a flurry of excitement in the scientific community because it flew in the face of the then "Central Dogma" of biology, which stated that the flow of information was only possible in one direction from DNA to RNA to protein. Reverse transcriptase overturned this idea, challenged many of the biological theories that were based on this assumption, and also helped explain phenomena that appeared to contradict the central dogma. Although the route outlined by the central dogma still holds good in many cases, we now know that other pathways of information are possible in biology.

See also: Polymerase, Transcription; DNA Synthesis/Replication and Repair (Ch. 4).

REVERSION

The term reversion is typically used to describe a mutation that appears to reverse or correct the effects of mutation in a gene that occurred during its previous cycle of replication. In other words, a gene undergoing a mutation in one generation reverts to its wild-type form in the successive generation. Technically, reversions may occur much more frequently than we realize, simply because of the occurrence and subsequent correction of invisible point mutations. For practical purposes, however, the term is applied only to episodes involving visible changes in phenotype.

See also: Mutation, Phenotype.

RIBONUCLEIC ACID (RNA)

RNA, or ribonucleic acid, is the second important type of nucleic acid that is present in living organisms. This molecule, like DNA, encodes information in the form of linear nucleotide sequences, but with the exception of certain families of viruses, RNA is not the carrier of genes in living organisms. Whereas DNA stores the genetic information safely in the chromosomes, it is RNA that makes this information available to and usable by an organism during its lifetime.

The structure of RNA is similar to that of DNA in many ways but has some crucial differences that relate to its function. Its backbone contains the sugar ribose instead of deoxyribose (hence the names of the respective nucleic acids) and the nitrogen base uracil (designated U) replaces thymine (T) as one of the four building blocks. Like DNA, RNA stores its information in the form of linear nucleotide sequences. But whereas DNA's double helix is important for its function as a stable storage device for information, this duplex is a hindrance in accessing this information. So RNA is, for the most part, a *single*-stranded polynucleotide. However, because uracil, like thymine, can form hydrogen bonds with adenine, base pairing is possible in RNA in much the same manner as in DNA. This often results in very extensive secondary structure—or the folding of an RNA strand into complex three-dimensional shapes. Another distinction between DNA and RNA arises from the sugar ribose, which contains a reactive hydroxyl group that forms a center for chemical action.

Living organisms contain different species of RNA molecules, which perform specialized functions in the translation of information from DNA to proteins. The most important types of RNA molecules in living systems include the following.

Messenger RNA (mRNA)

Messenger RNA is often considered the "essential link" between the information in a gene (i.e., in DNA) and the final sequence of amino acids in a protein. In order to be expressed in the cell, a gene must first be transcribed into mRNA. This RNA transcript, or a modified version thereof, is then the physical template on which the protein's polypeptide chains are built.

The structure of an mRNA molecule is prototypical of RNA (i.e., a single-stranded polynucleotide), with a base sequence complementary to the DNA strand from which the gene was transcribed. In prokaryotes, the mRNA transcript is directly taken up by the protein synthesis machinery for translation. The process is a little more complicated in eukaryotes, where the newly made mRNA is separated from the protein-synthesizing enzymes by the nuclear membrane. Before they are transported from the nucleus to the cytoplasm, these mRNA molecules need to undergo a number of modifications to produce a mature mRNA that is ready for translation. Key steps in the processing of mRNA include the addition of stabilizing features such as several adenine bases (known as a polyA tail) to its 3' terminus and a trinucleotide "cap" at the 5' end of the molecule, as well as the very important process of removing (splicing) the introns. The mature mRNA contains a copy of the uninterrupted gene sequence flanked on either side by certain sequences that remain untranslated. After it is transcribed and processed, the mRNA is transported through the nuclear membrane into the endoplasmic reticulum, where it is ready for translation.

Ribosomal RNA (rRNA)

As its name indicates, ribosomal RNA is contained within the ribosomes of a cell, which are the sites of protein synthesis in a cell. It has become increasingly clear that it is the RNA component of the ribosomes, not the protein, that is responsible for most of their protein synthesis activities, including key catalytic functions. (This fact has prompted the suggestion that RNA predates DNA and proteins as a molecule that can both store information and catalyze reactions in the so-called RNA world.) During translation, the ribosome holds the mRNA transcript "in place" as it were, enabling the tRNAs to read the codons and bind with the transcript. It also furnishes the enzyme peptidyl transferase, which catalyzes the formation of the peptide bond between the consecutive amino acids of the protein (see the entry on protein synthesis in Chapter 4 for more details).

There are several species of rRNA, typically named for their size (as derived from their sedimentation coefficients). In eukaryotes, these rRNA species are the 18S rRNA of the small subunit of the ribosome and the 28S, 5.8S, and 5S rRNAs that together with proteins make up the large ribosomal subunit. Because ribosomes are present in all living organisms and have preserved their function in evolution, the sequence of various species of rRNA has also been quite highly conserved. These characteristics of rRNA have made it an extraordinary molecular clock (namely a tracker of evolution and evolutionary distances among different organisms). An analysis of its sequence conservation has revolutionized our ideas about the course of evolution. Taxonomists used to divide the living kingdom into two main groups, the prokaryotes and eukaryotes, but evidence from rRNA sequences has changed our view so that prokaryotes are now divided into two groups, bacteria and archaea, as different from each other as they are from eukaryotes.

Transfer RNA (tRNA)

Elsewhere we likened the process of protein synthesis to an assembly line at a factory, where designated workers are responsible for the proper placement of pre-assigned parts of a product according to a master plan. In that analogy, transfer

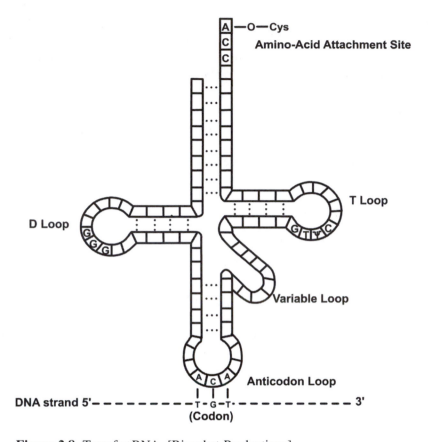

Figure 2.8. Transfer RNA. [Ricochet Productions]

RNA (tRNA) plays the role of the workers, who assemble a protein one amino acid at a time based on the instructions on the mRNA.

As discussed earlier (see the entry on the genetic code), there is a specific relationship between the sequence of bases in a gene (and its mRNA transcript) and that of the amino acids in the cognate polypeptide. Each of the 20 possible amino acids in a protein is encoded by a trio of nucleotide bases—a codon—in the mRNA. During protein synthesis, each tRNA reads the mRNA and plugs in the appropriate amino acid. How does a tRNA read and recognize a codon in mRNA? Once again, complementary base pairing comes to the rescue. Contained within the sequence of the tRNA (see Figure 2.8) molecule is a specific sequence called the "anticodon," a nucleotide triplet that is complementary to the codon on the mRNA. The binding of the anticodon to the codon is a sign for the tRNA to release the amino acid it is carrying on one end. The specific amino acid carried by a tRNA depends on the anticodon it contains; each molecule may carry only one specific amino acid. Because of the degeneracy of the genetic code, many amino acids may be carried on more than one tRNA type, and many cells contain as many as 32 different tRNA molecules.

Unlike the mRNA molecules, whose linear sequence is all-important, the tRNAs have a specific three-dimensional structure that plays a role in their functioning. Each tRNA is between 73 and 93 ribonucleotides long and is folded into a "clover leaf" structure with four main loops due to internal complementarity (see Figure 2.8). The central loop contains the anticodon. Although this schematic three-leaf clover figure is still used to depict tRNAs, X-ray crystallography and other 3D techniques show that the molecule is actually folded into a rough L shape with the anticodon exposed on one end and an amino acid attached to the other. Peppered throughout the nucleotide chain are unusual modified nucleotides that are not seen in any other RNA species. Before a tRNA enters the protein synthesis arena, it must be activated. Activation involves attachment of the appropriate amino acid to the 3' end of the tRNA molecule, which is catalyzed with high accuracy by specific enzymes called amino-acyl tRNA synthetases. Once a tRNA gives up its amino acid to a growing polypeptide, it is released back into the cytoplasm, where it can be reactivated or recharged with another amino acid.

In addition to these three main RNA species, the cell contains other RNA molecules such as the small nuclear RNAs (snRNAs), which play a role in the processing of other RNA molecules; the small nucleolar RNAs (snoRNAs), which play a number of roles, including the modification of bases in RNA and in telomere synthesis; extremely tiny microRNAs, which appear to regulate the expression of mRNA; and XIST RNA, a special molecule encoded in the X chromosome and involved in the inactivation of one of the two X chromosomes in female vertebrates.

RNA has occupied a rather volatile position in the history of molecular biology, waxing and waning in popularity among scientists as its various properties have come to light. RNA, not DNA, was the first nucleic acid that scientists seriously considered as a candidate for being the primary carrier of genetic information. With the confirmation of DNA's role in this regard and the discovery of its "double-helix" structure—which RNA obviously lacked—RNA's importance faded to the background. The discovery of the messenger and transfer functions secured its place in biology, albeit as a shuttle between proteins and DNA, but the discovery of its catalytic abilities in the 1980s brought it under the spotlight once again. Today RNA is recognized as a vital player in living organisms, with important functions in all major life activities!

See also: Ribosome, Ribozyme, Transcription; Protein Synthesis: The Translation Apparatus (Ch. 4).

RIBOSOME

A ribosome is an organelle found in the cytosol of the cells of all living organisms—whether prokaryotic or eukaryotic—and functions as the site of protein synthesis. Chemically it is composed of both RNA and protein complexed with each other to form three subunits that act in concert during the process of translation. The ribosome provides both the physical scaffolding that holds the mRNA in place and binds to appropriate tRNA molecules during protein synthesis as well as the enzymatic functions for mediating synthesis. Ribosomes engaged in active synthesis are often found in clusters known as polyribosomes. Sometimes these structures may be found bound to the endoplasmic reticulum (ER), and as the protein chain is formed it is extruded into the intermembrane space of the ER.

There are two main species of ribosomes that are found in prokaryotes and eukaryotic cells, respectively. The major difference between these two types of organelles is size—the prokaryotic organelles are smaller than their eukaryotic counterparts. The mitochondria in eukaryotes present a special case, as they have their own autonomous ribosomes that correspond in size and composition to the prokaryotic ribosomes.

Because ribosomes are found universally in the living kingdom and have retained their function in evolution, they are good indicators of the evolutionary relationships among organisms, especially those very far apart in evolution.

See also: Cell, Ribonucleic Acid (RNA), Translation.

RIBOZYME

The term ribozyme was initially coined in the early 1980s to describe RNA molecules capable of enzymatic activity (thus, *ribo*nucleic en*zymes*), which until that time was an activity believed to be restricted to proteins alone. The first RNA molecules reported to be capable of catalytic activities were synthesized by Thomas Cech in the test tube, but very soon thereafter Sidney Altman and other scientists demonstrated the presence of naturally occurring ribozymes as well. We now know that many biological reactions, particularly those involving RNA processing and protein synthesis, are catalyzed by ribozymes.

The discovery of naturally occurring ribozymes caused an upheaval in molecular biological thought, as proteins had been held as the sole biological catalysts since the beginning of the twentieth century. They also suggested a possible solution to the chicken-and-egg dilemma of the origin of life: namely the question of whether it was protein or DNA that was present first in Earth's first living beings. As we know, DNA encodes all of the proteins needed for life, but without proteins this information is inaccessible, as DNA replication, transcription, and translation all require proteins. However, proteins, in turn, can only be synthesized from their DNA templates. So which came first in living things? The answer, say some scientists, is neither. Because RNA can perform the functions of both DNA and enzymes, namely information (as it does to this day in the case of many viruses) *and* catalysis (ribozymes), it could have preceded both in living systems.

In addition to solving theoretical problems, ribozymes also present exciting prospects for human therapy. Vascular endothelial growth factor (VEGF) is a major stimulant of angiogenesis, which is the formation of new blood vessels in cancerous tissue. A synthetic ribozyme that blocks the action of the VEGF receptor by destroying its mRNA can potentially help starve cancers of their blood supply.

See also: Deoxyribonucleic Acid (DNA), Enzyme, Protein, Ribonucleic Acid (RNA).

SEGREGATION

The separation of homologous chromosomes into separate gametes during meiosis is called segregation. The segregation of each chromosome is independent of that of the others. So the source of a chromosome in any given generation, whether maternal or paternal, has no bearing on the particular combination of chromosomes in the gametes. As we can see from the hypothetical example shown in Figure 2.9, even three diploid chromosomes can segregate after meiosis in eight

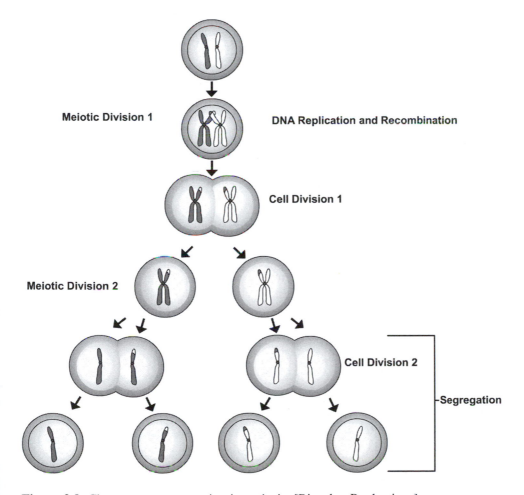

Figure 2.9. Chromosome segregation in meiosis. [Ricochet Productions]

(2^3) possible ways. This means a single diploid individual has the potential to segregate in a staggering 8,388,608, or 2^{23}, different ways and produce correspondingly different types of gametes. And this is discounting any mutations that may have occurred during meiosis. As one may well imagine, segregation is yet another mechanism to ensure genetic diversity over generations. It is the underlying basis for the observations that led Mendel to propose his laws of independent assortment of various traits.

See also: Chromosome, Heredity, Independent Assortment, Meiosis.

SELECTION

Selection refers to the force that favors the transmission of a particular phenotype (and hence genotype; i.e., a gene or allele) through generations over other phenotypes. It is the major driving force in evolution and was first described by Darwin. The well-known phrase "survival of the fittest" is the observation of selection at work. In any generation, the survival of an organism is determined by how it responds to its environment. Say an animal survives a certain environmental condition, such as a cold winter. If it reproduces in the spring it will contribute its genes to the gene pool of the next generation, whereas an animal that has died from the cold will not do so. The genes associated with cold survival are said to be positively selected. Any genes unique to the animal that was killed by cold weather will not be transmitted to the next generation and are negatively selected. Mutations (or alleles) that are associated with lethal genetic diseases are negatively selected, as they would kill the individual and thus not be passed along. In the case of recessive diseases, this selection operates very slowly because of heterozygosity. Heterozygotes carrying the recessive allele would pass along the disease allele from one generation to the next. Only homozygotes for the mutation that manifest the disease would actually stop transmitting the gene. Any mutation that interferes with the fertility of an organism is automatically negatively selected, even when it does not cause any disease fatal to the organism itself. Such mutations are termed "genetically lethal" mutations.

See also: Evolution, Inheritance, Mutation.

SIGNAL TRANSDUCTION

An important characteristic of a complex, multicellular organism is for the different parts to cooperate with one another so that the organism may function properly as a coordinated whole. For instance, the function of breathing, which occurs in the lungs, must be coordinated with the process of oxygen–carbon dioxide exchange occurring in other parts of the body. The liver and pancreas must supply digestive enzymes to the gut at the appropriate times so that incoming food is digested and not in the absence of food, because the latter might result in the breakdown of the gut's normal contents. This type of coordination is important down to the level of single cells; that is, two cells in a particular tissue must be able to communicate with one another in order to respond to changes in their immediate environment.

The process through which cells of a large organism communicate with one another is called signal transduction. This may involve the communication between individual cells or the response of many cells (e.g., in one tissue site) to a single signal. Every cell has many pathways in place whereby external signals are transferred to the inside, where they "tell" the cell what to do. The first step in the pathway is the binding of the signal—usually in the form of some chemical—to appropriate receptors on the cell's surface. This triggers a series of reactions, whose ultimate function is to stimulate the cell to a certain specific activity (e.g., the expression of a particular gene or an increase in the rate of metabolism).

Signal transduction is a very complex system, with many interacting components. One advantage of this complexity is that a defect in any one part may often be compensated for by various homologous proteins. However, defects in the system can also lead to disastrous complications—many diseases, including several types of cancer, can be traced to defects in some signal transduction pathway in a cell.

See also: Organism, Receptor; Signal Transduction (Ch. 4).

SOMATIC CELL

A somatic cell is any cell of an organism that is not a gamete or a sex cell. For example, any human cell other than eggs and sperm is somatic. Human somatic cells are diploid (i.e., they have 23 pairs of chromosomes—22 pairs of autosomes and a pair of sex chromosomes) and multiply by a process of mitosis. Another property of somatic cells is that they are quite highly differentiated, even though they contain the same genes for the most part. Certain populations of somatic cells, such as the skin cells, continue to divide throughout the lifetime of the individual, whereas others, such as brain cells, do not have the capacity for regeneration. Somatic cells might undergo mutations during cell division. These mutations, known as somatic mutations, are not typically passed down to the next generation.

See also: Cell, Mitosis; Signal Transduction (Ch. 4).

SOUTHERN BLOTTING

In the 1950s, well before the arrival of PCR and other high-throughput techniques, a scientist named Edwin Southern invented a method for identifying DNA fragments in a complex mixture. Most, if not all, DNA analytical methods we use today are faster, more efficient, smaller-scale versions of this basic technique, which was named after its originator (and not any geographical context).

Genomic DNA from a person is extracted from a suitable sample and digested with specific restriction enzymes to cut the strands into shorter pieces. These fragments are separated by gel electrophoresis. The DNA on the gel is made single-stranded, and hence suitable for hybridization, by treatment with sodium hydroxide, and it is then transferred (immobilized) onto a nylon membrane or filter. It is this transfer step that gave the name "blotting" to the technique. The filter can then be probed with a radiolabeled sequence, which will bind to DNA fragments containing complementary sequences. The position of the annealed fragment can be determined by autoradiography—exposing the filter to a film on which the radioactive probe shows up as a dark band. The patterns generated by different individuals will be somewhat different because of variations in the positions of various restriction sites.

This technique may be used to identify the presence of mutations in specific genes. For example, the beta-globin gene in sickle-cell anemia has a mutation that results in the loss of a specific restriction site for the enzyme MstII. Although this loss does not directly cause the disease, it does change the number and size of DNA fragments generated when the gene is digested with the enzyme. A normal beta-globin gene generates a 1.15 kilodalton (kDa) fragment when digested with MstII, and a sickle-cell mutant gene yields a larger fragment of 1.35 kDa. Both of these fragments would be picked up by the same beta-globin probe.

Southern's basic technique of immobilizing DNA for probing was subsequently modified for other macromolecules as well. mRNA is analyzed using a similar technique in a "Northern" blot, and a "Western" blot analyzes protein mixtures using radiolabeled antibodies as probes.

See also: Probe.

SPLICING

Eukaryotic genes contain many introns—long DNA segments interspersed between exons—which have to be removed when the DNA is transcribed into mRNA. The process of removing the introns during transcription is called splicing. In a broader sense, splicing refers to any DNA manipulation that involves excision of certain segments. Strictly speaking, the splicing discussed here is called *pre-mRNA* splicing and takes place in the nucleus before the mature mRNA is extruded into the cytoplasm for protein synthesis (see Figure 2.10).

Splicing has to be an extremely accurate process because any mistakes could result in changes in the gene sequence and hence an incorrect reading of the code. Therefore, the ends of the introns have to be easily recognized, and the splicing scissors (a collection of specialized RNA and protein molecules) have to cut at a

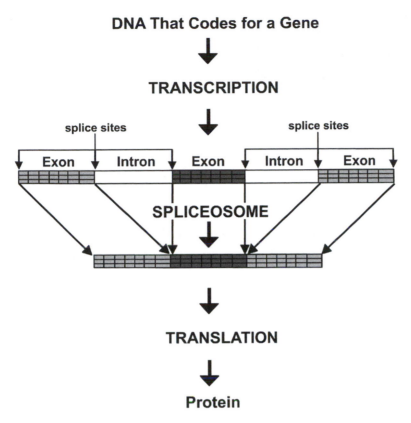

DNA That Codes for a Gene

TRANSCRIPTION

Figure 2.10. Splicing. [Ricochet Productions]

specific location. Introns almost invariably have GU at the 5' end and AG at the 3' end. The spliceosome, made up of small nuclear ribonucleoproteins (snRNPs), is responsible for the accurate excision of the intron and joining together of the free 3' and 5' ends of the exons.

See also: cDNA, Gene, Ribonucleic Acid (RNA), Transcription.

SYNTENY

Genes that occur together on the same chromosome are called syntenic genes. Genes that are grouped together in one species may or may not be similarly grouped in a different species—in general, more closely related organisms show higher degrees of synteny. When a number of genes occur together in similar groupings in species that are quite distant from one another, they are said to exhibit a conservation of synteny.

TANDEM REPEAT SEQUENCES

Any sequence of DNA that is found in repeated units in tandem (i.e., side by side) is called a tandem repeat sequence. The tandem sequences may be short (e.g., the trinucleotide repeats found in specific genes associated with Huntington's disease and myotonic dystrophy and those found in satellite DNA) or may in fact be entire genes. The ribosomal RNA genes, for example, are present in large clusters of repeated units, probably to enable the cell to synthesize the gene products rapidly in large quantities.

TRANSCRIPTION

Transcription is the process through which a cell synthesizes RNA molecules using the information encoded in the DNA template. It is always the first step in the expression of a gene, the final product being a messenger molecule—in which case it is followed by translation—or functional RNA species such as transfer RNAs and ribosomal components.

Like other synthetic processes in the cell, transcription is a complex, multistep process involving a number of enzymes. It is considerably simpler in prokaryotes than in eukaryotes, although the basic process is the same. First, there is the initiation of synthesis, which occurs by the binding of certain subunits of the RNA synthesizing enzyme—called RNA polymerase—to specific locations of DNA, known as the promoters. This is followed by the elongation of the new RNA chain, one nucleotide at a time, as specified by the DNA template. The newly forming chain is complementary to its DNA template, with uracil (U) bases replacing the T component of DNA in positions complementary to A. The final step is the termination of synthesis at the appropriate locations, where a special protein called "rho" binds to the polymerase and causes it to fall off the template.

As mentioned earlier, transcription in eukaryotes is more complicated than in prokaryotes. Initiation in this case involves the recognition of two sites rather than a single promoter. Another difference is that eukaryotic mRNA molecules undergo several modifications, both during and after the chain elongation step itself. One important post-transcriptional modification entails splicing the introns to join the exons and form the template for the translation of a functional protein.

See also: Gene Expression, Polymerase, Ribonucleic Acid (RNA).

TRANSGENIC

The term transgenic refers to an experimentally produced organism whose genome has been altered by the inclusion of foreign genetic material stably into the germ line of the organism. This foreign genetic material may be derived from other individuals of the same species or from wholly different species. The genetic material may also be of an artificial nature. The foreign genetic information may be added to the organism during its early development and incorporated into the cells of the entire organism. For example, mouse embryos have been given the gene for rat growth hormone, allowing mice to grow into large adults. Genetic information can also be added later in development to selected portions of the organism. Experimental genetic therapy to treat cystic fibrosis involves the selective addition of genes responsible for lung function and is administered directly to the lung tissue of children and adults. Transgenic organisms have been produced that provide enhanced agricultural and pharmaceutical products. Insect-resistant crops and cows that produce human hormones in their milk are just two examples.

See also: Biotechnology, Chimera.

TRANSLATION

Translation is the second step in the process of the expression of a gene into its protein product (i.e., protein synthesis). It is a complex process involving several different participants, where the ultimate product is a peptide chain whose amino acid sequence is encoded in a messenger RNA molecule.

The analogy of translation from one language to another is perhaps most aptly applied to the synthesis of proteins from RNA because this is the stage when the information from the genes—encoded in the form of a sequence of four possible bases—is transformed into an entirely new form or language, this one consisting of sequences of 20 rather than four basic units. In contrast, transcription—from DNA to RNA—may be regarded as a simple variation of a theme, a dialect containing the same recognizable patterns. The basis for the translation of genetic information to protein lies in the genetic code, as already described.

The main components participating in translation include an mRNA transcript of a gene properly processed, transfer RNA molecules that supply specific amino acids to the nascent protein, and the ribosomes, which play two important roles in this process—first by providing the important enzymes to mediate various steps of protein chain elongation, and second by acting as the physical anchor for the mRNA template during peptide construction. Energy for the various steps in the process is provided by intracellular ATP.

Translation takes place in the cytoplasm and begins when a ribosome encounters an mRNA molecule, newly extruded from the nucleus. Upon binding to the mRNA, the system is activated to receive tRNA molecules. Based on the codon on the mRNA, a matching tRNA bearing the appropriate amino acid attaches to a specific site on the ribosome (called the P site). A second tRNA with the next amino acid also attaches onto a second site (called the A site) in close proximity to the P site. Enzymes on the ribosomes then catalyze the formation of a peptide bond between the two amino acids and release the now empty tRNA from the P site. The dipeptide bearing tRNA is now transferred from A to P, while a new tRNA with the next amino acid attaches to the vacated A site, ready to form a new peptide bond. Elongation proceeds in this fashion until the ribosome reaches a stop codon for which there is no cognate amino acid–bearing tRNA. At this stage, translation is complete, and the peptide is released into the cytoplasm for additional modifications (see Figure 4.10).

See also: Gene Expression, Genetic Code, Protein, Ribonucleic Acid (RNA).

TRANSLOCATION

The transport of proteins from the cytosol, where they are synthesized (i.e., translated from mRNA into peptide) into the endoplasmic reticulum (ER), is an important step in the synthesis of many proteins. This transfer process is called translocation. Not only secretory and membrane proteins but also proteins that are destined for organelles of the secretory pathway, such as the Golgi apparatus, lysosomes, and the endosomal compartment, undergo translocation. The process is triggered by a signal sequence, normally located at the amino terminus of the newly translated peptides. For many membrane proteins, the signal sequence is identical to the first membrane anchor (called the signal–anchor sequence). Protein translocation can occur either during translation, with the ribosome tightly bound at the membrane, or post-translationally (i.e., after the peptides have been released from the ribosomes). Note that protein translocation is in no way related to the process of chromosome translocation.

See also: Protein, Recombination, Translation; Protein Synthesis: The Translocation Apparatus (Ch. 4); and also Chapter 3.

TRANSPOSABLE ELEMENT

Also known as transposons, transposable elements are DNA sequences that can move or "jump" from one chromosomal site to another. This phenomenon of jumping was first demonstrated in the 1950s by Barbara McClintock in the genes of the maize plant. She found that the transposons could create chromosomal insertions and deletions (typically as coupled events) and gene translocations, which resulted in changes in the appearance of the maize plants. Nearly 50 percent of the total genome of this plant consists of transposons! The importance of these findings garnered McClintock a Nobel Prize in Physiology or Medicine in 1983.

Scientists have since demonstrated the ubiquity of this phenomenon. Transposons are believed to be a means for a cell to respond rapidly to environmental stress. In the process of jumping from one site to another, they can cause mutations and change the amount of DNA in the genome. Some of the mutations may be advantageous in a given environment, whereas others could have harmful effects. Diseases such as Duchenne muscular dystrophy and severe combined immunodeficiency (SCID) are caused by the disruption of normal genes by transposable elements (see the entries for muscular dystrophy and primary immune deficiencies in Chapter 5).

See also: Mutation.

TWINS

Although multiple births are quite common in other species, humans most commonly have one baby at a time. However, the occurrence of multiple simultaneous births is possible, and when two babies are born at the same time, they are called twins. (Higher numbers are termed triplets, quadruplets, etc., and are progressively less common.)

Human twins may be the result of two independent fertilizations occurring in the womb at or around the same time, or a single fertilization event. The former are known as heterozygotic or fraternal twins, and they may or may not be of the same sex, depending on chromosomes of two different sperms. In the case of the latter, the zygote begins to undergo mitotic division as usual but is then split into two blastocysts, the splitting being triggered by certain unknown factors. Each of these blastocysts develops independently into a complete organism. Such twins are called monozygotic or identical twins. By virtue of their origin from the same egg and sperm, such twins are natural clones, at least at the beginning of development. Because of their genetic identity, homozygotic twins have been used as models for studying the relative effects of genetics and the environment on various characteristics, such as disease susceptibility and behavioral traits. Their similarities also make twins very good potential donors (for blood, organs, etc.), with low risks for rejection.

See also: Cell Cycle, Development, Mitosis.

VARIATION

Genetic variation refers to the differences in DNA sequence from one individual to another. Except for clones, all individuals, even within a single species, show some degree of variation from another. Variation is what makes each one of us unique. Even identical twins, although they start out with the same genotype, have a few minor differences in their DNA that arise during the course of embryonic development. Humans differ from one another in only 1 in 1,000 nucleotides or so, which may seem small but is actually quite a huge difference when we consider that the total genome contains 3 billion base pairs. This variation is clearly sufficient to create the enormous diversity we see among us, both with respect to such obvious traits as color and height as well as to less obvious ones, such as the ability to smell freesias or the ability to roll one's tongue. Differences in RFLP patterns are also caused by genetic variation.

See also: Population Genetics, Restriction Fragment Length Polymorphism (RFLP).

VECTOR

In the world of biotechnology, a vector is a vehicle for moving foreign genes into an organism. It is made of the same substance as the genes (i.e., DNA). Naturally occurring DNA molecules (e.g., retroviral genomes and plasmids) are often used as vectors for inserting new genetic information into the genome of human cells. Synthetic vectors, which are also known as artificial chromosomes, are recombinant DNA molecules that are not found in nature but are constructed in the laboratory by ligating together different genes of choice. Examples of artificial chromosomes that were used in mapping the human genome are bacterial and yeast artificial chromosomes (BACs and YACs).

Bacterial artificial chromosomes are made from bacterial plasmids and maintained and propagated in bacteria such as *Escherichia coli*. The length of the foreign DNA fragments—which despite the bacterial origins of the vector may come both from prokaryotic and eukaryotic sources—in a BAC typically ranges from 100 to 300 kb (with an average of 150 kb). Yeast artificial chromosomes are maintained and propagated in yeast cells. A YAC is capable of independent replication and also has other features, such as restriction sites and selection markers, that allow easy selection of yeast cells that have taken up the DNA of interest. YAC cloning allows the insertion of DNA fragments of up to 1 million base pairs.

See also: Biotechnology, Cloning, Plasmid, Vector.

VIRUS

Derived from the Latin word for "poisonous fluid," the term virus was first used as a label for the filterable, submicroscopic (and hence invisible) causative agents of certain diseases that were not caused by bacteria or other microscopic pathogens. Although the properties in that definition still apply to the viruses, we now use the term for a very large group of intracellular parasites found in virtually all forms of life, ranging from prokaryotic bacteria to multicellular beings (both plants and animals). Most often consisting of little more than a piece of nucleic acid (either DNA or RNA but never both) covered in a protein sheath, a virus occupies a special niche in the biosphere and is considered to exist at the "threshold of life." Incapable of independent existence, a virus behaves much as an inert chemical when outside a cell. However, when introduced into an appropriate host cell, the same virus can harness the cellular machinery to perform such living activities as replication—namely the synthesis of new proteins and nucleic acid to produce several new viruses. Viruses are typically host-specific and have not been seen to infect across different living kingdoms under natural circumstances (e.g., a plant virus does not infect a human, and a bacteriophage would not thrive in a duck). Many, but not all, viruses induce diseases in their host organisms. Among some well-known examples in humans are polio myelitis, smallpox, and influenza. Because of their small genome size, they lend themselves to easy genetic manipulation and as vectors for foreign DNA. This has resulted in many practical applications of the viruses, for instance as cloning vectors in genetic libraries and more recently in gene therapy to introduce foreign genes into target cells.

See also: Bacteriophage, Gene Therapy, Vector.

WILD TYPE

Wild type refers to the genetic form of an organism that occurs most commonly in nature. For example, fruit flies most commonly have black eyes, but occasionally it is possible to find a fly with red eyes. The black-eyed fly is considered as the wild type and the red-eyed one as a mutant.

See also: Mutation.

X-LINKAGE

The X chromosome exists in all of the diploid cells of both males and females. However, unlike the other chromosomes (autosomes), it does not have an equal counterpart, with an allele-for-allele correspondence in males. This is because the Y chromosome is much smaller and has much fewer genes. Indeed, the difference between the X and Y chromosomes is so great that there are very few genes on the X chromosome with analogs on the Y chromosome. As a consequence, virtually every gene on the X chromosome is expressed in males as if it were a dominant gene. This phenomenon is called X-linkage, and the traits or diseases expressed in males are correspondingly referred to as X-linked traits or diseases. X-linkage is obviously absent in females, where the X chromosome follows a normal dominant/recessive gene-expression pattern. A well-known example of an X-linked disease is Hemophilia A, a blood-clotting disorder caused by a mutation in the gene for the blood-clotting factor VIII. Red–green color blindness, or daltonism, is another example.

Because of their uneven distribution, X-linked traits and diseases have very characteristic inheritance patterns (see Figure 2.11). Whereas a mother may pass an X-chromosomal gene to progeny of either sex, fathers can only transmit these genes to their daughters. Although the daughter may herself be heterozygous and thus safe from the X-linked disease, she can pass the gene along to her son, who will then show symptoms. The only way a female shows signs of an X-linked disease is when she is homozygous for that trait, for instance if she is the daughter of an affected father and a carrier mother and inherits the defective gene from both parents.

See also: Chromosome, Inheritance.

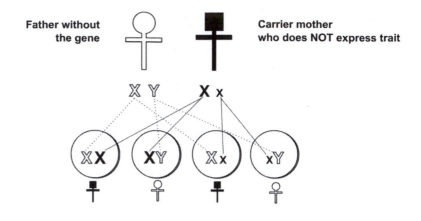

Father without the gene

Carrier mother who does NOT express trait

$\mathbb{X}\mathbf{X}$ = Daughter who does NOT carry gene and does NOT express trait

$\mathbf{X}\mathbb{Y}$ = Son who does NOT carry gene and does NOT express trait

$\mathbb{X}$ x = Daughter who is a carrier but does NOT express trait

x $\mathbb{Y}$ = Son who has the gene and expresses the trait or condition

Figure 2.11. X-linkage. [Ricochet Productions]

ZYGOTE

A zygote is the cell produced by the fusion of an egg and a sperm during fertilization. Zygote formation is a crucial step in the reproduction of a diploid organism. The zygote is diploid (i.e., it contains two copies of each chromosome, one each from the egg and from the sperm). After fertilization, the zygote rapidly undergoes mitosis and develops into the embryo and eventually an adult organism.

See also: Fertilization, Mitosis, Organism.

Chapter 3

The Chromosomes

A chromosome, as described in the previous chapter, is the physical structure in a living cell that contains an organism's genes or DNA. This chapter describes the organization and working of human chromosomes in some detail.

STRUCTURE AND ORGANIZATION

The human genome (actually the diploid genome) consists of 46 chromosomes—22 pairs of autosomes and two sex chromosomes—the entire set of which is present in the nucleus of nearly every functioning cell in our bodies. The only exceptions are the sperm and egg cells, which, being haploid, have just one set of 23 chromosomes: all the autosomes and either an X or Y chromosome. For identification purposes, the autosomes are numbered 1–22, more or less in order of decreasing size. The sex chromosomes are called X and Y chromosomes: a female cell contains a pair of X chromosomes, and a male cell contains one X and one Y.

The Human Chromosomes

Chromosome 1	263 million base pairs	Chromosome 9	145 million base pairs
Chromosome 2	255 million base pairs	Chromosome 10	144 million base pairs
Chromosome 3	214 million base pairs	Chromosome 11	144 million base pairs
Chromosome 4	203 million base pairs	Chromosome 12	143 million base pairs
Chromosome 5	194 million base pairs	Chromosome 13	114 million base pairs
Chromosome 6	183 million base pairs	Chromosome 14	109 million base pairs
Chromosome 7	171 million base pairs	Chromosome 15	106 million base pairs
Chromosome 8	155 million base pairs	Chromosome 16	98 million base pairs

Chromosome 17 92 million base pairs	Chromosome 22 56 million base pairs
Chromosome 18 85 million base pairs	Chromosome X 164 million base pairs
Chromosome 19 67 million base pairs	Chromosome Y 59 million base pairs
Chromosome 20 72 million base pairs	Mitochondrial plasmid 16.5 kilobases
Chromosome 21 50 million base pairs	

Each chromosome is essentially a single molecule of double-stranded DNA. The smallest chromosome, number 21, contains about 50 million nucleotide pairs (stretched to its full length, this molecule would extend 1.7 cm), whereas the largest, chromosome 1, contains up to 263 million base pairs (which would stretch to 8.5 cm). Imagine that, if it were stretched out, the chromosomal DNA in a single human diploid cell would be over 2 meters long!

How does such a long molecule fit, so compactly as to be practically invisible, into a microscopic cell? This is where the spiral structure of DNA plays an important role. The double-helical structure of DNA enables it to coil around itself, much like a telephone cord, and form compact supercoiled packages that can fit into a nucleus. The DNA strands are actually wrapped (like thread on a spool) around tiny proteins called histones that help them to stay supercoiled. Approximately 146 base pairs worth of DNA is wound in a double loop around each histone molecule to form a unit known as a nucleosome. These nucleosomes are then tightly packed alongside each other to form the chromatin fibers that make up a chromosome (see Figure 3.1).

The highly coiled structure just described is the state of the chromosomes in the resting or nondividing cells of an organism. In a dividing cell, however, this structure is slightly loosened in order to allow various enzymes access to the DNA so it can replicate itself. When chromosomes in this loosened state are treated with a special stain known as the Giemsa stain, they can be seen under a normal light microscope. Each stained chromosome possesses a uniquely identifiable pattern of light and dark bands. The lightly stained regions, known as euchromatin, contain single-copy, genetically active DNA (namely, DNA that is actively being converted into RNA; see the entry on transcription in Chapter 4), and the darkly stained regions, known as heterochromatin, contain repetitive sequences that are genetically inactive.

The gross structure of a chromosome has certain features that are essential for its activity and survival. The *centromere*, which is visible as the constricted region, is the point at which two strands of a replicating chromosome are held together during cell division. It is involved in distributing the chromosomes between the progeny cells during mitosis and meiosis. The relative position of the centromere (i.e., how close it is to the center) is different in different chromosomes. The other important chromosomal structures are the *telomeres*, long stretches of repeating units of DNA found at the ends of the chromosomes. Telomeres form

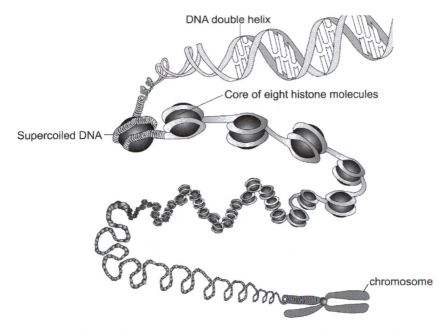

DNA double helix

Core of eight histone molecules

Supercoiled DNA

chromosome

Figure 3.1. How DNA is packaged into chromosomes. [Ricochet Productions]

complexes with proteins that help prevent the DNA molecule from unraveling and contribute to the stability and longevity of the chromosome.

Less than 2 percent of the human chromosome encodes the information of our genes. There are some estimated 30,000 genes in the human genome. They vary greatly in length from a few hundred base pairs (e.g., certain cell surface receptors) to about 2 million base pairs, which is the size of the largest known human gene for dystrophin. As mentioned earlier, gene organization within human and other eukaryotic chromosomes is considerably more complex than in prokaryotes. For instance, eukaryotic genes are made of sequences called *exons* interspersed by fairly long stretches of DNA called *introns*. Only the DNA of the exons actually contributes to the sequence of the gene. Introns are transcribed into RNA but are then spliced out before the gene is translated into proteins. However, they play important roles in regulating various aspects of gene expression.

The remaining 98 percent or so of human genomic DNA is not known to code for any genes. Up to 5 percent may be accounted for by promoters and other gene regulation functions, but the role of the bulk of the noncoding DNA is still unknown to us. Certain areas of the chromosome, such as the DNA near the centrosomes and telomeres and in the Y chromosome, are seen to be composed of multiple tandem repeats of shorter sequences. This DNA goes by the label of *satellite DNA*, and although its biological function remains unknown, it has proven useful for fingerprinting purposes.

If there is any functional or logical significance to the distribution of different genes (i.e., patterns of synteny) among different chromosomes, it has not yet been discovered by scientists. Certain genes of similar function are found close together in clusters on the same chromosome (e.g., the alpha-globin genes on chromosome 16). But at the same time, related genes are found in completely different locations; for example, the alpha-globins themselves are just one component of hemoglobin, which contains a second protein, beta-globin, that is encoded (also in a cluster) in chromosome 11. These clusters very likely arose by a process—taking place over several generations—of changes in the DNA such as gene duplication and the accumulation of random mutations.

THE SEX CHROMOSOMES

The sex chromosomes deserve special mention because of their somewhat unique status in the genome. Unlike the autosomes (the 22 other chromosome pairs), which are homologous, containing very similar sister chromosomes, this pair can consist of two very distinct types of chromosomes, named the X and Y chromosomes. Whereas all female organisms have two X chromosomes, males have one X and one Y chromosome. This means that the Y chromosome is the decisive factor in determining the sex of an individual.

The X chromosome is nearly three times as large as the Y chromosome and consequently has the capacity for carrying many more genes. Curiously enough, researchers have found that among the hundreds of genes in the X chromosome are nearly half of the genes related to early stages of sperm production, a strictly male function! As it happens, these genes are silenced (or not expressed) in females.

The Y chromosome may be considered as a shrunken version of the X chromosome. Although it lacks most of the genes on the X chromosome, it does possess a small region, called the pseudoautosomal region, that shows sequence homology with a part of the X chromosome. In addition, it contains a set of housekeeping genes that are essential for survival, as well as genes responsible for sex determination (called the *SRY* genes) (see Figure 3.2).

The inequality between the sex chromosomes gives rise to a difference in the expression of the genes on these chromosomes in the two sexes. In females, the genes of the X chromosome follow the normal patterns of expression, which means that the phenotype is related to the dominant allele, just as in the autosomes. In males, however, the situation is quite different because most of

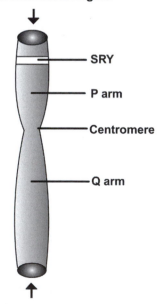

Pseudoautosomal Region

SRY

P arm

Centromere

Q arm

Pseudoautosomal Region

Figure 3.2. Y chromosome. [Ricochet Productions]

the genes on the X chromosome have no corresponding alleles on the Y chromosome. The phenotype in males always corresponds to the genes on the X chromosome whether these genes are dominant or recessive. The phenomenon of the X-linkage of certain genetic diseases (see Chapter 2) is a result of this inequality.

The inheritance pattern of the sex chromosomes is also related to their inequality, although in this instance it is the distribution of the chromosomes rather than differences in size that is the controlling factor. Every meiotic event in a man leads to the production of two sperm cells, one with an X chromosome and the other with a Y chromosome. Meiosis in females, on the other hand, will always produce gametes (egg cells) with an X chromosome. So the sex of the offspring is determined by the sex chromosome in the fertilizing sperm. Contrary to the folk wisdom of certain cultures, then, it is the father who determines the sex of the baby!

MITOCHONDRIAL DNA

A second species of human DNA that deserves special mention is the mitochondrial plasmid. As described in Chapter 2 (see the entry on mitochondria), mitochondrial DNA is autonomous from the nuclear chromosomes both for expression and replication. Although it accounts for a very small fraction of the total cellular DNA—the mitochondrial plasmid is about 1,000 times smaller than the smallest chromosome—it is nevertheless essential for the well-being of the organism because it contains genes for vital respiratory proteins. The importance of this DNA is also evident from the diseases that can develop when mitochondrial genes are mutated.

Whereas the Y chromosome is passed on only from fathers to their sons, the mitochondrial plasmids follow a strictly maternal inheritance (i.e., they can only be passed on from one generation to the next via the mother). This is because mature sperm contains practically no cytoplasmic organelles. During fertilization, only the DNA of a sperm cell enters the ovum, where it fuses with the egg's DNA in the nucleus. So all the cytoplasmic contents of the zygote—such as its mitochondria and other organelles—come only from the egg or the mother. And because the mitochondria are completely independent of nuclear DNA, only they will give rise to more mitochondria as the zygote develops into a new multicellular organism.

This gives rise to a special kind of non-Mendelian inheritance pattern of the mitochondrial genes. For example, whereas a man might have a disease due to a mitochondrial mutation (having inherited it from his mother), he cannot pass it on to any of his children.

CHROMOSOMAL ABERRATIONS

As with all other molecules and structures in a living cell, the chromosomes are subject to various defects. The simplest, most basic types of defects that are possible in a chromosome include the different types of mutations (deletions, insertions, and substitutions), which, as described earlier, cause a change in the base sequence in a DNA molecule. Like individual mutations, chromosomal abnormalities can occur either in the germ line or in specific somatic cells. This plays a role in the nature and inheritance of the genetic defects that result. *Constitutional* abnormalities result from changes in the germ line and give rise to heritable diseases. *Somatic* abnormalities in contrast are usually specific to an organ or tissue and thus give rise to a mosaic pattern of chromosomes in the individual. The deleterious effects of different mutations in individual genes on our chromosomes are discussed in detail in Chapter 5. In this section, we shall concentrate on aberrations and abnormalities that involve larger portions of the chromosome.

ANEUPLOIDY

Aneuploidy refers to the presence of an abnormal number of chromosomes in the nucleus of the cell. It should be noted that this is distinct from the different phases in the life cycle of the organism, where cells contain a multiple of their normal chromosomal complement. For example, a haploid human cell (sperm and egg cell) has 23 chromosomes, whereas a diploid cell (e.g., a skin cell) has 46. Aneuploidy refers to aberrations in one normal set, usually of the diploid cell (because this is the phase in which humans exist). The presence of 45 or 47 chromosomes instead of 46 is an aneuploidy.

The most common cause of aneuploidy is *nondisjunction*, which is the failure of chromosomes to separate during cell division. Nondisjunction seems to happen most often during meiosis, when homologous chromosomes do not segregate properly into the daughter cells during division. This results in an unequal distribution of the chromatin material into the gamete, with one cell receiving an entire pair of chromosomes and the other lacking a chromosome altogether. When such abnormal gametes (with 22 or 24 chromosomes) participate in fertilization, they give rise to abnormal zygotes or embryos with conditions known as *trisomies* (47 chromosomes) and *monosomies* (45 chromosomes).

The survival rate of most aneuploid zygotes or embryos in humans is quite low, and these zygotes are often spontaneously aborted so soon after fertilization that the pregnancy goes unnoticed. When the embryos do survive to term, as in the examples cited herein, the babies are often born with various birth defects, including both physical abnormalities and mental retardation. The most frequently observed aneuploidies in humans involve the sex chromosomes. Examples include *Klinefelter syndrome* and *Turner syndrome*. Perhaps the best-known example of a trisomy in humans is *Down syndrome*, which results from the nondisjunction of chromosome 21. (See Chapter 5 on genetic diseases for details on these diseases.) As people grow older, errors in meiosis increase in frequency, increasing the likelihood of producing aneuploid, often nonviable embryos.

TRANSLOCATIONS

Whereas aneuploidy is a numerical abnormality of the chromosomes that results from defects in meiosis, translocations are structural abnormalities in which the DNA in the chromosomes is significantly rearranged and the continuity and integrity of individual chromosomes are disrupted. Such abnormalities typically result from the recombination between fragments of DNA derived from sequences that were not contiguous with one another in the original genome. They may involve one or more breaks in one or more chromosomes. In general, these translocations are said to be balanced if there is no net gain or loss of genetic material (DNA) from the chromosome and unbalanced if there is such a change.

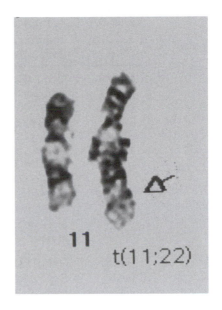

Chromosome translocation (between human chromosomes 11 and 22). [© J. Siebert/Custom Medical Stock Photo]

Translocations within a single chromosome often take the form of *inversions*. That is to say, a piece of DNA that is broken off from within the chromosome is rejoined after flipping 180 degrees. Depending on the location of the breakpoints, this can cause the interruption of genes, or the juxtaposition of genes to new promoters. The recombinational event that is most relevant to our discussion of genetic diseases is a *reciprocal translocation*, in which pieces of DNA from two different chromosomes trade places, so to speak. An example of a reciprocal translocation is the *Philadelphia chromosome*, which results from the reciprocal translocation of pieces of chromosomes 9 and 22 and results in the activation of an oncogene because of its juxtaposition with a new promoter. (See the section on chromic myelogenous leukemia in the entry on Cancer in Chapter 5.)

GENES AND DISEASES: A PHYSICAL ROADMAP

A complete catalog of all of the human genes is well beyond the scope of this volume, and indeed is not even possible yet, but Tables 3.1–3.24 list some well-known examples of genes in the various chromosomes. Individual genes are discussed in further detail in Chapters 4 and 5.

Table 3.1. Chromosome 1: Genes and diseases at a glance

Gene	Protein product/function	Disease	Nature of mutation
C1q (*A*, *B*, and *C*)	C1q subunit of complement: Binds Fc portion of antibody and initiates complement cascade	Systemic lupus erythematosus[1]	Binding capacity of Fc with immunoglobulin is changed
GBA	Enzyme (glucocerebrosidase) involved in fatty acid metabolism, specifically in recycling blood cells and myelin	Gaucher disease (Types 1–3)	Enzyme deficiency
GLC1A (also called *MYOC*)	Cytoskeletal protein (myocilin) produced in specialized eye tissues, where it regulates intraocular pressure	Glaucoma	Amino acid substitutions that affect structural integrity of protein
HFE2 region	No specific gene identified	Juvenile hemochromatosis	Specific variant at locus 1q associated with disease phenotype
LEPR	Membrane receptor that transmits leptin signals into cells	Obesity	Missense mutation leads to a single amino acid substitution in external, leptin-binding portion of the receptor
LMNA	Nuclear matrix protein (lamin) involved in maintaining the stability of the nucleus and chromatin material	Charcot–Marie–Tooth disorder (Type 2B1 or axonal type)	Defects in protein reduce ability of axons to conduct electrical impulses
		Emery–Dreifuss muscular dystrophy	Dysfunctional protein
MPZ	(Myelin protein zero) Structural protein found in the myelin sheath of peripheral nerves	Charcot–Marie–Tooth disorder (Type 1A)	Defective protein leading to demyelination
NRAS	Oncogene	Melanoma	Gene is highly susceptible to cancer-causing mutations if exposed to UV light

Table 3.1. (continued)

Gene	Protein product/function	Disease	Nature of mutation
PSEN2	Presenelin 2: regulation of amyloid precursor proteins	Alzheimer's disease	Improper regulation of amyloid proteins results in formation of plaques
SELE	Selectin E: adhesion molecule produced by endothelial cells, which mediates sticking of leucocytes to blood-vessel walls at sites of inflammation	Atherosclerosis	Polymorphism associated with high susceptibility: A to C base substitution resulting in serine to arginine change
UROD	Enzyme (uroporphyrinogen decarboxylase) produced in liver and RBCs catalyzing the final step of heme synthesis	Porphyria cutanea tarda	Enzyme deficiency

[1]Polygenic diseases linked with certain alleles of this gene in addition to other genes elsewhere on genome.

Table 3.2. Chromosome 2: Genes and diseases at a glance

Gene	Protein product/function	Disease	Nature of mutation
CNGA3	Nucleotide-gated ion channel of the light-induced signal transduction pathways in cone cells	Achromatopsia (specific to European populations)	Protein function disrupted
CTLA4	T-cell specific serine esterase, involved in putting the brakes on an immune response	Autoimmune disorders (e.g., Graves disease)	Point mutations reduce the ability of CTLA4 to keep the immune system in check
CYP1B1 (or GLC3)	Cytochrome P450 subunit involved in metabolizing signaling steroids of the eyes	Glaucoma	Functional disruption of gene product
MSH2	DNA mismatch repair enzymes	Colorectal cancer	Heterogeneous group of mutations disrupting DNA repair activities

Table 3.2. (continued)

Gene	Protein product/function	Disease	Nature of mutation
MSH6	DNA-binding protein specific for GT mismatches	Colorectal and endometrial cancers	Heterogeneous mutations disrupting DNA repair activities
PAX3	Paired-box transcription factor: development of the face and ears	Waardenburg syndrome	
		Type 1 WS	Point mutations within gene
		Type 3 WS	Deletion of part of chromosome that includes portions of the PAX3 as well as other genes
POMC (Proopiomelanocortin)	Pituitary hormone involved in stimulating the adrenal cortex and melanocytes	Obesity	Base substitutions/ point mutations leading to early truncation of normal protein
SCN2A1	Voltage-gated sodium channel subunit expressed primarily in the brain	Epilepsy	Structural mutations disrupting proper ion passage
SCTR	Secretin (hormone) receptor in brain cells	Autism	Mutations affect binding to secretin
SLC11A3	Ferroportin: iron transport	Hemachromatosis (type 4)	

Table 3.3. Chromosome 3: Genes and diseases at a glance

Gene	Protein product/function	Disease	Nature of mutation
CLCN2	Chloride ion channels especially up-regulated in glioma cell or the brain; involved in membrane depolarization	Epilepsy, juvenile absence type	Dominant mutation that produces defective channels that interfere with the depolarization of membrane
MITF	Transcription factor involved in the regulation of melanocyte development and melanin biosynthesis in epithelial cells of the retina	Type 2 Waardenburg syndrome	Deletion in one of the exons, which alters the regulatory properties of mutation; dominant expression

Table 3.3. (continued)

Gene	Protein product/function	Disease	Nature of mutation
MLH1	Mismatch repair enzyme	Colorectal cancer	Heterogeneous group of mutations often associated with this cancer
SCLC1 region	No specific gene at locus	Small cell lung carcinoma	Deletion of this locus, seen to occur especially with prolonged cigarette smoking
SCN5A	Sodium ion channel expressed specifically in cardiac muscles	Long Q-T syndrome	
Transferrin	Ion-transport protein that transfers iron ions to growing cells	Atransferrinemia with severe microcytic anemia	Mutations leading to defective gene expression
ZNF9 (Zinc finger protein 9)	RNA binding protein with specific affinity for cellular retroviral sequences	Myotonic dystrophy type 2 (dystrophia myotonica 2)	Expansion of a tetranucleotide sequence (CCTG) in the first intron

Table 3.4. Chromosome 4: Genes and diseases at a glance

Gene	Protein product/function	Disease	Nature of mutation
ANK2	Neuronal ankyrin (ankyrin B) that links integral membrane proteins to cytoskeleton and plays roles in cell motility, activation, proliferation, and contact of brain cell	Long Q-T syndrome (type 4)	Amino acid substitution leading to defective protein
EVC	Ellis–van Creveld protein: possible membrane protein of unknown function, containing a leucine zipper	Weyers acrofacial dysostosis	Heterogeneous mutations in single copy of gene
		Ellis–van Creveld syndrome	Homozygous or heterozygous mutations resulting in two defective copies of *EVC*

Table 3.4. (continued)

Gene	Protein product/function	Disease	Nature of mutation
EVC-2	Function unknown; also known as limbin, this protein is expressed in the growth plates of long bone	Ellis–van Creveld syndrome	Single base-pair mutations resulting in defective expression of protein
FGA, *FGB*, and *FGG*	Fibrinogen subunits alpha, beta, and gamma	Bleeding disorder due to partial or complete fibrinogen deficiency	Mutations cause formation of one or more defective subunits, which leads to deficiency of fibrinogen in blood
FGFR3	Member of family of fibroblast growth factor receptors whose specific function is to bind specific factors to inhibit cartilage-forming cells in bony tissues	Achondroplasia	Amino acid substitution leading to a gain of function so that bone growth is limited
FSHMD1A	Function unknown	Facioscapulohumeral (FSH) muscular dystrophy	Deletion of a repeated nucleotide unit from gene in this region

Table 3.5. Chromosome 5: Genes and diseases at a glance

Gene	Protein product/function	Disease	Nature of mutation
APC	Tumor suppressor gene	Familial adenomatous polyposis	Mutations in area of gene called the mutation cluster region (MCR) that lead to truncated protein
Asthma susceptibility locus (candidate genes):		Asthma	Not fully characterized
BHR1	Unknown	Specifically associated with bronchial asthma	

Table 3.5. (continued)

Gene	Protein product/function	Disease	Nature of mutation
IL13	Interleukin 13: immunoregulatory peptide produced by helper T cells and involved in B-cell maturation and proliferation		
UGRP1	Secreted globulin of lungs similar to uteroglobin (hence called uteroglobin-related protein) that is the target of transcription factors		
DAT1	Dopamine transporter	ADHD	
ERCC8	Transcription factor involved in DNA excision repair	Cockayne disease (type I)	Deletions leading to dysfunctional protein product
GABRG2 or *CAE2*	Ligand (GABA) gated chloride channel functioning as inhibitory neurotransmitter	Epilepsy	Functional mutation
MATP	Membrane-associated transporter protein that mediates melanin synthesis	Oculocutaneous albinism type 4	Homozygous or compound heterozygous mutations leading to the formation of a defective transporter
PCSK1	Pro-insulin processing enzyme (Prohormone Convertase 1) involved in regulating insulin synthesis	Obesity (with impaired proinsulin processing)	Heterogeneous mutations leading to diminished enzyme activity
SLC26A2	Sulfate transporter critical in formation of cartilage	Dwarfism	
		Achondrogenesis	Mutations in regulatory portion of gene leading to developmental defects
		Diastrophic dysplasia	Mutations leading to structural/functional defects of gene product

Table 3.6. Chromosome 6: Genes and diseases at a glance

Gene	Protein product/function	Disease	Nature of mutation
EPM2A	Laforin: regulatory phosphatases of neuron cells	LaFora disease (epilepsy)	Deletion and missense mutations leading to defective proteins that form glycogen-like masses in cells
HLA-D locus	Unknown genes or proteins	Susceptibility locus for insulin-dependent diabetes mellitus (IDDM)	
HFE	Membrane protein that regulates iron uptake by mediating interaction of transferrin with its receptor	Hemochromatosis	Recessive mutations that reduce protein activity
LAMA2	Subunit of laminin 2 (merosin), a basal membrane protein that mediates cellular organization during embryonic development	Congenital muscular dystrophy	Missense and nonsense mutations
LPA	Apolipoprotein (a): cholesterol processing lipoprotein	Atherosclerosis/ thrombogenesis	Exact nature unknown
NFKBIL1	MHC class I protein	Rheumatoid arthritis	Polymorphism, not a functional mutation, associated with RA susceptibility
NHLRC1	Malin: zinc-finger protein with function not fully characterized but seems to act in same pathway as laforin (see *EPM2A*)	Epilepsy (LaFora type 2)	Heterogeneous mutations leading to defective, non-functional protein

Table 3.7. Chromosome 7: Genes and diseases at a glance

Gene	Protein product/function	Disease	Nature of mutation
BRAF	Oncogene: homolog of mouse sarcoma virus with kinase activity	Lung cancer	

Table 3.7. (continued)

Gene	Protein product/function	Disease	Nature of mutation
CFTR	Chloride ion transporting protein (cystic fibrosis transmembrane conductance regulator) expressed in epithelial cells	Cystic fibrosis	Loss of function mutation; recessive inheritance
GPRA	G-protein coupled receptor 154: signaling protein expressed in bronchial epithelial and smooth muscle cells; forms part of susceptibility locus	Asthma susceptibility	
MUC3A		Ulcerative colitis	
PDS	Member of a family of sulfate carriers (called pendrin) expressed specifically in the thyroid gland and cochlea and appears to be involved specifically in chloride and iodide transport	Pendred syndrome	Recessive, loss of function mutation
OPN1SW	Blue-light absorbing opsin (photoreceptor) in retina	Blue-yellow color blindness	
GCK (or *HK4*)	Glucokinase enzyme	IDDM (type I diabetes mellitus)	
TFR2	Transferrin receptor 2	Type 3 hemochromatosis	
KCNH2		Long Q-T syndrome	

Table 3.8. Chromosome 8: Genes and diseases at a glance

Gene	Protein product/function	Disease	Nature of mutation
c-myc	Transcription factor with an essential role in mitosis (normal cell division)	Burkitt lymphoma	Translocation of gene to other chromosomes (2, 14, or 22) where expression of *c-myc* is deregulated

Table 3.8. (continued)

Gene	Protein product/function	Disease	Nature of mutation
CNGB3	Nucleotide-gated ion channel in retinal photoreceptors	Achromatopsia (frequent in population of Pingelap—rare elsewhere in world)	Defective protein lacks proper signaling capability
GDAP1	Ganglioside-induced differentiation-associated protein: nerve cell development	Charcot–Marie–Tooth disorder	Loss of function mutations leading to recessive form of disease. Severity varies according to extent of mutation
HT locus	Unknown	Hashimoto syndrome	
RECQL2	Protein product is homologous to a bacterial helicase (involved in unwinding and stabilizing DNA) but with no detected activity	Werner syndrome	Point mutations that result in the formation of an altered gene product

Table 3.9. Chromosome 9: Genes and diseases at a glance

Gene	Protein product/function	Disease	Nature of mutation
ABL1	Proto-oncogene that encodes a tyrosine kinase involved in cell differentiation and development and stress response	Chronic myeloid leukemia (CML)	Translocation within gene that forms new "Philadelphia chromosome" in which *ABL1* activity is deregulated
FCMD	Fukutin: controls the migration and assembly of the neurons of the brain's cortex during development	Fukuyama congenital muscular dystrophy	80% of all cases have an inserted tandem repeat sequence within gene
FRDA	Frataxin: mitochondrial protein involved in regulation of iron transport and respiration	Friedreich's ataxia	Expansion of GAA triplet sequence

Table 3.9. (continued)

Gene	Protein product/function	Disease	Nature of mutation
TMC1	Transmembrane protein, possible ion channel, expressed specifically in hair cells of the cochlea	Hereditary deafness as well as progressive deafness	Mutations in different sites of gene cause either dominant or recessive forms of deafness
TYRP1	Tyrosinase-related protein involved in melanin production	Type 2 oculocutaneous albinism	Mutations affect maturation of melanocytes leading to drastic reduction in insoluble type of melanin
		Rufous type oculocutaneous albinism, also known as xanthism	Unrelated class of mutations interfere with proper melanin production

Table 3.10. Chromosome 10: Genes and diseases at a glance

Gene	Protein product/function	Disease	Nature of mutation
ALOX5	5-lipoxygenase: enzyme involved in generating leukocyte-specific cytokines	Atherosclerosis	Deletions or insertions in transcription factor-binding region of promoter (i.e., regulatory rather than structural defect)
ERCC6 (also *CSB*)	DNA-binding protein important in transcription-coupled excision repair	Cockayne disease	Truncating or missense mutations interfering with protein activity
OAT	Ornithine ketoacid aminotransferase enzyme in mitochondira: converts ornithine into the amino acid glutamate	Gyrate atrophy	Amino acid changes resulting in defective enzyme
OPTN	Optineurin: interacts with transcription factors to prevent cell lysis	Glaucoma	Amino acid substitutions

Table 3.11. Chromosome 11: Genes and diseases at a glance

Gene	Protein product/function	Disease	Nature of mutation
ATM	Phosphatidylinositol-3-kinase enzyme participating in signaling pathways to repair damaged DNA	Ataxia telangiectasia	Recessive mutations affecting ability of enzyme to act under conditions of DNA damage (e.g., radiation)
CNTF	Ciliary neurotrophic factor; neurotransmitter synthesis promoting protein specific to nerve cells	Amyotrophic lateral sclerosis (ALS)	Modifier mutation only; does not cause disease by itself but accelerates disease caused by *SOD1* mutation
HBB	Beta-globin: subunit of hemoglobin that transports oxygen in red-blood cells	Beta-thalassemia	Gene deletion
		Sickle cell anemia	Point mutation leading to amino acid substitution and change in structural properties of proteins
HMBS (or *PBGD*)	Hydroxymethylbilane synthase (more commonly called porphobilinogen deaminase): enzyme involved in heme-group metabolism	Acute intermittent porphyria	Mutations decrease enzyme activity
INS	Insulin: gene actually encodes a precursor protein called proinsulin that is processed further to make the functional molecule	Insulin-dependent diabetes mellitus (IDDM2)	Different alleles contribute to disease susceptibility rather than a single mutation causing the disease outright
KCNQ1	Voltage-gated potassium channel	Long Q-T syndrome	Dominant mutations causing a gain of function change in ion channel
SCT	Secretin: endrocrine hormone best known as a digestive hormone that stimulates liver and pancreas to send secretions to gut; also believed to have neuroendocrine functions because of specific receptors in brain cells	Autism	Mutations affect interaction of protein with receptors in brain cells

Table 3.11. (continued)

Gene	Protein product/function	Disease	Nature of mutation
SLC22A1L (aka *SLC22A18*)	Tumor suppressor gene encoding a carrier protein for organic actions	Breast, lung, ovarian, and other cancers	Mutations leading to uncontrolled activity of protein
SMPD1	Sphingomyelin phosphodiesterase: enzyme involved in metabolism of brain glycolipids	Niemann–Pick disease	
		Type A	Small deletions, nonsense mutations that result in truncated gene product of missense mutations that render the enzyme noncatalytic
		Type B	Missense mutations that produce defective enzyme with partial activity
TYR	Tyrosinase enzyme: participates in melanin biosynthesis	Oulocutaneous albinism (tyrosinase-negative type)	Recessive nonsense and missense mutations causing either complete (Type 1A) or partial (Type 1B) loss of enzyme function

Table 3.12. Chromosome 12: Genes and diseases at a glance

Gene	Protein product/function	Disease	Nature of mutation
ALK1	Activin-like kinase, receptor for transforming growth factor beta proteins with role in vascular development	Hereditary hemorrhagic telangiectasia	Mostly single-site mutations, resulting in improper development of blood vessels
AID	Activation-induced cytidine deaminase—RNA editing enzyme in B cells to splice out the constant chain genes of IgM antibodies	Autosomal recessive hyper-IgM immunodeficiency (type 2)	Mutations lead to dysfunctional enzyme and antibodies

Table 3.12. (continued)

Gene	Protein product/function	Disease	Nature of mutation
AQP2	Aquaporin—regulates the passage of water in and out of cells	Nephrogenic diabetes insipidus	Mostly point mutations that result in dysfunctional protein
CDKN4	Cyclin-dependent kinase 4, helps in regulating cell division	Cutaneous (skin) melanomas	Germ-line mutations result in abnormal protein
COL2A1	Alpha1(II) chain of type II collagen, which adds structure and strength to connective tissues	Achondrogenesis, type II (Langer-Sardino)	Several types—substitutions, deletions that interfere with normal formation of collagen
HNF1	Hepatic nuclear factor—regulates transcription of the glucagon gene	Type 1 diabetes mellitus	Structural defects that interfere with HNF dimerization and hence protein function
IBD2	Region on chromosome 12 linked to inflammatory bowel disease	Susceptibility to ulcerative colitis, minor effect on Crohn's disease	Unknown
KRTHB6	Member of keratin family—found in hair and nails	Monilethrix	Point mutations in coding regions prevent normal formation of keratin
PAH	Phenylalaninhydroxylase—enzyme in phenylalanine metabolism	Phenylketonuria	Several types of mutations resulting in enzyme with low or no activity
RECQL3	Helicase-like (DNA-unwinding) like protein with unknown function	Bloom syndrome, pre- and post-natal growth deficiency	Mutations have an impact on DNA replication
vWF	Complex protein that functions both as an anticoagulant factor and as a mediator of adhesion between platelets to the collagen exposed on the walls of blood vessels	von Willebrand disease—common inherited bleeding disorder, especially among women	Heterogeneous disorder—different types of mutations, resulting in defective von Willebrand factor

Table 3.13. Chromosome 13: Genes and diseases at a glance

Gene	Protein product/function	Disease	Nature of mutation
BRCA2	Tumor suppressor	Susceptibility to breast cancer	Mostly deletions of short DNA sequences that can lead to loss of tumor suppressor function
Cx26	Connexin-26: gap junction protein that regulates flow of potassium ions in cells of inner ear	Hereditary deafness	Various mutations lead to dysfunctional protein
EDNRB	G-protein-coupled receptor for endothelin-3, a growth factor in the embryo	Hirschsprung's disease; Waardenburg–Hirschsprung disease	Point mutations disrupt ability to bind endothelin
IRS2	Insulin receptor substrate involved in signaling pathways of sugar metabolism	Type 2 diabetes mellitus	Deletion leads to disease; certain polymorphisms are protective
PHF11	Putative transcription factor that regulates IgE-producing B cells	Susceptibility to asthma	Mutations cause splicing at different sites, which influences levels of IgE

Table 3.14. Chromosome 14: Genes and diseases at a glance

Gene	Protein product/function	Disease	Nature of mutation
NPC2	Glycoprotein involved in cholesterol transport	Niemann–Pick disease type C (small proportion of cases)	Single amino acid changes prevent cholesterol binding and the restoration of normal cholesterol levels in cells
PS1	Presenilin-1—membrane protein that may be involved in apoptosis	Alzheimer's disease (type 3)	Alternative splicing influences secretion of beta-amyloid
TCL1	Oncogene	T cell lymphomas	Translocation of gene within chromosome removes brakes on gene control

Table 3.14. (continued)

Gene	Protein product/function	Disease	Nature of mutation
TCRA	Alpha subunit of T cell receptor A	Burkitt and T cell lymphomas	Translocation of genes (*c-myc* from chromosome 8 and *TCL1*) to enhancer sequences in this gene

Table 3.15. Chromosome 15: Genes and diseases at a glance

Gene	Protein product/function	Disease	Nature of mutation
FBN1	Fibrillin—essential component of elastic fibers of connective tissue	Marfan syndrome	Mutations inherited in autosomal dominant manner; mutated FBN1 gene has variable expression
HEXA	Encodes part of an enzyme called beta-hexosaminidase A, which breaks down a lipid of nerve cells called ganglioside GM2	Tay–Sachs disease	Mutations in HEX A result in inability to break down GM2, which then accumulates in nerve cells
OCA2	Homolog of P-protein in mouse involved in tyrosone and melanin transport	Autosomal recessive oculocutaneous albinism	Several types (missense, nonsense, deletions, insertions), lead to inability to distribute melanin

Table 3.16. Chromosome 16: Genes and diseases at a glance

Gene	Protein product/function	Disease	Nature of mutation
CARD15	Caspase recruitment domain—intracellular protein involved in apoptosis	Crohn's disease	Mutations in leucine-rich repeat domain of CARD15 interfere with signal transduction
CD11	CD11 integrin cluster—involved in microbacterial cell adhesion	Crohn's disease	Unknown
CD19	Cell-surface protein involved in B-lymphocyte function; belongs to Ig superfamily	Crohn's disease	Mutations may interfere with growth of B-lymphocytes

Table 3.16. (continued)

Gene	Protein product/function	Disease	Nature of mutation
CYBA	Peptide chain of cytochrome b-245, part of NADPH complex responsible for respiratory burst in phagocytes	Chronic granulomatous disease (CGD)	Impairment in the function of NADPH oxidase complex
Dnase1	Enzyme that catalyzes breakdown of DNA	Systemic lupus erythematosus	Several mutations disrupt function; accumulation of undigested DNA
HBA1	Alpha chains of hemoglobin	Alpha-thalassemia	Mutations interfere with normal expression of hemoglobin alpha chains
IL4R	Interleukin-4 receptor	Crohn's disease	Mutations affect IL4-mediated regulation of mononuclear phagocyte function
MC1R	Melanocortin hormone receptor	Albinism	Mutations lead to defects in regulating production of melanin
PKD1	Integral membrane protein; participates in signal transduction pathways especially in cilia of kidney tubule cells	Polycystic kidney disease (PKD)	Defects in polycystins prevent their interaction
SPN	Sialophorin—involved in leukocyte adhesion	Crohn's disease	Unknown
TSC2	Tuberin; regulates GTPase-activated signaling pathways	Tuberous sclerosis	Deletions disrupt signaling pathways

Table 3.17. Chromosome 17: Genes and diseases at a glance

Gene	Protein product/function	Disease	Nature of mutation
AXIN2	Controls expression of genes in cell division process	Associated with colorectal cancer	Mutations may result in uncontrolled cell division
BRCA1	Tumor suppressor gene	Breast and ovarian cancer susceptibility	Mutations disrupt tumor suppressor function

Table 3.17. (continued)

Gene	Protein product/function	Disease	Nature of Mutatiion
GCGR	Glucagon receptor—G-protein-coupled receptor involved in sugar metabolism	Type 2 diabetes mellitus	Heterozygous missense mutations affect ability of receptor to bind glucagon
p53	p53 tumor antigen	Several types of cancer	Mutations lead to uncontrolled cell division
PMP22	Peripheral myelin protein—major component of myelin	Charcot–Marie–Tooth disorder type 1A	Gene duplication leads to excess PMP22; abnormalities in myelin structure and function
Rb	Expresses protein involved in regulating cell cycle in retina and bone cells	Retinoblastoma	Recessive mutations lead to retinoblastomas and osteosarcomas

Table 3.18. Chromosome 18: Genes and diseases at a glance

Gene	Protein product/function	Disease	Nature of mutation
FECH	Ferrochelatase—catalyzes last step of heme synthesis	Erythropoietic protoporphyria (EPP)	Mutations lead to low or no enzyme activity, which leads to the accumulation of protoporphyrins in body
MADH4	Tumor suppressor gene	Hereditary hemorrhagic telangiectasia (HHT) along with juvenile polyposis	Defects disrupt tumor suppressor function of protein
NPC1	Protein believed to help regulate intracellular transport of cholesterol from lysosomes to other parts of cell	Niemann–Pick disease—Type C	Mutations in protein seem to affect transport function; accumulation of cholesterol within lysosomes

Table 3.19. Chromosome 19: Genes and diseases at a glance

Gene	Protein product/function	Disease	Nature of mutation
APOE4	Apolipoprotein E: lipid/cholesterol transporting protein of the chylomicron	Alzheimer's disease type 2 Atherosclerosis	Defects can lead to inability to bind receptors, increased blood cholesterol, and risk of atherosclerosis as well as formation of amyloid plaques
DM1	Protein kinase expressed in skeletal muscle	Myotonic dystrophy type 1	Amplification of CTG triplet nucleotide; disease severity varies with repeat number
HAMP	Antimicrobial peptide produced in liver	Juvenile hemochromatosis	Deletions, point and frameshift mutations, and truncations lead to inactive protein
JAK3	Signaling molecule in interleukin-2 mediated pathway for lymphocyte development	Severe combined immunodeficiency (SCID)	Mutations disrupt lymphocyte development
SCN1B	Voltage-gated sodium channel	Epilepsy	Unknown

Table 3.20. Chromosome 20: Genes and diseases at a glance

Gene	Protein product/function	Disease	Nature of mutation
ADA	Adenosine deaminase—metabolism of adenosine	Severe combined immunodeficiency (SCID)	Deficiency, low activity results in accumulation of adenosine; immature lymphocytes are sensitive to adenosine and fail to develop properly
		Hemolytic anemia	Mutation leading to excess ADA causes hemolytic anemia due to destruction of red blood cells

Table 3.20. (continued)

Gene	Protein product/function	Disease	Nature of mutation
ADAM33	A disintegrin and metalloprotease (ADAM) involved in cell to cell-surface interactions	Asthma susceptibility	Unknown
AVP	Arginine vasopressin—pituitary hormone/growth factor	Neurohypophyseal diabetes insipidus	Mutations affect activity of anti-diuretic hormone vasopressin
EDN3	Endothelin-3—components of signaling pathway in neural crest colonization process	Hirschsprung's disease

Waardenburg syndrome | Mutations interfere with signaling pathways and disrupt proper development |
| *TNFRSF5* | CD40 receptor for the tumor necrosis factor | Hyper Ig-M immunodeficiency syndrome | Truncations and point mutations lead to inability to express the receptor |
| *SNAP-25* | Synaptosomal-associated protein 25; expressed specifically in nerve cells; helps trigger release of neurotransmitters | Attention deficit hyperactivity disorder (ADHD) | Unknown |

Table 3.21. Chromosome 21: Genes and diseases at a glance

Gene	Protein product/function	Disease	Nature of mutation
AIRE	Autoimmune regulator—transcription factor found in cytoplasm and nucleus of cells	Autoimmune polyglandular syndrome (APS)	Unknown
APP	Amyloid precursor protein	Alzheimer's disease type 1	Mutations in exons 16 and 17 lead to formation of beta-amyloid, which gives rise to plaques
SOD1	Superoxide dismutase—housekeeping enzyme; removes supercharged oxygen molecules from cells	Amyotrophic lateral sclerosis (ALS)	Unknown

Table 3.21. (continued)

Gene	Protein product/function	Disease	Nature of mutation
20 to 50 genes of chromosome 21	Several	Trisomy 21 (Down syndrome)	Presence of an extra copy of chromosome 21 in cells may disrupt normal gene-expression patterns

Table 3.22. Chromosome 22: Genes and diseases at a glance

Gene	Protein product/function	Disease	Nature of mutation
BCR	Breakpoint cluster region—normal function still unknown, may be signaling protein	Chronic myeloid leukemia (CML)	Reciprocal translocation between chromosomes 22 and 9; results in short chromosome 22 ("Philadelphia chromosome") and BCR-ABL1 fusion protein (see Table 3.9)
NF2	Putative tumor suppressor protein	Neurofibromatosis 2	Mutations interfere with function of protein (as yet unclear)
Sox10	Transcription factor involved in regulation of embryonic development and in determination of cell fate	Waardenburg–Hirschsprung disease	Mutation leads to extension of the peptide, which alters its function

Table 3.23. Chromosome X: Genes and diseases at a glance

Gene	Protein product/function	Disease	Nature of mutation
ATP7A	Enzyme that transports copper ions into cells of brain and bones during development	Menkes disease	Point mutations leading to deficiency or complete loss of enzyme activity
BTK	Bruton tyrosine kinase: signal transduction member involved in maturation of B cells	X-linked agammaglobulinemia	Various point mutations leading to loss of function of kinase

Table 3.23. (continued)

Gene	Protein product/function	Disease	Nature of mutation
COL4A5	Alpha chain of type IV collagen, which is a structural component of basement membranes	Alport syndrome	Formation of abnormal proteins, which causes a dysfunctional basement membrane
DMD	Dystrophin: cytoskeletal protein in muscle fibers that bridges the internal cell surface with its extracellular matrix	Muscular dystrophy	
		Duchenne type (severe)	Small deletions leading to a shift of reading frame and complete loss of function of protein
		Becker type (mild)	Large deletion without shift in frame resulting in the production of a partially functional dystrophin
EMD	Emerin: nuclear membrane protein of skeletal and cardiac muscles that anchors membrane to cytoskeleton	Emery–Dreifuss muscular dystrophy	Missense or nonsense mutation leading to formation of a non-functional protein product
HEMA or *F8*	Blood-clotting factor VIII	Hemophilia A	Heterogeneous group of point mutations
HEMB or *F9*	Blood-clotting factor IX	Hemophilia B	
		Classical hemophilia B	Mutations in functional gene
		Hemophilia B Leyden	Point mutations in promoter
IL2RG	Component (gamma subunit) of an interleukin (IL2) receptor present on lymphocytes	Severe combined immunodeficiency (SCID)	All types; more than 250 different mutations have been found to cause gamma chain deficiency
MAGEB gene family	Melanoma-associated antigens: normal function unknown	Melanoma	Expressed in melanoma patients

Table 3.23. (continued)

Gene	Protein product/function	Disease	Nature of mutation
MeCP2	Nuclear protein that inhibits the expression of specific genes during embryonic development by specifically binding methylated DNA in the promoter regions	Rett syndrome	Lethal mutation in males; mosaicism in females allows them to survive because protein in normal cells partially compensates for the mutation
PIG-A	Enzyme in biosynthesis of cell membrane proteins that act as anchors for various molecules	Paroxysmal nocturnal hemoglobinuria	Point mutations that inhibit enzyme leading to destabilization of RBC membranes
TNFSF5	Marker on T cell surface for specific B cell marker CD40	Hyper Ig-M immunodeficiency syndrome	Point mutations that interfere with ability of molecule to bind to B cells

Table 3.24. Chromosome Y: Genes and diseases at a glance

Gene	Protein product/function	Disease	Nature of mutation
CSF2RA	Granulocyte-macrophage colony-stimulating factor, alpha subunit	Acute myeloid leukemias of the M2 subtype	Mutations may lead to inability to bind colony-stimulating factor
SRY	Transcription factor involved in sexual differentiation	Gonadal dysgenesis (in XY females)	Point mutations leading to impaired transcription of genes under the control of the SRY transcription factor
		XX male syndrome	Translocation of region to X chromosome followed by nondisjunction

Chapter 4

Genes for Normal Functions

The human genome consists of tens of thousands of genes that contain the information for making proteins and RNAs. These, in turn, perform the myriad functions that give us our shapes, forms, and abilities. They affect virtually all of our activities from the invisible intracellular events that keep our genes intact and read them into proteins to the things we do to live—eat, breathe, and sleep—as well as all of our physical traits, both seen and unseen. We have genes for basic sub-cellular activities such as DNA and protein synthesis and maintenance, and for the key cellular components that help conduct these processes. Our cells are organized into specialized compartments and organelles to perform specific activities—and so at this next level we have information for the genetic basis of these structures and functions. Genes also control cell division, ensuring that all cells have the same genetic information. Our bodies have many different types of cells even though they all contain the identical genetic information. How does this happen? Again, the answer is in our genes—they provide the control mechanisms that enable different parts of the genome to be switched on and off as needed. And finally there are genes that influence the physical and behavioral features that make each one of us who we are.

The preceding description should not be interpreted as a statement of genetic determinism—namely the idea that "genes are us" and determine who we are or what we can be. Our genes do not have that power. On the other hand, it would be very difficult, if not impossible, to think of any area of our lives where our genes do not have some effect. In this chapter, we look at different genes in the human genome to see how they exert their influence on different aspects of our lives.

Obviously, a complete catalog of all of the human genes and their functions is beyond the scope of a single volume. Instead, in this chapter, we have chosen some representative examples of genes from a broad spectrum of activities. We have organized these functions in what we hope is a user-friendly fashion—alphabetically by traits or functions. Wherever possible, we have used the common terms for structural or functional characteristics rather than the names of the genes to identify an entry. Of course, we realize that various people might refer to the same trait by different names, so we encourage frequent use of the subject index to pinpoint the gene you are looking for. Readers are also urged to use the "see also" notes at the end of each entry because these are intended to point you toward related entries not only in this chapter but also in other chapters. Occasionally, you might find that brushing up on some of the terms in Chapter 2 will help you to understand the material about specific genes more easily.

ALCOHOL METABOLISM

Everyone knows about the intoxicating effects of drinking alcohol. And while it is true that this is the inevitable consequence of drinking to excess, it is also true that the definition of "excess" varies from one individual to another. Some people can drink several glasses of wine or beer without any noticeable effects, whereas others start to feel intoxicated after just one glass. Often, one feels the effects more quickly when drinking on an empty stomach, as compared with having a drink with a meal. In general, women have been found to be more susceptible than men to the physiological effects of alcohol. And finally, intoxication and associated ills notwithstanding, some researchers have claimed that the daily consumption of a moderate amount of alcohol can actually be *beneficial* to one's health, helping for instance in staving off heart attacks!

Why are the effects of alcohol so varied? At least part of the explanation lies in the way the human body processes alcohol. Intoxication is the result of the accumulation of alcohol and intermediate products of its metabolism in the blood. How someone reacts to alcohol depends on how efficiently the alcohol is broken down and removed from his or her system. When a person drinks, alcohol is absorbed from the stomach and intestines into the bloodstream. The rate of absorption depends on other contents of the stomach and intestines—food, especially fats and proteins, slows down the rate of alcohol absorption considerably. Blood transports the alcohol to sites where it is metabolized, but metabolism usually occurs at a slower rate than absorption so that once the alcohol is in the blood, it stays there for some time. The majority of alcohol metabolism occurs in the liver, but some alcohol is broken down in the kidneys and lungs and excreted in urine or breath, respectively. Alcohol that is not broken down is also excreted via the urine, sweat, and breath.

It should be noted that we are using the word alcohol to mean ethanol (chemically CH_3CH_2OH). Ethanol is the product of fermentation of sugars by yeast and bacteria, and we frequently encounter it in our guts, not only when we drink alcoholic beverages but also more regularly because of the metabolic activities of bacteria that normally live in our intestines—studies have shown that our intestinal flora produce anywhere from 12 to 40 grams of alcohol per day. In fact, the alcohol-degrading enzymes probably evolved to detoxify this alcohol, and our ability to process alcoholic drinks is just a side benefit! Although alcohol has no nutritive value, it does provide energy: about 7 kilocalories per gram consumed. Meanwhile, there are several alcohols besides ethanol, many of which also have physiological functions in the body and must be processed appropriately. An example is retinol, a precursor to retinoic acid, which is a hormone important in cellular differentiation.

Alcohol Dehydrogenase (ADH)

Alcohol metabolism is a process of gradual or stepwise oxidation, which is the replacement of hydrogen groups with oxygen. The first step is the conversion of an alcohol to aldehyde, such as ethanol (CH_3CH_2OH) to acetaldehyde (CH_3CHO). By and large, this reaction is catalyzed by a group of enzymes called the *alcohol dehydrogenases* (ADHs).

ADHs are produced by a 380 kb cluster of seven genes (*ADH1–7*) located on chromosome 4. Based on patterns of distribution and variation of these genes, scientists have deduced that these distinct ADHs evolved from a single common ancestor and differ from one another with respect to the particular alcohol on which they act and the tissues in which they are produced. An individual's sensitivity and susceptibility to alcohol is determined to a large extent by the way in which these enzymes function.

ADH 1, 2, and 3 together form the class I alcohol dehydrogenases, whose chief function is to catalyze the conversion of ethanol to acetaldehyde. ADH1, which is produced in the liver, is most active in the fetus, particularly during the early stages of development. Its activity falls off in adulthood. In the adult liver, it is ADH2 that is responsible for most of the alcohol metabolism. ADH2 is also produced in the lungs—where it appears quite early in development—and in the adult kidneys. ADH3 is most active in the intestines and kidneys of fetuses and newborn infants. Most of us have a "cocktail" or mixture of ADHs, the exact composition of which is dependent on several factors, including age, location, and, most importantly, the specific *ADH* alleles that are present. This heterogeneity plays a role in determining the variations in alcohol susceptibility of different people.

Epidemiological studies have shown that consumption of moderate quantities of alcohol is associated with lowered incidence of heart disease. This association is related to the specific *ADH3* genotype of an individual. The two allelic forms of this gene code for a rapid-acting and a slow-acting enzyme, respectively. The studies showed that men who were homozygous (i.e., have both copies of the gene) for the slow-acting form of ADH3 had the most reduced risk for heart attacks with regular but moderate alcohol intake. (These men also had the highest levels of high-density lipoproteins [HDLs] in the blood, another factor shown to be linked to lowered risk for heart attacks.)

Scientists believe that ADH4 is called into action when a person consumes very large amounts of alcohol. In these instances, this enzyme accounts for up to 40 percent of the total alcohol breakdown activity of the liver. ADH5 has virtually no effect on ethanol but acts upon high-molecular-weight alcohols. It also plays a role in eliminating formaldehyde, a strong irritant in cells. *ADH5* is the only gene from the group that is expressed in the brain, although its exact physiological function there is as yet undiscovered. Even less is known about the physiological activity of ADH6, although it is thought to have a function distinct from that of the other ADH enzymes. Messenger RNA transcripts of this gene (but not the protein) have been detected in the stomach and liver. Finally, ADH7 is one of the major ADH species produced in the stomach but not in the liver. It does not act on ethanol but is specific for retinol, which suggests a function for this enzyme in retinoic acid synthesis.

Besides ADH, liver cells sometimes use two additional enzymatic pathways to convert ethanol to acetaldehyde, albeit at much lower levels of activity. The endoplasmic reticulum, which is a cellular site with known detoxification functions, contains a cytochrome enzyme (called P450IIE1 or CYP2E1, encoded on chromosome 10) that is induced in response to episodes of chronic drinking. The liver enzyme catalase (*CAT* gene on chromosome 11) can also catalyze alcohol oxidation, although it does not appear to do so efficiently under physiological conditions.

Acetaldehyde Dehydrogenase (ALDH)

The next step in alcohol breakdown is the conversion of acetaldehyde to acetate or acetic acid (CH_3COOH), which is catalyzed by enzymes called *acetaldehyde dehydrogenases* (or ALDHs). There are two major forms of ALDH in liver cells, which differ mainly with respect to where they are present within the cells. The gene *ALDH1* (also known as retinal dehydrogenase 1), located on chromosome 9, encodes a form of the enzyme that is distributed throughout the cytosol of liver cells. The second gene, *ALDH2*, is on chromosome 12, and its protein product ALDH2 is found specifically inside the mitochondria, for which reason it is also known as liver mitochondrial ALDH. Population studies have shown that about 50 percent of East Asian people do not have a functional ALDH2. In these people, the mutation responsible for this lack of function is associated with a lower alcohol tolerance. This low tolerance in turn correlates with a lower frequency of alcohol abuse and addiction in these populations compared with others.

See also: Respiration; Alcoholism (Ch. 5).

APOPTOSIS

A cell's cycle consists of many phases such as cell division, growth, and finally, inevitably, death. Programmed cell death, also called apoptosis, is very important in keeping an organism healthy and functioning normally. Apoptosis prevents the overgrowth of tissues in the adult animal. During embryonic development, it takes place in specific cells of the body at specific times in order to correctly form different body parts. As you can imagine, apoptosis has to be a very tightly controlled process. On the one hand, old cells must die in order to make room for the newly forming cells of an organ or tissue. On the other hand, if too many cells are killed, the organ will not be able to function properly. Consequently, apoptosis is a multistep process that involves many participants acting in different ways to coordinate the events in one cell with other events going on in the body. Activators initiate apoptosis, effectors carry out the actual process of killing the cell, and finally inhibitors halt the process when enough cells have died.

Scientists first studied apoptosis in detail while investigating the nature of development in the nematode *Caenorhabditis elegans*, and so the first apoptosis genes to be identified and characterized in detail were from this tiny worm. As it turns out, these genes are fairly well-conserved in the evolution of animals, and nearly all of the worm genes have functional counterparts in human as well as other animal genomes. Much of this early work of apoptosis gene identification was carried out in the laboratories of Sydney Brenner (1927–) and Sir John Sulston (1942–) in Cambridge, England, and later at that of H. Robert Horvitz (1947–) at MIT in Cambridge, Massachusetts. These three researchers received the 2002 Nobel Prize in Physiology or Medicine for this work.

Apoptosis is the outcome of different signals from the cell's environment. Consequently, many of the mechanisms involved in this process coincide or intersect with other cellular pathways (called signal transduction pathways) that transmit messages either from the cell surface to different locations inside the cell or from one part of the cell to another. In the next section, we look at some members of these pathways that are involved in initiating or effecting cell death. For a more complete discussion of signal transduction, see the corresponding entry later in this chapter.

Nuc-1

In 1976, John Sulston was studying the function of apoptosis in the embryonic development of the nervous system in *C. elegans* when he discovered the first apoptosis gene. He found certain mutants that were unable to get rid of fragmented DNA in dying cells. He named the mutated gene *nuc-1* (for nuclease defective) because affected cells were deficient in the activity of a DNA endonuclease (i.e., an intracellular DNA-degrading enzyme).

The human analog of *nuc-1*, also known as *PPARD*, is found on chromosome 6. This gene encodes a nuclear hormone receptor that controls the activity of the endonuclease. The receptor belongs to a large class of proteins called the peroxisome proliferator-activated receptor (PPAR) superfamily. Peroxisomes are membrane-bound, enzyme-containing compartments found in most eukaryotic cells, where the breakdown of complex molecules such as fatty acids and nucleotides takes place. These compartments prevent the accumulation of various harmful compounds in the cell by ensuring their degradation and also protect other cellular components from these wastes and the enzymes that process them. Peroxisomes are typically formed in response to chemicals—also called their proliferators—that tell the cells when they are needed. (Their importance is evident from the severity of genetic diseases associated with peroxisomal defects such as

Refsum's disease and Zellweger's syndrome.) PPARs, acting in response to a large number of chemicals, mediate the size and number of peroxisomes that are formed in a cell. The human *Nuc-1* controls the activity of peroxisomes that digest DNA fragments formed during programmed cell death events. The activity of this gene product is therefore a later step in the process of apoptosis. Defects in human *Nuc-1* have been associated with colon cancer.

Ced-1 and Ced-2

Following the discovery of *nuc-1*, Edward M. Hedgecock, a postdoctoral researcher in Sulston's lab, identified mutations in two new genes that affected apoptosis during development. As in the case of *nuc-1*, these genes did not seem to be essential for causing cell death per se. Instead, they interfered with the postdeath processing of cellular material. Hedgecock found that the mutations in these genes (which he named *Ced-1* and *2* for *ce*ll *d*eath abnormal) blocked the phagocytosis (engulfment) of cells that died during development by macrophages and other phagocytes.

Ced-1 encodes a receptor protein on the cell membrane that allows an engulfing cell to recognize a dead cell. This protein is similar to human SREC (scavenger receptor from endothelial cells), a protein of endothelial cells (i.e., cells that make up the walls of blood vessels) that mediates the binding and degradation of lipoproteins. When a cell dies, it releases lipoproteins. This activates the scavenger receptors on endothelial cells and prompts them to move to the site of the dead cells and engulf the lipids. SREC is the product of the *SCARF1* gene on chromosome 17. Scientists have found evidence to suggest that a gene, called *CD91*, on chromosome 12, may be the human homolog of *Ced-1*. *CD91* expresses a protein that is believed to act as a receptor involved in engulfing apoptotic cells in mammals.

The nematode gene *Ced-2* encodes a protein similar to the human protein CrkII. This protein is the product of a gene on chromosome 17 that is also homologous to certain avian sarcoma virus sequences—in other words, it is an oncogene (see the entry on oncogenes and proto-oncogenes in this chapter). The CrkII protein takes part in signal transduction mechanisms involving tyrosine kinase proteins. Both Ced-2 and CrkII are believed to act via a pathway that regulates the reorganization of a dying cell's cytoskeleton during its engulfment by phagocytes.

Suicide Genes

Although the discovery of *nuc-1* and *ced* genes opened the way to studying the process of apoptosis in greater detail, researchers found that these genes were not in fact directly involved in controling the process of cellular suicide. Robert Horvitz identified key genes for the actual cell death process when he went to work with Brenner and Sulston in 1974. It was after this that scientists began to see apoptosis as a systematic process controlled by several genes. As it turned out, there were three worm genes—named *egl-1*, *ced-3*, and *ced-4*—required for activation of programmed cell death and a fourth, called *ced-9*, that prevented cell death.

- *Egl-1* encodes a protein that activates cell death by binding to ced-9 and releasing ced-4 from the ced-9/ced-4 complex. The expression of *egl-1* is controlled by other proteins (e.g., transcription factors) that decide the fate of different cells in a tissue or organ, namely which cells get to live and which must die. In the human genome, there are several genes called "BH3 only" killer genes that are very similar to *egl-1*. Examples include *HRK* (for "hara-kiri") on chromosome 12 and *BIK* on chromosome 22. One of the *C. elegans* transcription factors that controls *egl-1* is produced by the gene *ces-2*, which is similar to the gene for a human transcription factor called E2A-HLF. This protein is the product of a leukemia-associated oncogene that scientists believe acts by interfering with apoptosis.

- *Ced-3* encodes an aspartate-specific cysteine protease, also known as a caspase. Caspases are enzymes that catalyze the hydrolysis of peptide linkages within oligopeptides or polypeptides. In other words, they mediate apoptosis by degrading cellular proteins. The *CASP2* gene on chromosome 7 in humans has enzymatic properties and apoptotic functions similar to those of *ced-3*.

- *Ced-4* encodes an apoptotic protease activating protein, which binds and activates ced-3 proteins at special sites on the molecule called caspase-associated recruitment domains (CARDs). The homolog in humans is *Apaf-1* (apoptotic protease activating factor-1) on chromosome 12.

- The gene product of *ced-9* binds to and inhibits *ced-4* activity and is in turn inhibited by the *egl-1* encoded protein. In humans, a gene called *Bcl-2* (on chromosome 18) is the functional equivalent of *ced-9* and blocks the action of *Apaf-1*.

See also: Cell Cycle, Signal Transduction (Ch. 2).

BEHAVIOR

The relationship between genes and behavior is a subject of considerable discussion, confusion, and sometimes even outright denial. There is no simple answer to the question of whether or not genes control behavior, and any response must be qualified with several explanations. Certainly genes do not "determine" behavior in absolute terms—there is no one gene for being good, bad, intelligent, polite, or ill-mannered—but, on the other hand, they play a significant role in influencing such traits. The environment also plays a significant role, as does the interaction between different genes and the environment. In other words, behavior is a multifactorial trait. The relationship between genes and behavior is far from deterministic, but there is no doubt that our genes influence our behavior both in direct and in subtle ways.

At least part of the confusion in a discussion of this subject comes from the very broad range of functions and activities described by the word "behavior." We can think of behavior in terms of an organism's interaction with its surroundings. An instinctive response, such as the drive to evade a perceived danger—either by fighting it or by escaping from it—is a type of behavior that almost all animals exhibit. Also known as the "fight-or-flight" response, this type of behavior is the outcome of the action of the hormone adrenalin (for details, see the section on adrenalin under the entry for hormones and the endocrine system). Here the genetic link is relatively simple to understand—any gene involved in the production and regulation of the hormone, as well as the production and regulation of its receptors, will have an impact on the way in which the animal responds to a threat. Even in this relatively straightforward case, however, we should be cautious in our assessments because hormones have no more of a deterministic relationship with behavior than do genes.

When we think of human behavior, however, what usually comes to mind are actions associated with higher cognitive abilities, such as our ability to speak and communicate with others, display emotions, form social relationships (fall in love, choose friends or enemies), learn new facts, and form memories. Traits that give each of us our individual personalities also fall under the umbrella category of behavior. And it is in abilities such as these that the explanation of behavior in terms of genes and proteins becomes tricky. To attribute these types of behaviors to genes is not just an oversimplification but is in fact grossly misleading and possibly even harmful under certain circumstances. Many racial, ethnic, and cultural stereotypes are often based on exaggerated or false correlations between genes and behaviors. All that genes can do is encode and express proteins, which then carry out specific functions. Behavior is the result of complex and reticulated ways in which these proteins interact with each other and with the environment. Like all complex traits, even a simple behavioral trait involves many genes, environmental factors, and the interaction between these two sets of factors.

Sir Francis Galton (1822–1911) was one of the first scientists who attempted to examine the relationship between heredity and human behavior. He studied behavioral correlations within families and developed research techniques that are still used today, such as twin and adoption studies. Although Galton was convinced that criminal behavior has a hereditary basis, he did observe the "difficulty of distinguishing that part of [man's] character which has been acquired through education and circumstance, and that which was in the original grain of his constitution."

In many instances, it is easier to understand a system when it goes wrong rather than under normal conditions. Therefore, studies on patients with behavioral disorders can shed some light on the complex interplay between our genes and environment that results in the development of these disorders. In recent years, scientists have found evidence from such studies to link specific behaviors with genetic loci. Examples of behavioral disorders in which genes have been implicated include attention deficit hyperactivity disorder (ADHD), autism, Parkinson's disease, and schizophrenia. (See the next chapter for more details about these diseases.)

Many of the genes implicated in behaviors are linked to the neuroendocrine system. This is perhaps not so surprising considering the extent to which we use this system to respond to our surroundings. Attention deficit hyperactivity disorder, Parkinson's disease, and schizophrenia, for instance, have been linked to mutations in the genes of the *dopamine* system. (See the individual entries for diseases in the next chapter.) Dopamine is a chemical neurotransmitter that is produced in the brain and plays an important part not only in motor and endocrine functions but, as scientists discovered more recently, in higher-order cognitive functions as well. The intricate patterns and networks through which this system modulates cognitive functions are not well-understood, but investigations of the aforementioned diseases have helped to fill in parts of the puzzle.

One of the best-studied genes in the dopamine system is *DRD4* on chromosome 11. The gene, which has been implicated in all three of the disorders mentioned in the preceding paragraph, encodes a dopamine receptor that inhibits adenyl cyclase and triggers a specific cascade of signal proteins (see the entry on signal transduction) in the nerve cells where it is expressed. Polymorphisms in the coding region of *DRD4* have been found to be associated with what psychologists define as the "novelty-seeking trait." Individuals who score higher than average on the novelty-seeking scale are characterized as impulsive, exploratory, fickle, excitable, quick-tempered, and extravagant, whereas those who score lower than average tend to be reflective, rigid, loyal, stoic, slow-tempered, and frugal. (Note that these characteristics are only tendencies and not immutable traits.)

Autism is a developmental/behavioral disorder in which individuals are unable to form proper social and emotional relationships with others. Interestingly, this disorder has been linked to mutations in the gene (on chromosome 11) for secretin, which was first identified as a digestive hormone. However, further work on this hormone system revealed that receptors for secretin were present not only in the cells of the gut but in certain areas of the brain as well, indicating an as-yet-unidentified neuroendocrine function for this hormone. Autism has also been linked to mutations in the *SCTR* (secretin receptor) gene on chromosome 2— which is expressed specifically in the brain cells—but not to mutations in secretin receptor genes expressed in the gut.

In 1999, scientists showed that the learning ability of mice was boosted when a gene coding for a protein known to be associated with memory was inserted into the brain cells of these animals. The popular media quickly dubbed this gene "the smart gene" or the "IQ gene," even though improved memory is only one of many criteria for defining intelligence. The gene in question—*NR2B*—encodes N-methyl-D-aspartate receptor 2B, found in humans on chromosome 12. In the experiment, overexpression of N-methyl-D-aspartate (NMDA) receptor 2B (NR2B) in transgenic mice led to increased activation of NMDA receptors. NMDA receptors increase the efficiency of synaptic transmission of the nerve impulses thought to underlie certain kinds of memory and learning.

In 1993, there was considerable excitement in the media over the apparent identification of what some people called the "gay gene" (i.e., a gene that seemed to predispose for homosexuality). However, an examination of the literature shows that the researcher—Dean Hamer at the NIH—has only found a correlation between homosexual men and a certain genetic locus, and no causal or mechanistic links have been found so far. Researchers in the field of behavioral genetics have reported similar links between genetic loci and several other physical behaviors, including aggression, femininity, and nurturing. Most of these studies have not been replicated, and the nature of these associations is still not clear. Indeed, the picture remains as complicated as ever—namely, there is no deterministic relationship between genes and behavior, and any behavior is most likely a combination of genetic, environmental, and other interactive factors.

See also: Environment (Ch. 2); Attention Deficit Hyperactivity Disorder (ADHD), Autism and Pervasive Developmental Disorders (PDD) (Ch. 5).

BLOOD

Long before humans understood the functions of blood, they had recognized how essential this substance was for sustaining life. Ancient medical treatises dating back to Hippocratic times, and perhaps even earlier if we consider other cultures such as those of China and India, talk about the vitality of blood, and William Harvey's (1578–1657) discovery of blood circulation is marked by historians as one of the key events of the "scientific revolution" in the seventeenth century. Enduring metaphors such as "blood is thicker than water" or "blood brothers," or some trait being "in the blood," stand as testimony to the importance of blood in popular culture as well. Indeed, for centuries before the subjects of our book—genes—were discovered, it was blood that was held to be the carrier of our traits and identities.

So what is it about blood that is so important? Reaching virtually every nook and cranny in the body and constituting about 8–10 percent of the total body weight, blood is one of the most active contributors to the integrity of the organism. It supplies our tissues and organs with the many nutrients essential for their functioning and helps maintain a constant environment (in terms of temperature, moisture, or acidity, for instance) in the body. Blood serves as the "highway" or transport network through which cells and soluble substances produced in one part of the body reach other sites where they are needed. It is also responsible for protecting the body against various harmful agents, such as bacteria, viruses, pollen, dander, and dust. Last but not least, blood is the medium through which a pregnant mother nourishes her fetus until birth.

Blood is a complex substance composed of both cells (about 45 percent by volume) and a fluid portion called the plasma. The cellular components of blood include red blood cells (RBCs), or erythrocytes, a mixture of immune cells collectively called the white blood cells, or leucocytes, and platelets. The plasma consists of various salts and soluble proteins suspended in water. The proteins, which comprise nearly 10 percent of the total plasma volume, are responsible for the nutritive functions as well as providing blood with its volume and pressure. This section discusses some of the most important components of blood along with the genes that govern their formation and activity.

Albumin

The bulk of the plasma proteins—from 50 to 80 percent—are accounted for by the *albumins*, which play an important role in maintaining blood volume and pressure by providing the force needed by capillaries to draw water from tissues into the blood. The albumins also carry small molecules such as steroids, fatty acids, and hormones that need to be transported from one part of the body to another. Encoded by a gene on the human chromosome 4, the blood albumins are small proteins that are produced in the liver. Mutations in the albumin gene result in the formation of anomalous proteins but have not so far been associated with any serious diseases.

Globulin

Named for their globular shape, the *globulins* are a very widely dispersed group of blood proteins with a variety of activities in the body. The serum or plasma globulins comprise about 35 percent of the blood proteins, although actual levels fluctuate within a much broader range, depending on the circumstances under which the organism lives. There are three major types of globulins in human blood. They are designated as alpha, beta, and gamma globulins, depending on how they separate in an electrical field. The alpha and beta globulins are mainly involved in the transport of proteins, fatty acids, and vitamins from one part of the body to another. Monitoring the levels of these globulins is useful in diagnosing various diseases because their levels increase under conditions of acute inflammation, hepatitis, and malnutrition. Notable examples of serum alpha globulins include *thyroxine binding globulin* (TBG) and *retinol binding protein* (RBP4). TBG, encoded by a gene on the X chromosome, is important in transporting the hormone thyroxine from the thyroid gland to appropriate locations in the body. This is very important, as the thyroid hormone regulates key body functions such as the rate of metabolism and heartbeat in the body. Defects in the *TBG* gene are often associated with X-linked thyroid deficiencies. RBP4 is the specific carrier for retinol (vitamin A) in the blood and is expressed by a gene on chromosome 10. It delivers retinol from the liver, where the vitamin is stored, to peripheral tissues. Mutations in this gene cause persistently low levels of both the binding protein and vitamin A, leading to an accumulation of carotenes in the blood that is characterized by an orange discoloration of the face, nails, fingers, and toes. Vitamin A supplements seem unable to help correct this deficiency. Hereditary defects in the synthesis of retinol binding protein have also been associated with cases of night blindness (a consequence of vitamin A deficiency).

Perhaps the best-known example of a serum beta globulin is the protein *transferrin*, a glycoprotein encoded by a gene on chromosome 3. Its main function is to carry iron (in the form of Fe^{3+} or ferric ions) from the intestine, bone marrow, spleen, and liver to all proliferating cells in the body. Transferrin carries the iron via the blood to the recipient cells, enters the cells by means of specific receptors on the cell membranes, and then deposits the ions into a special acidic compartment. This intracellular iron is particularly important for the synthesis of enzymes that catalyze DNA replication, which is in turn required for proper cell division. In addition to iron transport, transferrin also seems to play an unrelated physiological role as a pollen-binding protein involved in removing certain types of allergy-inducing organic molecules from the bloodstream. Deficiencies in transferrin, due to improper gene expression, cause a genetic disease called atransferrinemia, which is characterized by severe anemia.

The gamma globulins, also known as *antibodies*, are important in fighting off disease, as they attach to and help destroy specific foreign substances in the body. Of all serum globulins, they are the most variable in level and increase quite dramatically when the body is subject to an infection or immunization. The gamma globulins are very complex in nature and activity and are discussed in greater detail under the entry on immunity.

See also: Immunity.

Fibrinogen

Fibrinogen is one of the important factors in mediating the clotting of blood. It comprises about 4 percent of the blood proteins. The gene for fibrinogen is also on chromosome 4 but at a different location from the plasma albumin gene. For more details about this protein, see the entry on blood clotting.

Hemoglobin

Unlike unicellular organisms such as the bacteria, which can easily obtain oxygen by diffusion across their cell membranes, multicellular organisms have many cells and tissues that do not have direct contact with the outside environment and so cannot get oxygen by simple diffusion. In the case of humans and many other animals, the function of transporting oxygen to different parts of the body is performed by the blood. More specifically, it is performed by hemoglobin, a protein that is carried in the red blood cells (RBCs) and that gives blood its characteristic red color.

There are some 20–30 billion RBCs in an average human, each containing about 280 million hemoglobin molecules. Hemoglobin picks up oxygen in the lungs and delivers it to the tissues so that cells can burn glucose and produce energy. It is therefore absolutely essential for the biochemical functioning of all the cells in the body. In addition to transporting oxygen from the lungs to peripheral tissues, hemoglobin also transports CO_2 in the opposite direction. In fact, hemoglobin acts by exchanging molecules at both ends: CO_2 is exchanged for O_2 in the lungs, while O_2 is exchanged for CO_2 in the muscles and other tissues.

Hemoglobin belongs to a class of proteins called the *metalloproteins*, so named because heavy-metal atoms such as iron, zinc, or copper form an integral part of the molecular structure of the proteins. Each hemoglobin molecule contains four iron atoms held in place by large organic chemical structures called "heme groups"—hence the name. These iron atoms are responsible for binding the oxygen and carbon dioxide. Indeed, the reason that carbon monoxide and cyanide are dangerous poisons is because these two molecules attach to the iron in hemoglobin so tightly that they block its normal ability to bind oxygen. The body is quickly starved of oxygen, and all biochemical functions stop.

The protein portion of hemoglobin in adults is a tetramer (i.e., with four protein chains) consisting of two similar subunits (called alpha and beta) that join together in developing red blood cells and remain joined for the life of the red cell. The protein is produced in bone marrow by red blood cells and remains within them until their destruction. It is then broken down in the spleen, and some of its components, such as iron, are recycled to the bone marrow. The heme groups are broken down, transported to the liver, and secreted with the bile into the intestine for eventual elimination from the body.

The alpha and beta subunits of hemoglobin are made by clusters of genes on chromosomes 16 and 11, respectively. The alpha cluster spans about 30 kb and has four functional genes. The beta cluster is about 45 kb and contains five genes, only two of which code for the proteins that form the normal adult hemoglobin. These two forms of the beta subunit are formed (i.e., the genes are expressed) only after birth. A third gene in the beta cluster makes a third type of protein, called gamma hemoglobin. This subunit is only made in the fetus and takes the place of normal beta protein in fetal hemoglobin.

Over the millions of years of human existence, the genes for hemoglobins have accumulated a number of mutations, which have resulted in several hundred variant genes in different populations worldwide. However, most of the mutations are "silent" and produce no change in protein sequence or function. As a result, there is a uniformity of protein composition of these hemoglobin variants throughout the world. Indeed, most variant hemoglobin genes have only been found through specialized research techniques. There are of course some mutations that dramatically change aspects of hemoglobin behavior. Specific mutations in the beta subunit gene, for instance, are associated with sickle-cell anemia. The absence of the beta protein altogether causes thalassemia.

See also: Sickle-Cell Anemia, Thalassemia (Ch. 5).

BLOOD CLOTTING (COAGULATION)

As we have seen in the previous section, blood is essential to the proper functioning of our bodies. All organs and tissues must be constantly supplied with blood in order to maintain their proper activities and will suffer severe damage and dysfunction if their blood supply is cut off for any length of time. The body ensures a proper blood supply to all of its parts by means of an extensive and intricate system of blood vessels, which includes the arteries, veins, and capillaries.

An injury to even a single vessel can cause the blood to spill out from the blood vessel at that location and disrupt the normal flow throughout the body. Excessive bleeding or blood loss will ultimately result in death. Consequently, the ability of the body to control the flow of blood following injury to a blood vessel is necessary for its survival. The process by which blood loss is controlled and the damaged blood vessel repaired is known as *hemostasis*. Basically this involves the coagulation of blood (or the formation of a clot) at the site of injury, followed by repair of the damaged tissue and the subsequent dissolution of the clot to restore normal blood flow. There are four major events in the clotting of blood:

1. The first step is vascular constriction in the vicinity of the injury. This shrinks the diameter of blood vessels in the area and reduces the amount of blood that can flow through the injured area, thus limiting blood loss.

2. Next, a protein called *thrombin* prompts the platelet cells to aggregate at the site of injury. This forms a loose plug of tissue—the beginnings of a blood clot—to temporarily patch up the injured site. Red blood cells may also get entrapped in this initial plug. Another protein, *fibrinogen*, stimulates these aggregated platelets to clump together by binding to the exposed collagen on the blood vessel walls. Upon activation, platelets release ADP, serotonin, phospholipids, lipoproteins, and other proteins important for the next step in the hemostasis process. The activated platelets also change their shape to accommodate the formation of the plug.

3. The aggregated platelets secrete proteins called clotting factors and set into motion a series or cascade of enzymatic events that convert *fibrinogen* to *fibrin*. The fibrin molecules become enmeshed between the clumped platelets (and RBCs when present) to form the fibrin *clot*.

   The formation of the fibrin clot is a very complex process and involves many different proteins that act one after another (hence the term cascade) to activate the successive protein to perform its function. In fact, there are two biochemical pathways known as the *intrinsic* and *extrinsic* blood-clotting cascades (see Figure 4.1) that lead to the formation of the clot. Although they are initiated by distinct mechanisms, the two eventually converge on a common pathway. The type of tissue injury described triggers the extrinsic pathway, whereas the intrinsic pathway is initiated in response to an abnormal vessel wall (e.g., in cardiovascular disease).

4. Finally, after the tissue repair is complete, the protein *plasmin* dissolves the clot so that normal blood flow may resume.

The Clotting Cascade

Details of the coagulation pathways are beyond the scope of this book, but essentially they are triggered by the protein thrombin binding to specific receptors on the surface of platelets, which initiates a signal transduction cascade involving a G-protein, a protein kinase, and several other enzymes and proteins. The

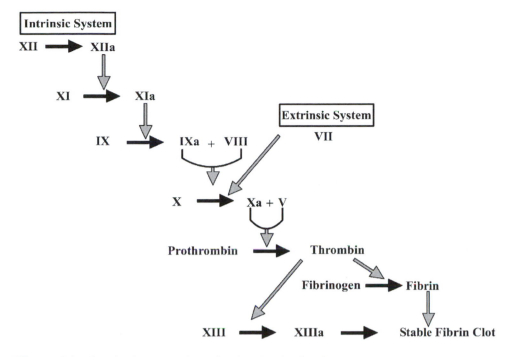

Figure 4.1. The clotting cascade. [Ricochet Productions]

different components of the coagulation cascade are summarized in Table 4.1, followed by details on selected proteins.

Fibrinogen (Factor I)

The principal material in a blood clot, *fibrinogen* is a plasma protein that makes up about 4 percent of the total protein content of circulating blood. It is a large molecule consisting of three pairs of polypeptides, called the alpha, beta, and gamma chains, that are linked together by chemical bonds. These peptides are made by three different genes (*FGA*, *FGB*, and *FGG*) that lie next to one another on chromosome 4. Normal circulating fibrinogen is inactive and cannot form clots. In order to initiate the clotting process, the protein must be broken down to release its component polypeptides or fibrin threads, which then begin to enmesh themselves into and around the platelet clot. The clot is further consolidated and stabilized when the fibrin threads are cross-linked by the formation of chemical bonds between specific amino acids; this step is catalyzed by the coagulation factor XIII (see Table 4.1).

Prothrombin/Thrombin (Factor II)

Normally circulating in the blood in the form of the cofactor *prothrombin*, factor 2 is activated to thrombin in response to the aggregation of platelets. Thrombin is a special type of enzyme, called a serine protease, that breaks down fibrinogen into the fibrin threads. It also activates clotting factor XIII and plays a role in maintaining

Table 4.1. The blood coagulation cascade

Clotting factor	Protein	Chromosome	Function/activity[1]
Factor 1 (Fibrinogen)	Plasma protein	4 (3 genes: *FGA*, *FGB*, and *FGG*)	Structural component of blood clot; initiates clot formation by binding damaged blood vessel walls and stimulating platelet aggregation
Factor II (Thrombin)[2]	Serine protease	11	Converts fibrinogen to fibrin threads; activates factor XIII
Factor III (Tissue factor)	Cell surface glycoprotein; cofactor	1	Initiates extrinsic coagulation pathway; receptor for clotting factor VII
Factor IV (Calcium ions)	Not a protein		Cofactor
Factor V	Cofactor (activated to Va/VI)	1	Cofactor in the activation of prothrombin to thrombin
Factor VII	Serine protease	13	Activates factors X and XI; accelerates prothrombin conversion
Factor VIII	Cofactor	X	Cofactor in the activation of factor X
Factor IX (Christmas factor)	Serine protease	X	Involved in the intrinsic blood coagulation cascade in response to endothelial defects; activates factor X
Factor X (Stuart–Prower factor)	Serine protease	13	Converts prothrombin to thrombin
Factor XI	Serine protease	4	Activates factor IX
Factor XII (Hageman factor)	Serine protease	5	Binds to exposed collagen in damaged blood vessel walls
Factor XIII			Stabilizes clot by cross-linking fibrin threads; B chain may serve as plasma carrier
A chain	Transglutaminase	6	
B chain	Non-enzymatic subunit	1	

Table 4.1. (continued)

Clotting factor	Protein	Chromosome	Function/activity[1]
Von Willebrand factor (VWF)	Regulatory protein	12	Carrier protein for factor VII; mediates adhesion between platelet wall and collagen on blood vessel walls

[1]With the exception of the tissue factor (Factor III), all coagulation factors are present in an inactive proenzyme form until activated at the appropriate time during the clotting cascade. The category in column 2 refers to the activity of the activated protein.
[2]The precursor to thrombin is prothrombin, itself a cofactor that is activated upon the aggregation of platelets.

vascular integrity during development and postnatal life. In its prothrombin form, factor II (F2) is a vitamin K–dependent proenzyme that is encoded on chromosome 11. Mutations in the F2 gene lead to the development of thrombosis. Conditions such as vitamin K deficiency and severe liver disease, and the use of anticoagulant drugs, cause acquired thrombin deficiencies called dyspothrombinemia.

Dissolution of the Fibrin Clot

The process of hemostasis cannot be complete until the injured blood vessels are repaired and normal circulation is resumed. Although the fibrin clot works admirably to patch up any injuries and staunch bleeding, it should not be present after the vessel wall is repaired because it impedes the normal flow of blood in the vessels. The function of breaking down the clot after tissue repair falls to an enzyme called *plasmin*.

The precursor to plasmin is a protein called *plasminogen* that is encoded by a gene on chromosome 6 and is a normal constituent of circulating blood. Plasminogen can bind to both fibrinogen and fibrin and gets incorporated into the clot as it is being formed. A serine protease protein known as tissue plasminogen activator (tPA—the gene for this is on chromosome 8) then converts plasminogen to active plasmin. (Inactive tPA is released from vascular endothelial cells following injury; it binds to fibrin and is consequently activated.) The active plasmin actually digests the fibrin mesh in the blood clot! The result is a soluble product to which neither plasmin nor plasminogen can bind, so the clot falls apart. We can see this happening slowly in a bruise that starts to turn yellow as the clot dissolves. What happens to the released plasmin after the clot is digested? If it is present in the bloodstream, it could potentially prevent other clots from being formed, so to avoid this it is quickly inactivated by a circulating protein that acts as an inhibitor of the enzyme.

See also: Bleeding Disorders; Blood (Ch. 5).

BLOOD GROUPS

Although people have carried out blood transfusions for hundreds of years—the first recorded human blood transfusion was performed in 1795 by Philip Syng Physick (1768–1837), a physician in Philadelphia—until the early 1900s, it was still a risky procedure, with many deaths following attempted transfusions. As it turns out, mixing blood from two individuals sometimes causes the clumping or agglutination of red blood cells. The agglutinated cells can clog blood vessels and stop the circulation of the blood to various parts of the body. These clumped cells break up, and their contents leak out into the body tissues. The hemoglobin that is so vital inside the red cells becomes toxic when outside the cell and has fatal consequences for the patient.

It was not until 1901, when Austrian physiologist Karl Landsteiner (1868–1943) discovered and identified the principle behind human blood groups, that blood transfusions became safer. He found that blood clumping was due to an immunological reaction that occurred when the receiver of a blood transfusion made antibodies against the blood cells from a donor. However, Landsteiner noted that not all individuals reacted to transfusions in the same way and that the survival of a patient was linked to the identity of the donor. Whereas certain donor–recipient pairs were compatible for transfusions, others were not. Investigating the phenomenon further, Landsteiner was able to determine the molecular or chemical basis for this compatibility, and from this he devised the method of classifying different individuals into specific blood groups. For this discovery, he was awarded the Nobel Prize in Physiology or Medicine in 1930.

The differences in blood from person to person are due to the presence or absence of antigens on the surface of the red blood cells. Different individuals have different types and combinations of antigens. When the blood of an individual with certain antigens is transfused into a patient with a different set—a person with a different blood group—the recipient's immune system sees the donor RBCs as "foreign" and mounts an immune reaction against them. This results in the production of antibodies and the harmful consequences just described.

Blood groups are genetically determined, which means that the blood group you belong to depends on what you have inherited from your parents. About 200 different blood group substances have been identified and placed within 29 known blood group systems. Of these, the ABO and Rh systems are the most common ones used to determine compatibility for blood transfusions, and these are discussed further.

ABO Blood Groups

In the ABO blood-typing system, there are four different groups, designated as A, B, AB, or O, based on the antigenic composition of the red blood cells. The genetic basis for these blood groups is the allelic variation in a single gene present on chromosome 9. There are three possible allelic variations of this gene—A, B, and O—whose normal protein product is a sugar-transferring enzyme that adds a specific sugar to proteins on the red cell surface. Whereas the A and B alleles produce enzymes with specificities for different sugars and lead to the synthesis of different surface enzymes, the O allele produces a nonfunctional enzyme. It is possible to have both the A and B alleles active in the same individual. A and B are codominant over the O allele. Your blood group ultimately depends on the combination of alleles you possess in your genotype. For instance, if you belong to blood group A, you might have the A allele on both chromosomes, or one A and one O allele. In either case, your RBCs will have only the A antigens, and hence you will produce antibodies against blood cells from a type B donor. Similarly, if you belong to the blood group B, you might have two B alleles or one B and one O allele. People with one A and one B allele belong to the AB group. These individuals will have both types of antigens on their blood cells and consequently do not mount an immune reaction against any blood donor. For this reason, the AB blood type is called the universal receiver. Because the O allele is nonfunctional, only people who are homozygous for the O allele will display the O phenotype. This blood group is also called the "universal donor" because O type blood can be transfused to any recipient without the fear of eliciting an immune reaction. However, people with this blood group cannot receive blood from any of the other donors.

Rh Factor Blood Grouping System

In addition to the ABO system, blood cells are also grouped according to the presence or absence of another antigen, called the Rhesus or Rh factor, on the surface of their red blood cells. This is a highly immunogenic (i.e., capable of eliciting a strong immune reaction) protein produced by a gene on chromosome 1. Those who have the antigen are called Rh+ (positive), whereas those who do not are Rh− (negative).

Because the protein is highly immunogenic, a person who is Rh-negative (Rh−) will produce antibodies to the Rh protein and develop a severe immune system reaction if Rh-positive (Rh+) blood gets into his or her bloodstream. (In contrast, a person with Rh+ blood can receive blood from a person with Rh− blood without any problems.) In addition to transfusions, exposure can occur during childbirth, when an Rh− woman gives birth to an Rh+ baby (who inherited the Rh+ gene from the father). When blood cells from the baby travel across the placenta, the woman's immune system regards the Rh+ cells as a threat. Specialized white blood cells will make antibodies designed to kill Rh+ blood cells. The first Rh+ baby is usually safe from attack because the mother is primed against the antigen but does not mount a secondary response. However, if she subsequently conceives another Rh+ baby, her immune system will mount a much stronger secondary response and flood her child with antibodies. These antibodies then destroy the baby's red blood cells. If left untreated, this can result in severe anemia, jaundice, or even death. The condition is called Rhesus disease. Rhesus disease is now rare because Rh− mothers who give birth to Rh+ babies are immunized within 72 hours of giving birth. The immunoglobulin preparation works by killing the baby's red blood cells inside the mother's bloodstream before her immune system has time to react.

The most common blood types in the United States are A+ and O+: about 72 percent of the population has one or the other. AB− is the rarest blood type (1 percent of the population). Because blood type is a genetic trait whose inheritance follows Mendel's laws, blood group typing is used legally to establish paternity. Anthropologists and population geneticists also use the frequency of occurrence of various blood groups as tools to study ethnic or racial origins.

See also: Immunity—Antibodies.

CARTILAGE

Cartilage tissue has many obvious functions in the body (as in giving shape to the nose and the ears), but do we realize how widely this kind of tissue is distributed in the body, or the variety of functions it performs? Just a sampler of some of its functions includes:

- *Shape and flexibility*—of the nose, ears, and other organs;
- *Support*—it is the c-shaped cartilaginous rings in the windpipes (trachea and bronchi) that help to keep these tubes open;
- *Flexible movement*—cartilage provides a smooth surface at the ends of bones in joints to reduce friction;
- *Shock absorption*—in the spinal cord, cartilage forms protective cushions between individual vertebral discs.

This versatile tissue is a specialized type of connective tissue whose main function is to bind different structures together and form a framework of support for organs and the body as a whole. Connective tissues have only a few cells held together in a matrix of extracellular ground substance. In cartilage, this matrix consists mainly of a network of sugar-rich proteins called *chondroitins*, produced by special cartilage cells called *chondrocytes*. These chondrocytes are present in fluid-filled spaces called the lacunae scattered throughout the matrix. *Collagen*, a ropelike protein, is woven around a network of strands of chondroitin. The collagen acts as scaffolding, giving the cartilage its strength, and the chondroitin traps water molecules and provides the cartilage with its cushioning properties. Cartilage is not served either by nerves or blood vessels and is often covered by a dense fibrous membrane called the perichondrium.

In some vertebrates, such as sharks, the entire skeleton is made up of cartilage. Although this is not true for adult mammals, cartilage is the material of which the skeleton is first constructed in the embryo. It then acts as a model and is gradually replaced by bone as the fetus develops. The process by which bone tissue follows the cartilage model and slowly replaces it is known as ossification. Not all cartilage is ossified, however—as we have already mentioned, structures such as the ears and nose as well as the trachea and larynx (voice box) are made of cartilage even in the adult.

There are three main types of cartilage in humans. *Hyaline cartilage* is the most common type. It is present in the trachea, the larynx, the tip of the nose, in the connections between the ribs and the breastbone, and also the ends of bones, where they form joints. Temporary cartilage in mammalian embryos also consists of hyaline cartilage. The bluish-white substance is rich in collagen and makes it extremely strong but at the same time flexible and elastic. It provides a smooth surface at the ends of various bones, which reduces friction and facilitates bone movement.

White *fibrocartilage* is an extremely tough and resilient tissue found between vertebral discs, between the pubic bones in front of the pelvic girdle, and in the shoulder joints. The matrix in this type of cartilage tissue consists of the chondrocytes sandwiched between bundles of collagen, aligned in parallel arrays. This structure strengthens the cartilage even further, allowing it to function as a shock absorber, and provides sturdiness in joints without impeding movement. It also deepens sockets in joints, for example in the ball-and-socket joints in the hip and shoulder regions, to make dislocation less possible.

The most pliable of the cartilaginous tissue, *elastic cartilage*, is similar to hyaline cartilage but contains an abundant network of a protein called elastin in its matrix in addition to collagen. These fibers run through the matrix in all directions. Elastin endows the tissue with a great deal of flexibility (elasticity), enabling it to return to its original shape after binding. This type of cartilage is found in the ear lobes, Eustachian tubes, epiglottis, and parts of the larynx.

Chondroitin

Chondroitin is the matrix of sugar-carrying proteins (also known as glycoproteins or proteoglycans) that makes up a substantial portion of cartilage. A characteristic feature of many of these glycoproteins is that they are attached to sulfate groups, because of which they are often called the proteoglycan sulfates. So far, we know of at least five different chondroitin components of cartilage encoded by genes on different chromosomes:

- The principal proteoglycan of chondroitin, *aggrecan 1* (also called chondroitin sulfate proteoglycan 1 or CSPG1), enables cartilage to withstand compression. The gene for this protein is on chromosome 15, and mutations in it may be involved in skeletal dysplasia and spinal degeneration.

- CSPG2, which is encoded on chromosome 5, is especially abundant in hyaline cartilage, where together with other matrix proteins it provides mechanical support.

- CSPG3, also called *neurocan*, is thought to be involved in the modulation of cell adhesion and migration. Its gene lies on chromosome 19.

- CSPG4, whose gene is on chromosome 15, represents an integral membrane protein that plays a role in stabilizing interactions between the cells and matrix. It is also expressed by human malignant melanoma cells.

- Also known as *bamacan* (BAM), CSPG6, which is encoded by a gene on chromosome 10, was originally isolated from embryonic yolk sac and is an abundant material in basement membranes of cartilage.

Collagen

If the chondroitin family, with five members, seemed complicated, collagen—the protein that imparts strength and elasticity to cartilage and other connective tissue—is even more so, consisting of a minimum of nine types of molecules whose constituent chains are encoded by at least 17 genes scattered throughout the chromosomes of the genome. The peptides encoded by the different collagen genes form bonds with each other and wind themselves into complex triple helices, much like three-stranded ropes, reinforced by bridges between the strands to lend additional strength and stability to the structure. These proteins are by no means confined to the cartilage tissue. In fact, collagen is the most abundant protein in the human body. Many collagens are intracellular proteins that make up the cytoskeletons that give individual cells their shape. Collagens are also found in the basement membranes that line different organs and tissues in the body.

Structurally, the collagen genes can be divided into two rough categories. "Tissue-specific" collagens, as their name indicates, are produced in specific tissues only. "Housekeeping" collagen genes are transcribed widely in many tissues and also have a higher GC content than the tissue-specific genes. Genes for the major fibrillar collagens, such as those of cartilage, belong to the tissue-specific gene category.

The key collagens of cartilage include collagens II, IX, and XI. These are found both in hyaline cartilage and in the intervertebral discs. Collagen II is the most abundant protein in hyaline cartilage. Because collagen IX is located on the surface of the collagen threads, scientists suspect that it mediates interactions between the collagen network and other proteins of the tissues. Collagen IV is an example of a basement-membrane collagen. Mutations in the various collagen genes are associated with a vast spectrum of diseases, including abnormalities in the growth and development of bone and cartilage (e.g., achondroplasia) as well as other abnormalities unrelated to these tissues (e.g., Alport syndrome).

See also: Alport Syndrome (Ch. 5).

Elastin

As mentioned, elastin is the protein that provides the soft cartilage in structures such as the ear lobes and windpipe (trachea) with flexibility and resilience. It is also a component of other tissues such as the skin, lungs, and large blood vessels (e.g., aorta). In fact, elastin is found throughout the vertebrate kingdom, with the exception of certain primitive fish, and performs much the same function in all of these animals (i.e., it imparts elasticity to their tissues). Readers may have noticed that many facial creams and moisturizers boast elastin as one of the ingredients, ostensibly to help keep the skin supple and wrinkle-free.

The structure and composition of elastin are peculiarly suited to its specific function in various tissues. It is a fibrous protein made up of polypeptides that are especially rich in hydrophobic amino acids such as glycine and proline, making it one of the most hydrophobic proteins in the human body. Individual chains are cross-linked by special chemicals that are derived from amino acids to form a rubber-like network that imparts elasticity to the tissues. The best-characterized elastin gene in humans is one on chromosome 7 called *ELN* that encodes the protein found in the extracellular matrices of blood vessels. This protein gives blood vessels their elasticity and strength. The *ELN* gene is one of the larger genes in the human genome, with a total of 34 exons and spanning approximately 45 kb of genomic DNA. Mutations in this gene cause supravalvular aortic stenosis, a heart defect characterized by the narrowing (stenosis) of the outlet from the heart immediately above the aortic valve. Another genetic disorder, known as Williams syndrome, is caused by a small chromosomal deletion on chromosome 7, including *ELN*. People with this syndrome lack the elastin protein and have disorders of the circulatory system.

Regulatory Genes and Cartilage

Much research has been done on identifying the regulatory genes involved in cartilage growth, particularly because of their relevance in painful conditions such as osteoarthritis. For example, scientists are seeking to identify transcription factors that are responsible for directing undifferentiated cells (or stem cells) to form cartilage. In the human embryo, some of these stem cells are pluripotent (i.e., they have the ability to develop along distinct pathways at different locations into connective or supporting tissues, smooth muscle, vascular endothelium, or blood cells).

The differentiation of stem cells into chondrocytes (cartilage cells) is an important molecular event that is essential for growth, repair, and regeneration of cartilage and bone. After receiving the signal to differentiate into cartilage, these cells stop expression of type I collagen and begin expression of other collagens of type II, IX, or XI, as well as a noncollagen protein called "cartilage oligomeric matrix protein" (COMP) and the chondroitin aggrecan.

A key component in this process is COMP, a large extracellular matrix protein whose specific role is still unknown. Mutations in the human *COMP* gene have been linked to the development of pseudo-achondroplasia and autosomal dominant forms of short-limb dwarfism (characterized by short stature and early-onset osteoarthritis). (See Chapter 5 for a discussion of specific diseases.) The *COMP* gene has been mapped to chromosome 19 and its promoter region identified. During embryonic and adult stages in several different species, including the mouse, rat, and human, *COMP* expression appears to be restricted primarily to chondrocytes. Scientists believe that *COMP* gene expression changes depending on how differentiated the chondrocytes are at each stage of their development. Work is now under way to identify the transcription factors that, by binding to the *COMP* promoter, direct cartilage cells' function and phenotype.

See also: Dwarfism—Achondroplasia (Ch. 5).

CHANNEL PROTEINS AND CROSS-MEMBRANE TRANSPORT

Have you ever noticed what happens when you try to mix oil and water, for example, when washing a frying pan containing oil in the kitchen sink? The two liquids do not mix with one another and the oil immediately forms a thin film on the surface of the water, spreading out over as much of the surface as possible. If you were to touch the surface of the oil film with a fingertip, it would not get wet—the oil forms a thin but effective barrier. This is precisely how the membrane behaves in a cell whether the cell is an entire organism by itself or part of a larger, multicellular entity. A double layer of phospholipids, the membrane (also called the plasma membrane), forms a hydrophobic and largely impermeable barrier between the cell and its exterior, both of which are aqueous (i.e., water-based) environments. The cell membrane keeps the cellular contents enclosed within a boundary and protects the cell from various threats in its surroundings.

But the same properties of the plasma membrane that make it so effective a barrier also pose a problem. Although a cell needs to preserve its individual identity, it also needs to maintain contact with its exterior environment. Water, that most essential of substances, must have a way of getting into and out of a cell, carrying with it the materials that a cell needs for all the activities that make up life: growth, metabolism, excretion, and so on. Cells of different organs need to know when to stop growing, and others, such as nerve cells, need to transmit messages from the extremities to the brain and outward to other cells such as the muscles for taking action. Communication can only take place if there is some way for different signaling molecules to get in an out of a cell.

How do all these aqueous molecules get past the seemingly impervious cell membrane? The cell membrane is not a continuum of impervious lipid but is like a sieve with pores or gaps that allow the passage of some molecules while holding back others. But whereas a sieve can only distinguish among substances on the basis of size, the pores have the ability to differentiate among different substances. This very important property, known as semipermeability, enables the cell to draw specific substances inside while keeping others out.

To understand how pores discriminate among different substances, let us return to the oil and water analogy for a moment. Imagine a plastic doughnut floating in the water. If you pour the oil gently onto the water surface, you will notice that the oil film forms around the doughnut, while the hole at the center of the doughnut gives you direct access to the water. The pores of the cell act in a manner similar to the doughnut hole. They are lined with special membrane-spanning proteins that form selective channels for different molecules or ions.

The channel proteins have complex structures that enable them to be embedded in a hydrophobic membrane and still allow the passage of hydrophilic or charged molecules. A single channel consists of multiple protein subunits—sometimes identical and other times different—that together form a single doughnut-shaped unit. Each protein consists of both hydrophobic (water-repellant) regions that are made of the nonpolar amino acids and hydrophilic amino acid residues,

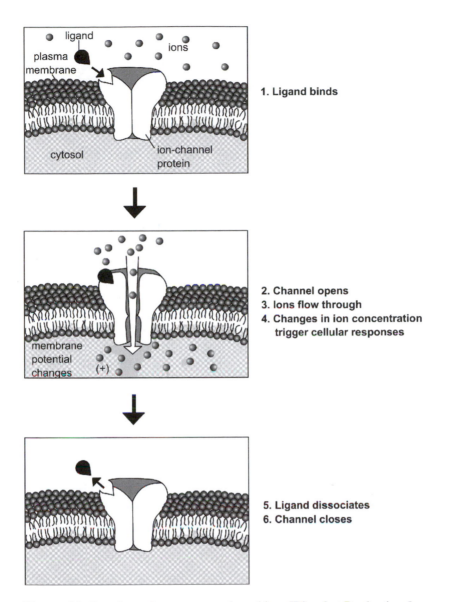

Figure 4.2. Ion channel: structure and working. [Ricochet Productions]

and is folded in such a way as to expose the hydrophilic residues in the central pore. The selectivity of the channel is determined by the nature and arrangement of these residues in the center (see Figure 4.2).

A very important aspect of membrane transport is its bidirectionality—the ability of different materials to flow either into or out of the cells. One way in which cells achieve this is by taking advantage of a concentration gradient of substances across the membrane. There is often a significant difference in the concentration of different ions between the interior and exterior of the cell—for example, potassium ions are generally present at higher concentrations in cytoplasm than in extracellular fluid, whereas sodium, chloride, and calcium ions have the opposite distribution. The membrane-embedded *ion channels* mediate the transport of ions in the direction along a concentration gradient (i.e., from higher to lower concentrations). Transport against a gradient, which is much more difficult and requires energy, uses *membrane pumps*, proteins that cannot only select for the appropriate ions but also harvest energy from ATP to transport them. Some representative examples of channels and pumps involved in membrane transport follow.

Ion Channels

In addition to the property of ion selectivity discussed earlier, channel proteins must also be able to regulate the flow of ions by opening and shutting their pores as needed. Depending on the mechanism they use for this function, there are two types of ion channels. *Voltage-gated* channels are controlled by the membrane voltage, which is generated by the gradient of electric charge across the membrane. *Ligand-gated ion channels*, on the other hand, are activated to open or shut by the binding of certain molecules, called ligands, to specific sites on the exterior portion of the channel proteins. Ligand-gated ion channels are therefore actually dual-function proteins, where the opening and closing of the pore may be regarded as the response to a signaling pathway that is set off by coupling a ligand to its receptor (see the entry on signal transduction in Chapter 2).

The human genome has many genes, scattered throughout the chromosomes, which code for different types of ion channel proteins. The structure of these channel proteins—especially the portion of the protein that is exposed in the pore—varies quite widely depending on the ion for which it is specific and its gating mechanisms.

Potassium (K^+) Channels

Of all ion channels, potassium channels have the simplest structure and are also the best characterized. Typically, a single channel consists of four identical subunits arranged in a petal-like fashion around the central pore. In some cases the channels are made of more than one type of subunit. So far, we know of at least 50 different genes on different chromosomes that encode for various K^+ channels, including both voltage- and ligand-gated types, as well as some that are gated by both mechanisms. Some examples are:

- The first gene found to encode an ion channel was for a voltage-gated K^+ channel isolated from a *Shaker* mutant of *Drosophila* (the fruit fly). Humans have more than 20 homologs of this gene, which are present on different chromosomes and show varying degrees of tissue specificity. *KCNA1* on chromosome 12, for instance, is specifically expressed in the neurons, heart, retina, and the islet cells in the pancreas, whereas *KCNA3* on chromosome 1 is expressed in the lymphocytes, brain, lungs, spleen, and thymus.

- The *KCNA10* gene on chromosome 1 encodes a K$^+$ channel that is both voltage-gated and also controlled by the ligand cyclic GMP. It is expressed in cells of the aorta, brain, and kidneys.

- A subset of K$^+$ channels called the MaxiK or large-conductance calcium-sensitive channels play important roles in controlling smooth muscle tone and neuron excitability. These channels—which are both voltage-gated and sensitive to calcium ion concentration—are made of two subunits, a pore-forming alpha unit encoded by the *KCNMA1* gene on chromosome 10 and a regulatory beta subunit. There are multiple genes for beta subunits on different genes, such as *KCNMB1* on chromosome 5 and *KCNMB3* on chromosome 3. Intracellular calcium regulates the physical association between the alpha and beta subunits.

- Small-conductance calcium-sensitive K$^+$ channels are voltage-independent ligand (calcium)-gated channels expressed predominantly in the brain and heart and are thought to control neuronal excitability in these organs. Examples include *KCNN1* on chromosome 19 and *KCNN2* on chromosome 1.

- The gene alternately known as *SUR1* (for *s*ulfonylurea-*u*rea *r*eceptor 1) or as *ABCC8* (for ATP-binding cassette C8), which is on chromosome 11, encodes the subunit for a ligand-activated K$^+$ channel that is found in the pancreas, neurons, and skeletal muscles. It is an ATP-sensitive channel that is also involved in insulin release. Different mutations in this gene have been associated with medical conditions characterized by defective insulin secretion such as type II diabetes mellitus.

Channels for Sodium (Na$^+$) and Other Cations

In general, channels for sodium, calcium, and other cations have more complex structures than the K$^+$ channels, with a large number of subunits that are organized more intricately.

- Voltage-gated sodium channels play important roles in conducting the action potential in nerves and muscles. In contrast with the K$^+$ channels, sodium channels are made up of more than one type of protein (i.e., they are heterodimeric proteins). The voltage-gated channels consist of alpha subunits that form the central pore and beta subunits that stabilize the channel in the membrane. *SCN1A* and *SCN2A1* are examples of genes on chromosome 2 that encode alpha subunits of these channels. *SCN1A* is especially abundant in brain cells. Mutations in these genes have been linked to epileptic seizures. An example of a beta subunit that combines with either of these subunits is produced by the *SCN2B* gene on chromosome 11.

- *CACNA1S* on chromosome 1 encodes an alpha (pore-forming) subunit for a voltage-gated channel for calcium (Ca^{2+}) ions that controls the release of these ions in skeletal muscles. The channel protein has over five subunits, and their corresponding genes are located on different chromosomes.

Chloride (Cl⁻) Channels

Chloride channels are relatively simple channels that usually consist of a single type of subunit. They have important functions in organs such as the kidneys as well as skeletal muscles.

- CLCN is a family of voltage-gated Cl⁻ channel proteins consisting of nine members whose genes have significant homology but quite diverse functions in different parts of the body. *CLCN1* on chromosome 7, for instance, regulates the electric excitability of the skeletal muscle membrane. Mutations in this gene cause two forms of inherited human muscle disorders: Becker-type muscular dystrophy and a dominant myotonia. Mutations on *CLCN2* on chromosome 3 are linked to different forms of epilepsy. *CLCN4* and *CLCN5* are both X-linked genes. The latter is expressed in the kidneys, and mutations in the gene result in Dent disease and other renal tubular disorders.

The idea that ion (salt) transport had important physiological functions was suggested as early as 1890 by the German chemist Wilhelm Ostwald (1853–1932) (who in 1909 won the Nobel Prize in Chemistry for work of a more fundamental chemical nature, namely the discovery of the principles of chemical equilibria). Examples of some common physiologically important ions include sodium, potassium, calcium, and magnesium, among the positively charged cations, and chloride, a negative ion (anion). In 1925, Leonor Michaelis (1875–1949) suggested that the cell membrane might contain narrow channels to allow the passage of such ions from cell to cell. Then, in the 1950s, neurophysiologists Alan Hodgkin (1914–1998) and Andrew Huxley (1917–) conducted electrophysiological experiments on the giant axons of squid—which garnered for them the Nobel Prize in Physiology or Medicine in 1963. Subsequently, others were able to devise models for the mechanisms of ionic transport across cell membranes and also demonstrate the existence of channels for potassium and sodium ions. Further advances in the field have continued to the present day, including the elucidation of the complete three-dimensional structure of a potassium channel, for which Roderick McKinnon (1956–) received his half of the 2003 Nobel Prize in Chemistry.

Water Channels

So far, we have talked about how ions—substances that are dissolved in water—pass through channels along a concentration gradient. But what about water itself? Simple diffusion is evidently ruled out because of the impermeability of the phospholipid membrane. The answer—namely, special channels earmarked for water—might seem obvious, and this in fact was postulated in the nineteenth century some decades before Ostwald proposed ion channels, but it was not until a century later in the mid-1980s that Peter Agre (1949–), recipient of the second half of the 2003 Nobel Prize in Chemistry, discovered these channels. And he did so quite by accident, while studying an unrelated issue—the Rh blood group factors of RBCs. In 1988, Agre isolated a new protein of unknown function and upon further investigation found it to be a water channel.

Agre's serendipitous discovery turned out to be an essential component of all living cells from bacteria to humans. Water channels play important roles in homeostasis, enabling cells to regulate volume and internal osmotic pressure. They also function in situations where water must be retrieved from a body fluid (e.g., when urine is concentrated in the kidney). Humans have at least 11 distinct water channel proteins, known as aquaporins (AQPs). They form relatively simple homomeric channels that are typically specific for different tissues. Defects in the genes for aquaporins have been linked to many different diseases.

- *AQP1* on chromosome 7 encodes a peptide that forms a homotetramer (four-subunit) channel for water in erythrocytes and kidney tube cells. Mutations in this gene show some links to imbalances in the movement of ocular (eye) fluids.

- *AQP2* on chromosome 12 forms water channels in the collecting tubules of the kidney. Mutations have been linked to diabetes insipidus (water diabetes—see the entry on diabetes in Chapter 5). This gene forms a cluster with three other aquaporin genes—*AQP0*, *AQP5*, and *AQP6*. The protein P0 is found in the fiber cells of the eye lens, where its exact function is unknown, P5 plays a role in the generation of saliva, tears, and pulmonary secretions, and P6 is specific for the kidneys.

- *AQP3* on chromosome 9 is localized at the basal lateral membranes of collecting duct cells in the kidney. In addition to transporting water, aquaporin 3 also facilitates the transport of other molecules such as urea and glycerol. The gene *AQP7*, which is found close to this gene, produces another aquaporin that is very similar in sequence and transport activity to P3. The latter is also believed to play an important role in sperm production.

- *AQP4* on chromosome 18 is the predominant aquaporin found in the brain.

- *AQP8* on chromosome 16 is expressed exclusively in the pancreas and colon.

- *AQP9* on chromosome 15 is quite similar in sequence to aquaporins 3 and 7. Like those channels, P9 allows the passage of nonionic substances and plays a role in urea transport as well as osmotic water permeability. Expressed abundantly in the leucocytes, this aquaporin is also believed to play a role in the immune response and in the bactericidal action of these cells.

- *AQP10* on chromosome 1 functions in the water-permeable channels in the epithelia of organs such as the small intestine, where water is both absorbed and excreted. This aquaporin also allows the passage of glycerol and urea.

Gap Junction Proteins

Both the water and ion channels discussed earlier are found at the interface between cells and their noncellular environments, such as the lumen of the organs or tubules (in kidneys). Cells that are adjacent to one another in organs or networks also have intermembrane channels to allow the passage of substances directly from the cytoplasm of one cell to another. The anatomical feature that constitutes the region of close proximity between two cells is called a *gap junction* and is made up of an array of channel proteins that form the link from one cell to the next. The proteins that form these links are called the *connexins*.

Connexins are products of a gene family that consists of many different members that have a high degree of sequence similarity in some portions of the molecule and also have highly variable regions. The variability contributes to the diversity of functional properties of the gap junction proteins. For the sake of convenience, the connexins are categorized into groups such as alpha, beta, and so forth (reflected in their gene nomenclature), which indicate their relative similarities and differences from each other. They may also be classified by molecular mass, which is indicated by a number (e.g., connexin *x*, where *x* is the mass in kilodaltons [kDa]). For example, *GJA1* (gap junction alpha 1) on chromosome 6 encodes a 43 kDa protein called connexin 43, which is the major component of the gap junctions in the heart. These gap junctions are believed to play a crucial role both during the development of the heart in the embryo and in synchronized contractions of the adult heart. A cluster of genes (*GJB3*, *GJB4*, and *GJB5*) on chromosome 1 encodes three different beta connexins that are expressed in different tissues. Mutations in the beta 3 gene for connexin 31 have been associated with congenital deafness.

See also: Signal Transduction; Epilepsy (Ch. 5).

DEVELOPMENT

Have you ever wondered how a fertilized egg grows and organizes itself to become a perfect human being, complete with arms, legs, and internal organs? Each one of us begins life as a single-celled zygote with all of the instructions for life contained in the 23 pairs of chromosomes, half of which were received from each parent. And within those 23 chromosomes are the instructions that enable us to make all the cells and organs that we possess as adults, in all their amazing diversity of function and appearance. The formation of these cells, namely the development from a single-celled entity to a multicellular individual, is a complex and carefully orchestrated process—the growing fertilized egg somehow knows exactly how and when to direct cells to become the heart, the brain, the arms, and the legs.

Although much is still not understood about the process of embryonic development, we do know that certain genes help determine the destiny of the cells of

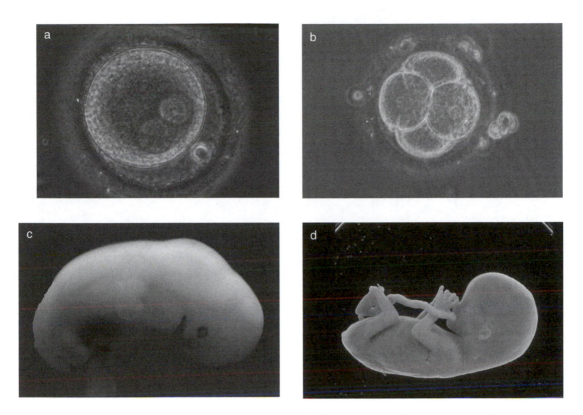

Stages in the development of a human fetus: (a) unicellular zygote; (b) embryo, 4-celled stage; (c) embryo, forty-eighth day; (d) fully developed fetus. [© R. Rawlins, Ph.D. (a and b), J. Siebert (c), and P. Birn (d)/Custom Medical Stock Photo]

the growing fertilized egg. At the very first stages of development, the mother controls the developmental processes, but very soon the genes of the growing zygote take over. After some time, the human embryonic cells organize themselves in layers and groups, a process called *gastrulation*. The cells arrange themselves to form a front and rear (along an anterior–posterior axis), back and belly sides (the dorsal–ventral axis), and left and right sides. At this point, the cells do not look outwardly different, but scientists have used probes for cell-specific proteins to show that specific genes of the zygote genome are being expressed and that in fact different groups of cells have already started to form. This *differentiation* creates three major "germ layers" in the gastrula: ectoderm, mesoderm, and endoderm. Each of these layers will give rise to specialized cells, tissues, and organs in the future. For example, some cells become the skin. Others organize themselves into a group of muscle cells that begin to expand and contract rhythmically—the heart. Still others, further along in time, become the arms and legs, and eventually a new human being is created.

Development is common to all multicellular organisms. In fact, scientists first unearthed the role of genes in development by studying fruit flies (*Drosophila*).

They correlated mutations in different genes with transformations in the flies' body patterns. These types of mutations, called *homeotic mutations*, cause structures in one body segment to be replaced by structures normally found in another segment.

One example of a homeotic transformation in fruit flies is the replacement of an antenna with a leg in the head. A group of scientists, led by biologist Ed Lewis (1918–2004), began to study mutant flies of this nature (i.e., flies that had legs on their heads instead of antennae) and traced the trait to mutations in a single gene, which they called *Antennapedia*. As it turned out, this gene was involved in development. The mutation changed the entire segment of the fruit fly's body and made it develop as if it were a different segment, one that sprouts legs instead of antennae. Over the years, scientists have discovered many more developmental genes with similar functions.

The study of the DNA sequences of genes such as *Antennapedia* that control body patterns revealed that each contains a similar stretch of about 180 nucleotides within its sequence. Scientists called this sequence a *homeobox* and all genes containing it homeotic or "*hox*" genes. The proteins expressed by these *hox* genes are special transcription factors. What is remarkable about the genes is that they are lined up along the fruit fly's chromosomes in a sequence that is parallel to the body part they control. In other words, the first *hox* genes control formation of the fly's head, the next ones in line control the thorax, and so on. At the other end of the cluster are genes controling abdomen formation. When all of these genes work correctly, the proteins they produce act together to ensure that each organism's body parts are made in correct locations.

Naturally, scientists were curious to know whether organisms other than fruit flies also had homeotic genes that regulated body patterning. It appeared that all animals seem to have homeobox sequences in their genomes! Even more incredibly, homeobox sequences found in most mammalian genes are very similar to those in fruit flies. In other words, these sequences have been conserved throughout evolution with hardly any change—indicating how ancient the process of development must be and how important these genes are in determining the fate of cells during development.

There are 39 human *hox* genes that are organized on four different clusters (A, B, C, D) on four separate chromosomal loci. These genes are closely related to the *Antennapedia* type of genes of the fruit fly. They are even arranged in the same order, and their action during embryonic development follows the same order in time and space as in the fly's development. Hox proteins form heterodimers with Pbx, another homeobox protein. When a Hox/Pbx heterodimer binds a gene, it activates transcription, whereas Hox homodimers tend to act as repressors. These transcription factors seem to act together in concert to determine the fate of developing tissue. For instance, if HoxA4, HoxA5, HoxB5, HoxA5, and HoxA6 are present together with no other *Hox* gene products, the seventh vertebra of a fetus will develop into a neck vertebra. If this delicate formula is disturbed, either through the absence or malfunction of a required Hox protein or through the presence of other transcription factors, the vertebra will develop ribs. This actually does happen in some people who grow up with an extra pair of ribs. Not surprisingly, most changes in development genes are not so benign. (Mutations in *Hox* genes often lead to developmental defects so critical that they lead to fetal death.) Polydactyly is a condition that arises from a mutation in the *HoxD13* gene. Mutations in *HoxA13* give rise to individuals with shortened thumbs, large toes, and genital malformations.

There are other homeobox genes that do not belong to the *Antennapedia* class. These genes form families of transcription factors such as *Pax6*, *Emx*, *Otx*, *Oct*, *Lim*, and *Msx*. *Pax6* is located on chromosome 11 and is a transcription factor important for the development of the eyes, the nose, the central nervous system, and the pancreas. Mutations in *Pax6* cause aniridia, a rare congenital condition in which iris is lost and the retina is underdeveloped in heterozygotes. In homozygotes there is complete loss of eyes, and usually these fetuses do not survive. The gene *Emx* is found on chromosome 2 and expressed in the embryo's cerebral cortex. It may be involved in the development of the central nervous system. The gene *Otx*, also on chromosome 2, probably plays a role in the development of the brain and the sense organs.

There are still other transcription factors, numbering perhaps in the thousands, that also participate in development, turning genes on and off to make cells follow a particular pathway to a particular fate. So, as you can imagine, the entire process of making a human being from scratch is incredibly complex and very carefully orchestrated.

In 1995, Edward Lewis, along with Christiane Nüsslein-Volhard (1942–   ) and Eric Wieschaus (1947–   ), received the Nobel Prize in Physiology or Medicine for their research on the genetic control of early embryonic development. Development biologists have also learned a great deal from the nematode worm *C. elegans*, which has a very precise pattern of development, leading always to the formation of 959 cells in the adult worm's body. Development once again came into the spotlight when *C. elegans* researchers John Sulston, Sydney Brenner, and Robert Horwitz won the 2002 Nobel Prize in Physiology or Medicine for their discoveries relating to "genetic regulation of organ development and programmed cell death"; in other words, apoptosis, an important developmental process (see the entries on apoptosis in this chapter and in Chapter 2 for details). The genetics of development continues to be a field of exciting and cutting-edge research.

See also: Apoptosis—Development (Ch. 2); Apoptosis, Transcription: RNA Polymerases and Transcription Factors.

DNA SYNTHESIS/REPLICATION AND REPAIR

In Chapter 2 we looked at the basic mechanism for DNA synthesis, a fundamental process that is required for all living things, both for maintaining the organism's genome during its own lifetime as well as passing on information to the next generation. In this section, we will be looking at the genes that encode some of the key enzymes involved in this process. For the sake of convenience, we will consider the enzymes under two main groups: first, the enzymes that direct the reactions of DNA replication, which occurs whenever a cell divides to form new progeny; and second, the enzymes that catalyze reactions involved in repairing damaged DNA both during replication and during the lifetime of a cell. This is to some extent an artificial classification because many enzymes involved in one are also involved in the other. Many of the early steps of DNA replication are also common to another fundamental cellular process, gene expression. In both cases, the compact DNA molecules, normally present as tightly wound packets in the nucleus, need to be opened up and made more accessible to enzymes that synthesize the end products, either mRNA or new DNA. A description of the enzymes of DNA synthesis will necessarily include a discussion of gene expression and RNA synthesis as well.

See also: Deoxyribonucleic Acid (DNA), DNA Synthesis/Replication and Repair (Ch. 2).

DNA Synthesis/Replication

Before discussing the main enzymes involved, we should look at some of the special characteristics that make DNA synthesis and replication noteworthy. Two features are particularly worth mentioning: (a) the process is dependent on a pre-existing *template* or model; and (b) replication is *semiconservative*. This means that every new or progeny DNA molecule contains a part of the parent molecule within it. These characteristics are closely tied to the nature of the DNA molecule. Because its nucleotide sequence is a fundamental aspect of the structure of a DNA molecule, the synthesis of a new molecule requires a template or model of this sequence. The fact that DNA is in the form of a double helix allows for the semiconservative mode of replication—each strand acts as a template for the synthesis of a new complementary strand with which it forms a new double helix when the strand is synthesized. Each progeny DNA molecule therefore consists of one parental strand and one newly synthesized strand (see Figure 2.3).

The main enzymes involved in DNA replication include the following.

Topoisomerases

The topoisomerases are a family of enzymes that alter the topology of DNA (i.e., the manner in which DNA appears on its surface). This particular activity is perhaps more important during transcription because it allows different genes to be exposed on the surface of the chromosome and be expressed at different times. Topoisomerases allow the temporary breaking and subsequent rejoining of DNA. Certain topoisomerases also play a role in unwinding the highly supercoiled DNA molecules to allow enzymes to gain access to the molecule.

Several genes located on different chromosomes encode topoisomerases. DNA topoisomerase 1 (Top1), for example, is encoded by a gene on chromosome 20. Topoisomerase 2 (Top2), which plays a role in unwinding DNA, has two forms: an alpha form, encoded on chromosome 17, and a beta form on chromosome 3.

The *Top2* genes are also targets for several anticancer drugs. Reduced activity of either Top2 form has also been linked to the disease ataxia-telangiectasia. Top3 also has alpha and beta forms, encoded on chromosomes 17 and 22, respectively. There is also a topoisomerase encoded on chromosome 8 that acts specifically on mitochondrial DNA.

Helicases

These proteins bind to double-stranded DNA and help to separate the two strands. This process, which is also known as "melting," is necessary for the two strands to function as templates for the new DNA molecules and to make genes accessible for expression. Melting requires energy, which is obtained by breaking down a molecule of ATP. There are many DNA helicase genes in the human genome, and they are found widely distributed among the chromosomes. Examples include a family of genes called *RECQL* genes—two forms may be found on chromosomes 8 and 12—and *HEL308* on chromosome 4. Defects in the *RECQL* genes have been associated with diseases such as Werner and Bloom syndromes. There are also enzymes with similar melting activity that are specific for RNA and play important roles in RNA processing, translation, and replication. An example is RNA helicase A, which is encoded by a gene on chromosome 1 (see Figure 4.3).

See also: Werner Syndrome (Ch. 5).

Single-Stranded DNA-Binding Proteins

Single-stranded DNA is quite unstable and has a tendency to try to revert to a helix by binding with its complement strand. The single-stranded DNA-binding proteins prevent this from happening by binding to single-stranded DNA generated by the helicases and stabilizing the single-stranded structure. Replication is accelerated by 100 times when these proteins are attached to the DNA strands.

Replication protein A (RPA) is the main nuclear single-stranded DNA-binding protein in eukaryotes. It is a key component of the DNA replication, repair, and recombination machineries. RPA binds single-stranded DNA and stabilizes the structure of the unwound DNA. It consists of three subunits (RPA70, RPA32, and RPA14) that are conserved in all eukaryotic cells. The gene for the largest and most important subunit (RPA70) lies on chromosome 17.

Eukaryotic Initiation Proteins

Following the unwinding and separation of DNA, a complex of proteins recognizes and binds to specific replication "start" sites on the DNA before the actual replication process begins at these sites. This DNA protein complex is known as the origin recognition complex (ORC) and involves six polypeptide subunits (designated as ORC1L–ORC6L) encoded by six separate genes. The ORC is then bound by a protein called Cdc6 (for cell division cycle 6), which is encoded by a gene on chromosome 17, followed by another protein, Cdt1, which activates Cdc6. At this point, another complex of six polypeptides, known as the MCM

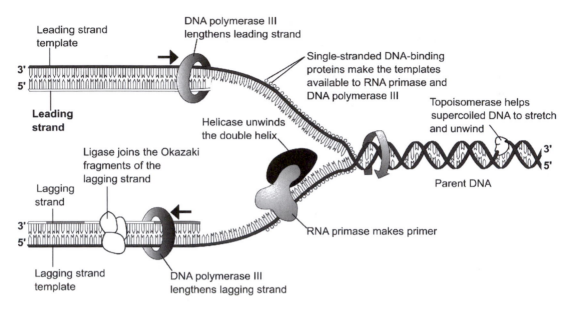

Figure 4.3. Enzymatic activity in DNA replication. [Ricochet Productions]

(mini chromosome maintenance) complex, binds to the ORC and triggers DNA replication. That these proteins are essential for DNA replication in eukaryotes is shown by the fact that all of the components have been highly conserved in nature from the yeast *Saccharomyces* to humans.

DNA Polymerase

DNA polymerase is the enzyme that catalyzes DNA synthesis, namely the stepwise building of the DNA molecule using the single-stranded parent molecule as the template to determine the specific nucleotide base (A, T, G, or C) added in each step.

The mechanism of DNA replication was first studied in the bacterium *E. coli*, which has three enzymes capable of catalyzing DNA replication. These are DNA polymerases (Pol) I, II, and III, each of which has a specific role to play. DNA polymerase III is largely responsible for replication or chain elongation, whereas the two other enzymes play a part in repair and proofreading. Repair is mainly achieved by excising wrongly incorporated nucleotides from the growing DNA strand (i.e., in the 5' to 3' direction). Proofreading refers to the excision of bases in the 3' to 5' direction and can occur anywhere on the molecule and not necessarily at the growing tip.

Several DNA polymerases have been identified in eukaryotes. Individual enzymes are identified on the basis of where they are present inside the cell and their primary function in replication. The human Pol-α, which is encoded on the X chromosome, is most closely related to Pol III of *E. coli*, but the enzyme that is primarily responsible for replication in eukaryotes in Pol-δ. This protein consists of two subunits that are specified by genes on different chromosomes. The gene *POLD1* on chromosome 19 encodes the catalytic subunit of the enzyme and *POLD2* on chromosome 7 makes a subunit that has regulatory functions. The Pol I equivalent in eukaryotes is Pol-β, which plays a role in DNA repair. The human version of this gene, *POLB*, is found on chromosome 8. Still another gene *POLG* on chromosome 15 encodes the enzyme Pol-γ, which is responsible for the replication of mitochondrial DNA.

See also: Polymerase (Ch. 2).

Primase

As discussed earlier, the presence of a short oligonucleotide primer with a free 3' hydroxyl group is an absolute requirement for DNA synthesis, whether in living cells or in the test tube. The enzyme primase catalyzes the synthesis of oligoribonucleotide primers at the replication initiation site. It is a complex protein made of three polypeptides encoded by genes found on chromosomes 1, 2, and 6.

See also: Primer (Ch. 2).

Ligase

We have already seen the uneven manner in which DNA synthesis proceeds off the two complementary templates. DNA synthesis off the lagging strand is not continuous but occurs in short stretches in the 5' to 3' direction. In order to produce a proper replica, these pieces need to be linked together with phosphodiester bonds between the 3'-hydroxyl and 5'-hydroxyl groups of the sugar phosphate backbone. The enzymes that catalyze this reaction are called DNA ligases. Five different DNA ligases have been found in mammals, of which LIG1 (encoded on chromosome 19) and LIG3 (on chromosome 17) are the most important in human cells. Both enzymes depend on ATP to catalyze the linking of strands. Of these, LIG1 functions both in DNA replication and in repair processes, whereas LIG3, which resembles a pox virus ligase more than mammalian forms of the enzyme, seems to play a role primarily in DNA repair and recombination (see the next section).

DNA Repair

Like all other molecules in the cell, DNA is subject to damage. This damage could be mechanical or chemical and may occur in cells that are either actively dividing or in the growth phase of their cycle. In any of these cases, the sequence of bases in the DNA molecules may be altered (see the entry on mutation). Some mutations may have no visible effect, or in a few cases might even prove advantageous to the organism, but others have adverse effects on the individual and must be repaired so that the organism can function properly and pass on the unaltered information to the next generation.

There are three main categories of "repair" mechanisms in human cells: damage reversal, damage removal, and damage tolerance.

Damage Reversal

Damage reversal refers to repair processes where enzymes simply help to restore the normal structure of the DNA without resorting to drastic chemical or physical changes in the molecule. Such repairs usually occur in response to stimuli such as X-rays or peroxides, which cause physical breaks in the sugar phosphate backbone of one or the other strand of a double helix. The enzyme *DNA ligase 3* (discussed in the previous section) is one of the main players in rejoining broken strands by catalyzing the formation of phosphodiester bonds between the 5' and 3' hydroxyl groups of adjacent sugars.

Damage Removal

Damage removal involves cutting out and replacing a damaged base or section of nucleotides from the DNA strand. A good analogy for DNA repair is modern medical practice, where the damage reversal corresponds to noninvasive treatments and damage removal is more analogous to surgery or amputation as the method of treatment. Three types of damage removal systems in mammalian cells are described below.

Base Excision Repair. This repair mechanism kicks in when a DNA molecule has a single inappropriate base at any location and occurs during DNA replication when the polymerase adds the wrong base into the growing DNA strand. The incorrect base is removed and replaced through the enzymatic action of enzymes called glycolyases, which break the sugar–base bond and remove the mismatched (noncomplementary) base. For instance, *uracil glycolyase* removes uracil from DNA. Uracil, as you might recall, belongs in RNA, not DNA. Sometimes, however, RNA primers are not removed properly following DNA replication, resulting in the presence of a uracil in a DNA molecule. Another, more common occurrence is the deamination of cytosine (a normal DNA building block) to uracil. The gene for human uracil glycolyase (UNG) is found on chromosome 12. There are also other specific glycolyases for DNA damage caused by radiation and chemicals.

Mismatch Repair. This proofreading process occurs after, not during, DNA replication. In *E. coli*, where this process has been studied the most, it adds another 100–1000-fold accuracy to replication. Following replication, a group of proteins scan the new DNA strand for any mismatched or unpaired bases. When they find any mismatched bases, they remove a short stretch of DNA around that base. DNA polymerase then returns to put in the correct base.

Human mismatch repair proteins that have been identified are very similar to those of prokaryotes. One family of mismatch repair genes, known as *MSH* genes, contains a number of proteins, including the one encoded by *BRCA1*, the gene found to be associated with breast cancer. Inherited mutations in some of the genes for mismatch repair have also been found to be associated with certain kinds of hereditary colon cancer.

See also: Cancer—Breast Cancer (Ch. 5).

Nucleotide Excision Repair (NER). This process is very closely related to mismatch repair. Here, the repair system recognizes different distortions in the DNA structure. The enzymes then excise the damaged portion and recruit DNA polymerase to fill in the resulting gap. Scientists have identified genes in simple eukaryotes (e.g., yeasts involved in these repair mechanisms, designated as *NER*). Organisms with mutations in these *NER* genes are UV-sensitive and have elevated levels of UV-induced mutation and recombination. People with the hereditary disease xeroderma pigmentosum are sunlight-sensitive, have very high risks of skin cancers on sun-exposed areas of the body, and have defects in genes homologous to those for *NER* in simple eukaryotes.

See also: Xeroderma Pigmentosum (Ch. 5).

Damage Tolerance

Damage tolerance is not strictly a repair system but a way for cells and tissues to cope with certain types of DNA damage while suffering the least amount of disruption to their normal functioning. Tolerance mechanisms are called into play when the cells are unable to correct the actual damage. For instance, exposure to UV light sometimes induces the formation of chemical bonds between adjacent pyrimidine bases (C and T) on a single strand. Most DNA and RNA synthesizing and processing enzymes cannot deal with such interpyrimidine bonds, and consequently such a dimer will act to block normal processes such as DNA replication. Tolerance mechanisms allow the rest of the DNA to replicate normally.

Recombinational (Daughter-Strand Gap) Repair. As described earlier, DNA replication in eukaryotes is initiated at multiple sites, and the different pieces are stitched together by ligases to form the whole molecule. If a polymerase encounters a pyrimidine dimer, it may stop chain elongation at that one site but replication continues at other sites, resulting in the formation of one or more incomplete pieces. When these pieces are subsequently stitched or ligated together, the final molecule will contain "gaps" of single-stranded unreplicated DNA, which are dangerous during cell division. The recombination repair mechanism leaves the pyrimidine dimer intact but works around it so that the new intact molecules are synthesized, although one of them still contains the dimer. If such a dimer is in a noncoding area, there will be no physical manifestations of the mutation. The proteins expressed by the human breast cancer susceptibility genes *BRCA1* and *BRCA2* seem to be involved in this kind of recombinational repair together with genes that are similar to the yeast genes *RAD51* and *RAD52*.

Mutagenic Repair (Trans-Lesion Synthesis). An alternative scenario for a DNA polymerase blocked at a dimer is for it to insert any nucleotide opposite the dimer and continue replication. (This is the "mutate or die" scenario.) This can happen in bacteria, and many scientists think it probably also happens in eukaryotes, although the mechanism is not well-understood. This is one reason why repair might actually cause mutations. The choice of base to incorporate at the site is more or less random and probably depends on other factors such as the relative abundance of the different bases. All other things being equal, there is a one in four chance that the right base is incorporated into the new strand!

See also: Cancer (Ch. 5).

FOOD AND DIGESTION

To live, we must eat. After water and air, food is perhaps the most essential requirement for all living beings, from the simplest to the most complex forms of life. It provides both the energy on which an organism sustains itself and the raw material for different components of cells, tissues, organs, and body parts. Small wonder then that so many organs and tissues in the body are devoted to the acquisition of food and its subsequent conversion into a form usable by the body.

Different forms of life derive nourishment from different sources—simpler organisms such as bacteria get their energy by absorbing chemicals directly from the environment, whereas humans and other multicellular animals have much more complex needs. Humans are heterotrophs, which means that we rely on other living organisms (plants, animals, fungi, etc.) as the primary source of our nutrition. The food we eat is mostly in the form of complex organic molecules. These molecules must be broken down to simpler chemicals (i.e., digested) that can be absorbed into the body (via the bloodstream) and used by its cells.

We take in food and digest it in the alimentary canal or digestive tract, which consists of a long and often convoluted, but continuous, tube that begins at the mouth and ends at the anus. The intermediate regions of the tract (in order) are the pharynx (throat), esophagus, stomach, duodenum, and small and large intestines. Food passes through the tract, where it undergoes enzymatic and chemical processes designed to extract all the useful materials from it. What cannot be used is excreted. A number of accessory organs feed into different parts of the alimentary canal and provide enzymes and other substances to digest the food and absorb the materials to be carried to the rest of the body. This section gives an overview of the mechanisms that the body uses for digestion, with specific attention to the genes that govern these mechanisms.

Appetite/Hunger

The *Oxford English Dictionary* describes appetite as both the desire and capacity for food, and hunger as a more extreme manifestation of this desire, namely the uneasy feeling in your stomach when it is empty. The human body has many different cues for hunger, including physical sensations (an "empty" feeling in the stomach, for example, or a feeling of dizziness or light-headedness), olfactory and visual stimulation—namely the smell and sight of certain foods—and internal signals such as hormones. An example is *ghrelin*, a recently discovered peptide hormone that stimulates the release of growth hormone from the anterior pituitary gland and has been found to have a stimulatory effect on appetite. *Leptin* (see the entry on homeostasis) has the opposite effect (i.e., of curbing appetite).

Food Ingestion

The first step in acquiring food is its ingestion through the mouth. It is here that food begins to undergo different changes that eventually result in the extraction of useful nutrients. Most of the changes to the food in the mouth are physical—the food is broken down into smaller pieces and emulsified for easier movement down the alimentary canal. Saliva from the *salivary glands* provides the moisture for this process. Composed mainly of water and mucus, saliva helps dissolve certain soluble chemicals in foods, which enables us to taste these foods. It also contains an enzyme called *amylase*. This enzyme, encoded by a cluster of genes on chromosome 1, begins the breakdown of complex carbohydrates such as starch and glycogen into smaller units of sugars called maltose and glucose. Two of the four genes of the amylase cluster are expressed in the salivary glands. Others, as we will see, are secreted by the pancreas and act in the small intestine. The action of salivary amylase (also called *ptyalin*) is what makes starchy foods such as bread and rice taste sweet when we chew them for a long time.

Taste

The sensation of "sweetness" that we just mentioned is one of four basic "tastes" that we can discern among different types of food. Taste is very important to us from a sensory perspective: we automatically like foods that taste better to us and are likely to avoid foods whose tastes we dislike. Our sense of taste often, but not always, comes to our rescue in avoiding poisonous food.

We are all familiar with the basic tastes: sweet, salty, bitter, and sour. Some maintain that there is a fifth taste—that of monosodium glutamate—which is called umami and roughly translates to pungent. Each of these sensations is the outcome of a particular signal transduction pathway (see entries on signal transduction in Chapter 2 and in this chapter) that is triggered when soluble or volatile chemicals in the food bind to specific G-protein-coupled receptors in cells of the tongue and olfactory receptors of the nose. Each one of us has some 30,000 to 50,000 taste receptors in clusters of cells called the taste buds, which are specific for the tastes. When a food chemical binds to its appropriate receptor, the cell responds to this signal by transmitting a nervous impulse from the taste buds to the brain. The brain processes the information and "tells" the tongue what the taste is.

There are many genes for the receptors of different tastes, scattered in different chromosomes. There are slight differences in the way each of these receptors binds to the chemicals and the sensory pathways that they stimulate. This variety gives rise to some of the subtle differences that we detect in food. For example, there are 40–80 different receptors, known as the type 2 taste receptors (TASR2s), that allow us to taste "bitter" foods alone. There is a specific receptor for the chemical 6-n-propyl-2-thiouracil on chromosome 5, whereas a cluster of four genes on chromosome 7 code for receptors that sense bitterness in compounds such as phenylthiocarbamide. Similarly, there are different receptors for the "sweet" tastes of sugar and saccharine.

After the preliminary processing in the mouth, food is swallowed and moves down through the pharynx (throat) into the esophagus. The esophagus is a narrow tube whose walls are made of smooth muscles that push the food toward the stomach by involuntary muscle contractions called peristalsis. The pharynx and esophagus do not produce any enzymes for digestion, but they do produce mucus to keep the food moist and the passages lubricated.

Gastric Digestion: The Stomach

Digestion begins in earnest in the stomach, a curved, baglike organ composed of a thick muscular wall that lies in the abdominal cavity just beneath the diaphragm. Embedded in the inner lining of epithelial cells of the stomach wall (known as the mucosa) are pits that lead to tubular gastric glands that produce enzymes and other fluids that make up the gastric juice. Although the gastric glands produce their secretions all the time, the amounts vary according to nervous and endocrine regulatory signals. Sensory (nerve cell) cues such as the sight, smell, and taste of food start what is known as the cephalic phase of gastric secretion. The next phase, called the gastric phase, is the period when the gastric glands are most active. The final phase, when the food passes on from the stomach to the intestines, is a signal to reduce the gastric secretions and is known as the intestinal phase.

Gastrin

Gastrin is a hormone encoded by a gene on chromosome 17 and produced by special endocrine cells in the gastric glands. Its secretion is stimulated by the entry of food into the stomach. It acts through special G-protein-coupled receptors (see the entry on signal transduction in this chapter) on other cells of the gastric glands—parietal and chief cells—to stimulate the production of the food-digesting components (hydrochloric acid and pepsinogen, respectively) of gastric juice.

Parietal Cells and Acid Production

One of the principal components of gastric juice is hydrochloric acid, which serves many digestive functions. It denatures and degrades food, provides a suitable environment (i.e., low pH) in which the stomach enzymes can act, and also activates pepsinogen to its active form (pepsin). The main producers of hydrochloric acid are the parietal cells of the gastric glands. These cells transport protons (H^+ ions) across the cell membrane into the stomach cavity in exchange for potassium (K^+) ions, using ATP as the energy source, and in the process concentrate the protons about 1 millionfold. They achieve this by the action of a transmembrane protein called H^+/K^+ ATPase. This enzyme or proton pump consists of two subunits: a highly catalytic alpha peptide, encoded by the *ATP4A* gene on chromosome 19, and a smaller beta-peptide, whose gene *ATP4B* resides on chromosome 13. The two subunits together form the channel through which the ions are exchanged. The alpha subunit is responsible for ATP hydrolysis. Parietal cells are full of mitochondria to provide the membrane ATPases with the energy they need to pump the protons into the stomach.

Pepsin

The main enzymatic activity in the stomach is the degradation of proteins to smaller peptides. The enzyme pepsin, which acts best at low pH, catalyzes the cleavage of peptide bonds near the amino acids tyrosine, phenylalanine, and tryptophan (called the aromatic amino acids). Proteins that have low amounts of these amino acids are therefore relatively resistant to peptic digestion.

Pepsin is produced as an inactive precursor or proenzyme called pepsinogen. Actually, there are many variants of this proenzyme that are coded by a cluster of similar genes on chromosome 11. The activation of pepsinogen to pepsin occurs by the removal of a 41-residue peptide from the amino end of the protein, which occurs under acidic conditions and hence the need for hydrochloric acid in the stomach cavity. Presumably the enzyme is secreted in this inactive form to prevent it from digesting cellular proteins in the gastric glands and the stomach walls. The mucus lining the walls of the stomach cavity also helps to protect the stomach from the action of pepsin.

Intestinal Digestion

By the time the food leaves the stomach, it is quite unrecognizable as the substances that went into the mouth. The combination of mechanical action, acid denaturation, and enzymatic degradation convert it into a semisolid form—known as chyme—consisting of partially degraded sugars (from starches) and proteins. These substances need to be broken down further into smaller subunits. This happens as the food travels through the long (about 3 meters), winding tube that is the small intestine, named for its diameter, *not* its overall mass. It is here that the digestion of another major food group—the fats (which have made it this far without any substantial changes)—occurs.

Although the small intestine is the site for the most intensive digestion, it produces very few of the digestive enzymes. The pancreas, which feeds its secretions into the duodenum (the first part of the intestine, immediately following the stomach), produces the bulk of these enzymes. Another accessory organ that plays an important role in digestion, especially of fats, is the liver. The largest gland in the body, the liver has many diverse functions in the body, the relevant one for digestion being the production of bile. Also called gall, bile is a greenish fluid consisting of water, salts, pigments, and cholesterol that is produced in the liver and stored in a small sac called the gall bladder until it is needed in the intestines. The bile salts emulsify the fats in food, physically breaking them down into smaller droplets that are more accessible to the fat-digesting enzymes in the small intestine. The process of fat emulsification also enables the body to absorb fat-soluble vitamins such as A, D, E, and K.

The Intestinal Hormones

The entry of chyme into the duodenum stimulates endocrine cells in the intestinal mucosa to produce certain hormones in much the same way as food entering the stomach stimulates the release of gastrin. The first hormone to be released is *secretin*, which then stimulates the pancreas to produce bicarbonates and the liver to secrete bile. This hormone, encoded by a gene on chromosome 11, belongs to the same family of hormones as glucagon. It binds to special G-protein-coupled receptors (the gene, called *SCTR*, is on chromosome 2) in the cells of the liver and pancreas. Interestingly, certain mutations in the genes for secretin and its receptor have been implicated in autism, which suggests that these proteins may have a neuroendocrine function apart from their role in digestion. Some kinds of pancreatic cancer are also caused by mutations in these genes.

The bicarbonates produced by the pancreas are alkaline and help to neutralize the stomach acids. This is important both for the activity of the pancreatic enzymes and to prevent the acid from corroding the intestinal lining. Neutralization also serves as a feedback inhibitory mechanism to stop secretin production.

The presence of chyme in the duodenum also stimulates the production of the hormone *cholecystokinin* (CKK) in the intestine. Acting through specific receptors in the cells of the gall bladder and pancreas, CKK induces the contraction of the gall bladder (which then pumps bile into the duodenum) and the secretion of digestive enzymes by the pancreas. This hormone is first made (by the gene *CKK* on chromosome 3) in an inactive pro-hormone form that is then activated by the removal of peptides. The receptors for CKK are the same molecules as those for pepsin, although on different cells. Curiously enough, the only other site in the body besides the gut where these receptors are found is the brain. The exact function of CKK in the brain is not clear yet, but the presence of its receptors at that site suggests that this hormone, like secretin, has some neuroendocrine function.

Digestive Enzymes

Pancreatic Enzymes. As mentioned, the pancreas is the gland that produces the greatest variety of digestive enzymes. These enzymes break down the larger, more complex molecules in food such as carbohydrates, fats, and proteins into their smaller component units, and their functions are summarized as follows.

- *Amylase*: The main source (more than three-fourths) of carbohydrates in the human diet is starch. (The rest comes from simple sugars, glucose and fructose—which are absorbed directly from the bloodstream—and the cellulose matter in plants, which is digested in our guts by the enzymes of intestinal bacteria.) As we saw earlier, the digestion of starches begins with the action of salivary amylases in the mouth but is then interrupted when the food reaches the acidic environment of the stomach. Digestion is resumed after the food reaches the intestine. The mechanical breakdown of food and emulsifying action of bile also free up more carbohydrates. The pancreatic amylases (produced by the same cluster of genes as the salivary enzymes) take on the task of digesting all of these carbohydrates in the intestines. They break down long chains into smaller molecules, typically disaccharides (i.e., carbohydrate molecules with two sugar units).

- *Trypsin* and *chymotrypsin*: These enzymes are the protein-digesting enzymes (proteinases) of the pancreatic fluid. Like pepsin in the stomach, they are first synthesized as inactive enzyme precursors (trypsinogen and chymotrypsinogen, respectively) that are activated in the intestine. The proteinases digest proteins by cutting them, usually at specific points, into smaller peptides. Trypsin is perhaps the most discriminating of all proteinases and can only cut proteins at arginine and lysine residues. Trypsinogen genes (*PRSS1* and *PRSS2*) are present on chromosome 7, very close to a locus for T cell receptors. Chymotrypsin, otherwise very similar to trypsin in structure and action, is somewhat less selective and attacks peptide bonds with the aromatic amino acids (tyrosine, phenylalanine, and tryptophan), as well as methionine and leucine. The gene for chymotrypsinogen (*CTRB1*) is on chromosome 16.

- *Lipases*: Lipases are enzymes that can break down fats. The pancreatic lipases are responsible for digesting dietary fats in the gut. Most of the fats that we consume belong to a class of chemicals called the triacylglycerols. These lipids are first emulsified by bile and then digested by pancreatic lipase to release the component molecules fatty acids and glycerol. *PNLIP*, the gene for human pancreatic lipase, is found on chromosome 10.

Enzymes of the Small Intestine. The final stages of digestion are carried out by sugar and peptide hydrolyzing enzymes that are produced by cells in the intestinal mucosa. *Maltase* completes the digestion of sugars by breaking down disaccharides such as maltose into single units (usually glucose) that are easily absorbed into the body via the gut. The gene for this enzyme is called *MGAM* (for maltase-glucoamylase) and is on chromosome 7. It belongs to a much larger family of sugar-digesting enzymes, called the glycosyl hydrolases, that have different specificities for disaccharides and are produced in a variety of cells and tissues. The *peptidase* that completes the final digestion of proteins is also called aminopeptidase N. The gene for this protein (*ANPEP* on chromosome 15) is expressed both in the cells of the intestinal epithelium, where its function is digestion, and in the epithelial cells of the kidneys, where its function is less clear. It acts by cutting peptide bonds in between the amino acids of small peptide chains to release free amino acids.

The process of digestion effectively ends in the small intestine, where the different breakdown products of food (i.e., sugars, fatty acids, amino acids, etc.) as well as other nutrients, such as vitamins, are absorbed via the intestinal epithelial tissues and distributed to the rest of the body. Undigested materials in the food (e.g., cellulose fibers from different plant foods) as well as other substances, such as the hemoglobin breakdown products and cholesterol that were processed in the liver and deposited in the gut via bile, are then discarded from the large intestine. The biochemical processes through which various molecules are assimilated into the body are considered in other sections of this book.

GROWTH FACTORS (CYTOKINES)

Growth is a fundamental property of all living organisms. It occurs at all levels from individual cells at the microscopic level to the whole organism. In the section on development, we talked about how genes influence cell differentiation and formation of the embryo. By the fourth month of life, the human fetus is practically fully formed, with all of its internal organs in place. The fetus spends much of its life in the uterus growing. This growth continues after birth as a baby grows up through stages of childhood, adolescence, and adulthood. Indeed, for the rest of our lives, even after we have stopped any visible physical growth, the cells in different organs and tissues (e.g., the skin) continue to undergo cycles of growth and division or death (apoptosis).

In other sections of this chapter, we talk about genes and proteins involved in processes such as DNA replication, protein synthesis (transcription and translation), and programmed cell death, which occur at different phases of the cell cycle and contribute to growth both at the level of individual cells and the whole organism. In this section, we look specifically at a special group of proteins called *growth factors* or *cytokines* that play an important role in regulating the growth of different cells. These proteins act by binding to specific receptors on the surface of cells (which allows them to communicate with the cell's interior) and setting off reactions—via signal transduction pathways—that ultimately stimulate cell division.

The key feature of growth factors is their versatility. Some growth factors are capable of stimulating cellular proliferation in many different cell types, whereas others act specifically on particular cell types. A discussion of a few examples follows.

Epidermal Growth Factor (EGF)

Epidermal growth factor is a growth factor that is active early in the life of the embryo. It binds specifically to receptors on the surface of embryonic cells of mesodermal and ectodermal origin. It continues to be active throughout life, particularly on keratinocytes (skin cells) and fibroblasts, and can also influence the expression of nuclear proto-oncogenes, such as *Fos*, *Jun*, and *Myc*.

The gene for EGF is located on chromosome 4. It produces an inactive, membrane-bound precursor protein, which is cleaved to generate the 53 amino acid peptide hormone that stimulates cells to divide. A single-nucleotide substitution, G to A, at position 61 of the *EGF* gene, has been found in patients with malignant melanoma. Epidermal growth factor stimulates its target cells by binding to a tyrosine kinase receptor encoded by the *EGFR* (EGF receptor) gene on chromosome 7. The kinase domain of the EGF receptor phosphorylates the EGF receptor itself (called autophosphorylation) as well as other proteins in signal transduction cascades.

Fibroblast Growth Factor (FGF)

Fibroblast growth factor is in fact a large family of growth factors, with at least 19 different members that are produced in a wide range of mammals. Studies of human disorders as well as gene knockout studies in mice show that FGFs are important for the development of the skeletal system and nervous system in mammals, as well as the generation of cells in the peripheral and central nervous systems. These factors are active in early embryonic development, apparently helping to induce differentiation in mesodermal cells. The FGF family of genes is widespread across the genome. *FGF-1* on chromosome 5 encodes a growth factor that controls the migration of endothelial cells during development and might also be angiogenic (i.e., involved in blood vessel formation). Other *FGF* genes have been identified on chromosomes 3, 13, 17, and the X chromosome. Kaposi's sarcoma cells that are found in patients with AIDS are known to secrete large amounts of FGF-4, which has oncogenic activity. The gene for this factor, also known as the Kaposi sarcoma oncogene, lies on chromosome 11, in close proximity to the *FGF3* gene, which is also known to be oncogenic.

At least four distinct receptors (FGFR1–FGFR4) have been identified for the FGFs. All have tyrosine kinase activity, like the EGF and PDGF receptors, which we will discuss. Like them, the FGFRs respond to the specific binding first by autophosphorylation. Following the activation of FGF receptors, several signal-transducing proteins associate with the receptor, and their tyrosine residues become phosphorylated to begin the signal cascade.

The FGF receptors are expressed in bone tissue, and many common autosomal dominant disorders of bone growth have been shown to result from mutations in the *FGFR* genes. The most common of these disorders is achondroplasia (ACH). Many people with this condition have a glycine-to-arginine substitution in the transmembrane domain of FGFR3, which corresponds to a guanine-to-adenine transition at nucleotide 1138 on the *FGFR3* gene on chromosome 4. The normal function of FGFR3 is to restrict the growth and proliferation of bone cells, and this mutation results in constant activation of the receptor whether or not FGF is bound to it.

Hematopoietic Growth Factors

These growth factors, also known as the hematopoietins, participate specifically in the process of hematopoiesis, namely the development of blood cells.

Erythropoietin (Epo)

This growth factor is the main regulator of erythrocyte (red blood cell, or RBC) production. It promotes the proliferation and differentiation of immature erythrocytes and initiates hemoglobin synthesis by the RBCs. The *EPO* gene, which is located on chromosome 7, is expressed primarily in the cells of the kidney. In fact, patients suffering from anemia due to kidney failure are often given Epo to induce a rapid and significant increase in red blood cell count. The gene for the Epo receptor, *EPOR*, is found on chromosome 19. Upon binding Epo, the receptor activates a tyrosine kinase, which goes on to activate many different intracellular signaling pathways.

Colony Stimulating Factors (CSFs)

This group of hematopoietins stimulates the production, differentiation, and function of different populations of white blood cells, or leukocytes. Humans have three distinct types of these proteins.

CSF1 is a growth factor specific for the production of macrophages, cells that play an important role in engulfing harmful substances in the bloodstream and also in various tissues. The precursor cells of macrophages have receptors that specifically bind to CSF1. The CSF1 receptor (CSF1R, encoded on chromosome 5) is actually a member of a much larger family of proteins that includes receptors for other CSFs as well as the PDGFs. Mutations in *CSF1R* have been associated with certain myeloid malignancies (cancers of the blood cells). CSF2, also known as the granulocyte-macrophage colony stimulating factor (GM-CSF), is less specific than the two other cytokines of this group because it affects the production of both granulocytes and macrophages. The gene for this protein is also on chromosome 5, and its receptor, *CSF2RA*, is encoded on the pseudoautosomal region of the X and Y chromosomes. CSF3 (also called G-CSF) is specific for granulocytes and is encoded on chromosome 17. The CSF3 receptor, which is encoded on chromosome 1, also seems to function in cell recognition and adhesion.

Lymphokines

These growth factors specifically stimulate the growth and development of lymphocytes, the white blood cells involved in mediating immune responses. See the entry on immunity for details.

Insulin-like Growth Factor (IGF-1)

Originally called somatomedin C, IGF-1 is the main protein involved in the response of cells to the growth hormone (GH). Cells that have the growth hormone receptor are stimulated by GH to secrete IGF-1, which then induces other cells, such as the bone cells, to grow. Structurally related to insulin, IGF-1 can bind both to its own receptor as well as to the receptor for insulin, albeit with a lower affinity. The gene for IGF-1 is located on chromosome 12 and has six exons, which are alternatively spliced in different tissues to make different forms of the protein. Deficiencies in IGF-1 caused by mutations in its gene are linked to growth retardation in combination with deafness and mental retardation. The *IGF1R* gene is found on chromosome 15. This gene plays an important role in cellular transformation and is overexpressed in many malignant cells.

Nerve Cell Growth Factors

Rita Levi-Montalcini (1909–) shared the 1986 Nobel Prize in Physiology or Medicine with Stanley Cohen (1922–) for their discovery and isolation of the nerve-growth factor (NGF), the first of many cell-growth factors discovered in animals. It plays an essential role in the differentiation and growth of nerve cells in the peripheral and central nervous systems. Nerve-growth factor is a protein that consists of three types of polypeptide chains—alpha, beta, and gamma—that interact to form the protein. Of these chains, the NGF beta chain (NGFB) is solely responsible for the nerve-growth stimulating activity of NGF. The *NGFB* gene is found on chromosome 1.

Platelet-Derived Growth Factor (PGDF)

A growth factor that is similar to EGF, PDGF is produced mainly by the blood platelets and promotes proliferation of connective tissue and glial and smooth muscle cells. Two genes on different chromosomes encode the two polypeptides of this protein: *PDGFA* on chromosome 7 expresses the alpha chain, and *PDGFB* on chromosome 22 encodes the beta chain. The receptor for this protein is encoded by the *PDGFRB* gene, which is situated on chromosome 5 between the genes for the two colony-stimulating factors mentioned earlier. This gene is involved in a chromosomal translocation between chromosomes 5 and 12 that results in chronic leukemia.

Transforming Growth Factors (TGFs)

Transforming growth factors are cytokines with the unusual ability to reversibly change cell types. There are two types of TGFs—alpha and beta.

TGF-alpha is a keratinocyte (skin cell) growth factor and may be important for wound healing. It binds both to its own receptor on cells and to the EGF receptor, and it is this interaction that is thought to be responsible for the growth factor's effect. Activated macrophages, keratinocytes, and possibly other epithelial cells secrete this protein, which is encoded by the *TGFA* gene on chromosome 2.

The TGF-beta family of proteins includes the activin and inhibin proteins. The Müllerian inhibiting substance that suppresses the development of female characteristics in an embryo (see the entry on sex determination) is also a TGF-beta-related protein, as are members of the bone morphogenetic protein (BMP) family of bone growth regulatory factors. The TGF-beta family may comprise as many as 100 distinct proteins. The *TGFB1* gene has been mapped to chromosome 19 and contains seven exons and large introns. Still other proteins, with distinct activities, have stretches of amino acid sequences homologous to the TGF-beta family of proteins, particularly at the C-terminal region.

Unlike the EGF, PDGF, and FGF receptors, the TGF-beta receptors all have serine/threonine kinase activity and therefore induce different cascades of signal transduction. TGF-beta proteins influence the proliferation of mesenchymal and epithelial cells.

HOMEOSTASIS: THE BODY IN EQUILIBRIUM

What is the first thing they do when you visit the doctor's office? Regardless of what you have gone in for, a nurse or assistant usually takes your temperature and checks your pulse rate and your blood pressure, even before asking you what is wrong.

Why are these things important? What possible relevance could your pulse rate, for instance, have to stomach cramps or a pain in the foot? What do body temperature, blood pressure, and pulse rate—the "vital signs" as they are called—tell the nurse and doctor? These signs tell them about your body's internal environment, which in turn gives them clues as to what the source of trouble might be.

More than a century ago, the French physiologist Claude Bernard (1813–1878) observed that the cells of warm-blooded creatures (mammals) such as humans were only capable of functioning properly, and surviving, if the temperature, pressure, and chemical composition of their environment remained relatively constant or balanced. If dramatic changes in these conditions occur, then the cells stop functioning properly. Depending on the extent of the disturbance, entire tissues, organs, organ systems, and ultimately the organism may stop functioning. A few decades after Bernard made his observations, another physiologist, the American Walter B. Cannon (1871–1945), coined the term *homeostasis* to describe the constant state of the body's internal environment. When doctors monitor our vital signs, they are looking to see how far the body has deviated from this normal state.

As Cannon stressed, despite its literal meaning ("homeo" means same and "stasis" means standing in Greek), homeostasis does not imply a rigid or unchanging body. It describes the body's attempt to remain as close to constant as possible within a range of conditions. Humans maintain a body temperature of around 37°C, an arterial blood pressure of 120/80 mm of mercury, and blood pH at 7.4. As we know from experience, the body is constantly subject to an enormous number of changes from the outside. But in addition to external factors such as the climate, virtually all of our activities, such as breathing, eating, walking, and talking, contribute in some way toward disturbing the balance within the body. Ingesting food, for example, brings more energy to the body. If this energy is not used up, it will accumulate in the body and raise the body's temperature. If the temperature rises too much, cellular proteins lose their activity and cells stop functioning. Consequently, the body needs to constantly counteract each perturbing effect with some activity to restore its inner balance. Major activities involved in maintaining homeostasis include responding to changes in the external environment, exchanging materials between the environment and cells, metabolizing foods, and integrating all of the body's diverse activities.

Homeostasis is based on a system of *negative feedback* (see Figure 4.4). The body detects deviations from normal through inbuilt sensors and responds to correct these deviations and restore the normal conditions. The principal detectors and messengers of change in homeostasis are the nervous system and the en-

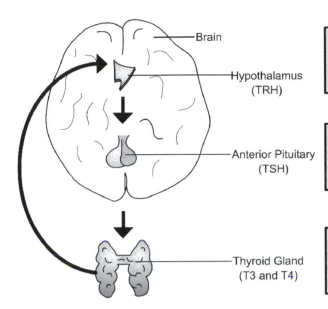

Step 1: Neurons in the hypothalamus secrete thyroid releasing hormone (TRH), which stimulates cells in the anterior pituitary to secrete thyroid stimulating hormone (TSH).

Step 2: TSH binds to receptors on epithelial cells in the thyroid gland, stimulating synthesis and secretion of thyroid hormones, which affect T3 and T4.

Step 3: When blood concentrations of thyroid hormones increase above a certain threshold, TRH-secreting neurons in the hypothalamus are inhibited and stop secreting TRH.

Figure 4.4. Negative feedback. [Ricochet Productions]

docrine system. The nerves use electrical impulses to transmit messages about change from one part of the body to another, and the endocrine system uses chemicals called hormones (see entry in this chapter). Hormones are secreted by endocrine glands located in different parts of the body. Glands important for maintaining homeostasis include the hypothalamus and pituitary glands at the base of the brain, the thyroid gland in the throat, adrenal glands on the kidneys, and islets of Langerhans in the pancreas. Whether the signals are electrical or chemical, homeostatic control at the level of individual cells comes down to the proper transmission of these messages. In other words, different signal transduction pathways (see entry in this chapter) are of key importance in homeostasis. Cells receive messages of the change and then respond with corrective measures, such as by increasing or decreasing their rate of metabolism (to regulate temperature) or changing the flow of water in and out of cells (to maintain pressure).

In the following section, we look at the genes for proteins that are involved in specific homeostatic mechanisms. For reasons of convenience, we have grouped the proteins according to the homeostatic parameters that they regulate (i.e., temperature, water, or sugar). However, this classification is a little contrived because many of these regulatory mechanisms are often linked with one another. Sugar utilization, for example, is a direct indicator of metabolism, which in turn affects temperature.

Temperature Homeostasis

If we think about the wide extremes of temperature that the human body encounters—from below freezing in the winter to more than 100° Fahrenheit (45° Celsius) in summer in many places—it seems extraordinary that it is able to maintain its internal environment within just 2–3 degrees (97–99°F or 37–37.5°C). Some of the external signs of the body's response to climatic changes are shivering and higher metabolic rates in cold weather and sweating in the summer to cool down the body. Still, these involuntary activities are just a few outward manifestations of a much more complex system of heat regulation within the human body.

The center for temperature control is the *hypothalamus*. Situated in the middle of the base of the brain, it serves as the physical junction for the nervous and endocrine systems of homeostatic control. The hypothalamus contains a sort of thermostat that is programmed to keep the body at a more or less constant temperature (37°C). It detects changes in temperature through cells called thermoreceptors and responds both by passing along these messages to the brain (which then initiates its own corrective responses) and by stimulating hormones that serve as messengers to other parts of the body to respond to the changes in temperature.

An example of a hypothalamic hormone involved in temperature homeostasis is *leptin*. Interestingly, this hormone was first discovered in the context of its involvement in appetite control, fat metabolism, and obesity (see the entry on obesity in Chapter 5), which points to the intricate links between different regulatory pathways in the body—in this case temperature, energy, fat storage and utilization, appetite, and body weight. Encoded by a gene on chromosome 7, leptin is secreted into the blood by fat cells called adipocytes in response to the accumulation of fat or other energy (e.g., heat). It carries a message to the hypothalamus, which has another control center for appetite (apart from the thermostat). Leptin activates both of these centers, which then respond with mechanisms to decrease food intake and energy production (both of which raise temperature). Like other hormones, leptin needs to bind a receptor in its target cells in order to pass on its message. The gene for the leptin receptor has been mapped to chromosome 1. Mutations in both genes (for leptin and its receptor) have been linked to obesity.

Water Homeostasis

Water, as we know, is essential for life. By weight, it accounts for more than 90 percent of the total human body, and chemically it is involved in virtually all of the biochemical pathways of life. Water also gives cells and tissues the ability to physically interact with their environment and with each other. Many nutrients are able to diffuse in and out of cells because they are dissolved in water. Water and certain ions (positively charged metal ions such as sodium and potassium and negatively charged ions such as chloride) provide cells with the internal pressure (osmotic pressure) they require for maintaining their integrity. Water homeostasis, also called osmoregulation, is very important for this. If there is too little water in the cells, their contents become too concentrated and their internal pH changes, which in turn affects the structure and function of intracellular proteins and nucleic acids. If there is too much water, the pressure builds up and cells can become leaky or even burst. The body exercises control over its water concentration by monitoring water in the bloodstream. By maintaining water pressure and concentration in the blood, the body determines how much water is available for absorption and usage by cells.

As in the case of thermoregulation, the negative feedback mechanisms of osmoregulation are headquartered in the hypothalamus, in this case the part of the gland next to the blood vessels. Changes in ionic concentration are detected by receptor cells in the hypothalamus, which relays these messages to the pituitary gland next to it. The pituitary gland then secretes a hormone called the *antidiuretic hormone*, or ADH. Also known as *vasopressin*, this hormone (whose gene is on chromosome 20) transmits signals to specific cells in the tubules of the kidneys, telling them to increase or decrease their permeability. This increases or decreases the ability of the kidney cells to release water. So far, we know of at least two distinct vasopressin receptor molecules in the kidney cells. Both are G-protein-coupled receptors that participate in a number of different signal transduction pathways. The genes for these receptors reside in the X chromosome and on chromosome 1. Mutations in any of these three genes can cause diabetes insipidus (see the entry for diabetes in Chapter 5).

Sugar Homeostasis

As the simplest and most direct source of chemical energy, sugar probably ranks next only to water and air in its importance to the human body. It also contributes a degree of viscosity and pressure to water, which means that the regulation of blood sugar levels impinges directly on many different aspects of homeostasis.

The pancreas is the gland that coordinates blood sugar regulation. Although it functions as an exocrine gland that produces digestive enzymes (see the entry on food and digestion) that are delivered via ducts to the gastrointestinal tract, the pancreas also contains special groups of cells called the islets of Langerhans that are entirely endocrine in function. This means that it can secrete hormones that are delivered directly into the bloodstream without the help of ducts or tubules. Responding to signals from pancreatic receptors that detect fluctuations in blood sugar levels, the islets of Langerhans produce the following two different hormones with roughly opposite effects.

Insulin

This hormone, which is produced by cells called beta cells in the islets of Langerhans, operates via the classical negative feedback mechanisms described earlier to effect sugar homeostasis. Acting through special membrane receptors—primarily in the liver, but also in other cells—insulin enables sugar to pass from the blood into the cells.

Glucose that is not needed immediately by the body is changed into glycogen for short-term storage in the liver, skeletal muscles, and skin, and fat for longer-term storage in the adipose tissue and liver. Insulin decreases the glucose level in the blood and increases glucose transport to skeletal, heart, smooth muscle, and fat cells, where it promotes the conversion of sugar to glycogen and fat. It also increases transport of amino acids into the cells and causes an increase in protein synthesis.

The discovery of insulin at the turn of the twentieth century was seen as a major breakthrough toward understanding and treating diabetes mellitus. In 1923, University of Toronto researchers Frederick Banting (1891–1941) and J.J.R. MacLeod (1876–1935) received the Nobel Prize in Physiology or Medicine for their part in this discovery. They shared the prize money with Charles Best (1899–1978) and James Collip (1892–1965), whose contributions had not been recognized by the Nobel committee.

The gene that encodes insulin lies on the short arm of chromosome 11. The gene for a specific insulin receptor that binds the hormone and signals the cells to absorb sugar is found on chromosome 19. Insulin can also bind to other receptors and set off other signal transduction pathways, such as amino acid uptake, as described earlier. Scientists have also found genes for transcription factors—such as *PDX-1*, *Neuro-D1*, and, in 2002, a gene called *RIPE3b1* factor—that can influence the ability of insulin genes to trigger insulin production in the beta cells of the pancreas. They are now exploring the possibility of inserting the three genes into stem cells, or undifferentiated pancreatic cells, to convert them into insulin-producing cells in patients with IDDM or type 1 diabetes mellitus.

Glucagon

Produced by a subset of islet cells (called alpha cells) that differ from those that produce insulin, the hormone glucagon acts as a positive control mechanism in controlling blood sugar. It responds to the decrease of sugar in the blood and acts on liver cells, stimulating them to release glucose into the blood. The glucagon gene is present in chromosome 2, and genes for glucagon receptors (expressed in muscle and hepatic cells) have been found on chromosomes 6 and 17.

Glucagon influences two metabolic pathways in the liver, in response to which the liver releases glucose for the rest of the body. In the first pathway, glucagon stimulates the breakdown of glycogen stored in the liver (synthesized earlier under the influence of insulin when blood glucose levels were high). Glucagon also activates a second metabolic pathway called gluconeogenesis, through which the liver converts nonsugar substrates such as amino acids to glucose. This pathway is usually a second resort for glucose production in humans and is activated primarily when the body is starved. Gluconeogenesis is the main mechanism of glucose release in animals such as cats and sheep that do not absorb much glucose from the intestines.

Although type 1 diabetes mellitus develops mainly from an insulin deficiency, abnormal glucagon secretion may also contribute to the disease, as indicated by the fact that the high blood sugar of many diabetic patients is accompanied by high levels of glucagon in the blood.

See also: Hormones and the Endocrine System; Diabetes (Ch. 5).

HORMONES AND THE ENDOCRINE SYSTEM

The endocrine system, along with the nervous and immune systems, constitutes one of the important communication networks that coordinate the activities in different parts of the body to keep it working as an integrated and balanced whole. This system is made up of glands that are located in different parts of the body and function by secreting chemicals called hormones that control physiological processes throughout the body. The characteristic feature of the endocrine glands is the absence of any ducts to carry their products (the hormones) to their destination. (This is where the system gets its name—"endo" in Greek means within and "krine" means to secrete.) Endocrine glands secrete hormones directly into the blood, which transports them to different parts of the body. At the receiving end of the endocrine signal are cells with special hormone-specific receptor molecules. Hormones can only deliver their messages to cells that have appropriate receptors (i.e., to the correct address). Cells without the specific receptors for hormones remain unaffected, at least directly.

Hormones are involved in four main physiological functions, namely homeostasis (see entry), energy management (i.e., production, storage, use), growth and development, and reproduction. They also influence behavior, often in very subtle ways. Although the endocrine system is diffuse and thus seems disjointed—its glands lack ducts and are quite distant from one another—the activities of its glands are, in fact, quite well coordinated and intricately interdependent. The main endocrine glands in the human body are the pituitary and pineal glands based near the brain, the thyroid and parathyroid glands in the neck region, the pancreatic islets of Langerhans, the adrenal glands in the abdominal region, and the gonads (ovaries in females and testes in males). Cells of certain other organs such as the stomach, intestines, heart, thymus, and placenta also produce endocrine hormones, but these usually act locally. For example, the thymus gland produces *thymosin*, which, among other activities, seems to control the maturation of lymphocytes (T cells). The thymus produces many types of thymosin, of which thymosin-beta is the best-studied. It is a small protein believed to be involved in wound healing, and its 5 kb gene is found on the X chromosome. The cells of the inner lining of the stomach respond to the presence of food by producing *gastrin*, which in turn stimulates the production of hydrochloric acid and pepsin to digest the food (see the entry on food and digestion). The gastrin gene has been mapped to chromosome 17.

Chemically, the hormones are classified into two main groups: those made up of amino acids and those derived from steroids. With the exception of the hormones produced by the adrenal cortex and the gonads (sex organs), all human hormones belong to the former category. The amino acid hormones may either be simple proteins or modified molecules such as amines. The former are encoded in genes and are synthesized by the normal mechanisms of protein synthesis. Amines are synthesized through enzyme-catalyzed biochemical pathways. The steroid hormones are derived from cholesterol, also by means of different enzymatic reactions. They consist of five major classes of molecules: progesterone, estrogens, androgens, glucocorticoids, and mineral corticoids. Protein and amine hormones act by binding to cell-surface receptors and initiating specific signal transduction pathways. Steroid hormones, on the other hand, enter the cell by diffusing across the cell membrane and bind to a specific receptor inside the cell's cytoplasm. The receptor–steroid complex moves to the nucleus to directly stimulate the translation of specific genes.

The endocrine system uses cycles and negative feedback to regulate physiological functions. Negative feedback regulates the secretion of almost every hormone. For example, as we saw in the entry on homeostasis, the hypothalamus maintains the water content of your body at a constant level through the antidiuretic hormone (ADH), also known as vasopressin. If you have too much water, your cells swell up and could burst. Too little could make your cells lose water by osmosis and shrivel up. When the hypothalamus senses too little water in the blood, it sends a message to the pituitary gland to release ADH, which stops the kidneys from removing water from the blood. If there is too much water in the blood, the hypothalamus stops signaling the pituitary gland to make ADH, and the kidneys remove the excess water. Cycles of hormone secretion maintain physiological and homeostatic control. These cycles can last from hours to months, and hormones can work at amazingly low concentrations (parts per trillion).

In the entry on homeostasis, we looked at some hormones active in different aspects of maintaining homeostasis. We now look at a few other examples of some hormones involved in some of the other functions mentioned earlier. For the sake of convenience, the sections are organized according to their source (i.e., the glands that produce them).

Pituitary Gland

The pituitary gland is a tiny pea-sized gland, less than 1 cm in diameter, that functions as a "master gland" in the endocrine system, secreting hormones that then control the activity of other glands. It is located at the base of the skull just below the hypothalamus, to which it is connected by a slender stalk called the infundibulum. The front of the pituitary gland is under the control of the hypothalamus and produces several hormones, including human growth hormone, hormones to control (stimulate) the activity of the thyroid gland, adrenal cortex, and the gonads, and a hormone called prolactin, the purpose of which is to promote lactation. The posterior part of the pituitary gland is controlled by the nervous system and produces vasopressin (see the entry on homeostasis) and oxytocin.

Growth Hormone (GH)

Human growth hormone (also known as somatotropin) is a hormone that promotes body growth. It acts by binding to receptors on the surface of liver cells and stimulating them to release cytokines such as the insulin-like growth factor-1 (IGF-1, also known as somatomedin), which then acts directly on the ends of the long bones to promote their growth. The GH-secreting cells in the pituitary gland are themselves under the control of signals by the growth hormone releasing hormone (GHRH) from the hypothalamus. Human growth hormone consists of a chain of 191 amino acids. It is expressed by a cluster of five genes located on chromosome 17. Most mutations in the *GH* gene give rise to hormone deficiencies, which usually result in short stature as well as more serious health complications.

Low levels of GH stunt growth and lead to dwarfism (see Chapter 5 for a detailed discussion of various forms of this condition). Dwarfism can also result from an inability to respond to GH. This can result from inheriting two mutant genes encoding the receptors for either GHRH or GH. Hormone-replacement therapy can be helpful for some cases. For several years, the only source of GH for therapy was that extracted from the glands of human cadavers (because animal GH does not work in humans). This carries some risk of causing a rare neurological disease attributed to contaminated glands. Luckily, recombinant human GH (rHGH) is now available for those who suffer from GH deficiency.

Prolactin

Prolactin is responsible for stimulating the production of milk in lactating mothers. It is produced by a gene 10 kb long, found on chromosome 6, that bears some homology for the growth hormone genes on chromosome 17.

Pineal Gland

The pineal gland is a small organ shaped like a pine cone (hence the name) at the base of the brain. It is also known as the epiphysis and functions mainly to synthesize and secrete the hormone *melatonin*, a very important hormone that controls biological rhythms (sleep–wake cycles) and also affects reproductive functions. It controls these activities because of its activity of responding to changes in environmental light and communicating this information to the retina and brain, which then relay the message to various other parts of the body. This ability to communicate information about surrounding light to parts of the body has led to the nickname of the "third eye" for the pineal gland.

Chemically, melatonin is a simple molecule, a modified version of the neurotransmitter serotonin, itself derived from the amino acid tryptophan. Serotonin is produced in the brain and transported to the pineal gland, where the successive addition of acetyl amd methyl groups converts it to melatonin. These two reactions are catalyzed by the enzymes serotonin-N-acyltransferase (SNAT) and hydroxyindole-O-methyltransferase (HIOMT), respectively. The *SNAT* gene spans approximately 2.5 kb in length and is on chromosome 17, and the *HIOMT* gene is on the X chromosome. The synthesis and secretion of melatonin are dramatically affected by light exposure to the eyes. In general, serum concentrations of melatonin are low during the daylight hours and increase to a peak during the dark. This pattern, in turn, is linked to the activity of the SNAT enzyme, which is also low during daylight and peaks during the dark phase.

Melatonin acts via G-protein-coupled receptors that are present on the membranes of the brain and retinal cells. Humans have two main types of these melatonin receptors: the melatonin receptor 1A (MTNR1A), encoded by a gene on chromosome 4 and expressed almost exclusively in the brain, and MTNR1B, the gene for which is on chromosome 11 and expressed both in the brain and the retina. The receptor binds melatonin and then transmits the signal through a G-protein-coupled signal transduction cascade.

Thyroid and Parathyroid Glands

These glands are found in the region of the neck over the trachea, or windpipe. Whereas most animals have a pair each of both the thyroid and parathyroid glands, the human thyroid gland is a single butterfly-shaped structure consisting of two lobes joined together by a narrow isthmus. The parathyroid glands exist as a pair of smaller clumps of cells, one on each lobe of the thyroid.

Thyroxine

The primary hormone secreted by the thyroid gland, thyroxine is an important physiologic modulator, with effects on the growth, development, and metabolism of cells. Of all the endocrine hormones, it has one of the largest and most diverse repertoires of target cells, which leads to a wide variety of manifestations of its activity, such as changes in the metabolic rate, the rate and strength of the heartbeat, body weight, and feelings of hunger and thirst, to name just a few.

Chemically, thyroxine is a small molecule consisting of two units of the amino acid tyrosine, to which are attached three iodide ions. It is first synthesized in the thyroid as a less active form known as T_4 that contains four iodide molecules. A globular glycoprotein called thyroglobulin (synthesized from the *TG* gene on chromosome 8) provides tyrosine residues for this enzyme and also acts as the scaffolding on which the hormone is synthesized. An enzyme called thyroid peroxidase (encoded on chromosome 2) catalyzes the addition of two iodide ions to each tyrosine molecule and the subsequent joining of these iodized amino acids. T_4 is released from its globulin scaffold into the gland by the action of lysosomal enzymes and subsequently converted to its active T_3 form by the removal of one iodide.

Thyroxine acts on its target cells by diffusing through cell membranes into cells and binding to receptors in the nuclear membrane. These nuclear receptors belong to a large family of receptor proteins that also includes steroid hormone receptors. An example of a gene for a receptor specific to T_3 in humans is *THRA* on chromosome 17. In the absence of the modulating hormones, these nuclear receptors are bound to DNA, which usually results in gene repression. The binding of the hormone to the receptor causes it to detach itself from the DNA and activate transcription. Genes that are regulated by T_3 include fatty acid synthetase and growth hormone. The increased activity of these genes raises the cell's basal metabolic rate. People who have mutations in the receptor beta gene (located on chromosome 3) that interfere with hormone binding suffer from a low response to the thyroid hormone.

Calcitonin

In addition to thyroxine, the thyroid gland also synthesizes and secretes calcitonin. This hormone, along with the parathyroid hormones, participates in controlling calcium and phosphorus in the blood and has significant effects on bone physiology. A pair of adjacent genes, *CALCA* and *CALCB*, on chromosome 11, encode polypeptides that exhibit calcitonin activity.

Parathyroid Hormone

Balanced calcium levels are very important for the human body. Even the smallest deviations can cause muscle and nerve impairment. The parathyroid hormone helps by stimulating bones to release calcium into the bloodstream and by the absorption of food by the intestines and conservation of calcium by the kidneys. Receptors for the parathyroid hormone are present on different cells, including cells of the blood vessels, brain, and pancreas. They are G-protein-coupled receptors that activate adenylyl cyclase to transmit its signal. Scientists have mapped a 32 kb gene for this type of receptor (*PTHR1*) to chromosome 3. Mutations in *PTHR1* cause a rare group of conditions known as metaphyseal chondrodysplasia, which are characterized by a short-limbed dwarfism associated with high calcium levels and low serum concentrations of the parathyroid hormone. A second gene, called *PTHR2*, expressed preferentially in the brain and pancreas, has been identified on chromosome 2.

Adrenal Glands

These glands, also called the suprarenal glands, are a pair of essentially identical structures that sit on top of each of the kidneys. Each gland is divided, both structurally and functionally, into an outer *cortex* and inner *medulla*. The cortex is the source of steroid hormones and is under the control of other hormones that are produced by the hypothalamus. The medulla secretes amino acid–derived hormones involved in regulating the body's energy and metabolism and is regulated by the nervous system.

Adrenalin

This hormone, also known as *epinephrine*, is very important to the body's metabolism. It is stimulated in response to different types of stress, including strong emotions such as fear or anger, as well as physical danger.

When one is faced with a stressful situation (e.g., suddenly coming face-to-face with a hungry lion, or something more mundane such as taking an exam), the adrenal glands release adrenalin. The hormone binds the beta-2-adrenergic receptor on the surface of cells (encoded on chromosome 5), which is coupled to G-proteins that go through a signaling cascade to activate adenylate cyclase. The ultimate effect of this signaling cascade is to get muscle cells to convert stored glycogen to glucose. Now the body is prepared for strenuous activity and has a readily available energy source to stand and fight or to turn around and run—the "fight or flight" response. Adrenalin also has a role in increasing heart activity, improving the power and longevity of skeletal muscle action, and increasing the rate and depth of breathing. Mutations in the gene for adrenalin receptors may be involved in errors in lipid metabolism that lead to obesity and hypertension. The diversity of adrenalin's physiological roles is reflected in its medical uses: physicians use it as a stimulant in cardiac arrest, as a bronchodilator to alleviate symptoms of asthma, and in the treatment of glaucoma.

Chemically, adrenalin belongs to a group of molecules called catecholamines. Catecholamines are synthesized in the body from molecules that are derived from the amino acids phenylalanine and tyrosine. Adrenalin is synthesized from noradrenalin (derived from tyrosine) in a reaction catalyzed by the enzyme phenylethanolamine N-methyltransferase (PNMT). The *PNMT* gene is located on chromosome 17 and spans about 2.1 kb.

Cortisone and Cortisol

Cortisone is a steroid hormone produced by the adrenal cortex. Its main job is to facilitate the conversion of proteins to carbohydrates, but it also has anti-inflammatory properties and helps regulate salt in the body. It is synthesized from cholesterol in the outer layer of the adrenal gland under the influence of adrenocorticotropic hormone (ACTH), itself produced by the pituitary gland. Cortisone is reduced to the more potent cortisol under the action of the enzyme hydroxysteroid (11-beta) dehydrogenase 1, the gene for which is found on chromosome 1. Mutations in this gene may lead to deficiencies in cortisol that cause symptoms such as excess hair growth, obesity, acne, and infertility.

Cortisol is also involved in the regulation of protein, fat, and carbohydrate metabolism, electrolyte balance, body water distribution, blood pressure, and immunosuppression. Like cortisone, cortisol is also synthesized and secreted in response to ACTH. In turn, cortisol regulates the production of ACTH by negative feedback inhibition. Cortisol helps to break down protein. The amino acids released are converted by the liver into glucose in a process called gluconeogenesis. People under stress or suffering from depression may have abnormally high cortisol levels. Not surprisingly, the hormone has earned a reputation as the "stress hormone."

Biochemist Edward C. Kendall (1886–1972) performed many experiments to isolate cortisone and other steroids from the adrenal glands starting in 1935. In 1948, Kendall and Philip S. Hench (1896–1965) used cortisone on patients suffering from rheumatoid arthritis. Their clinical trials were successful, and it became an established anti-inflammatory treatment, albeit with certain unpleasant side effects. These side effects are caused mainly by the fact that the therapeutic dose of cortisone is much greater than its normal levels in the body. This causes an exaggeration of its normal functions, leading to edema (swelling), increased gastric acidity, and imbalances in metabolism of sodium, potassium, and nitrogen. Nowadays, cortisone has been replaced by related synthetic compounds, called corticosteroids, that do not produce certain undesirable side effects.

Islets of Langerhans

Anatomically, the islets of Langerhans are collections or small islands of cells distributed somewhat randomly throughout the pancreas, itself a soft, rather shapeless organ lying in the abdomen in the space between the duodenum and intestines. As we mentioned earlier, the islets form the endocrine component of the pancreas and are composed of two distinct types of cells, called the alpha and beta cells, that secrete the sugar-controlling hormones *glucagon* and *insulin*, respectively. Details of the activity and genetics of these hormones have already been covered in the entry on homeostasis.

See also: Homeostasis: The Body in Equilibrium—Sugar Homeostasis; Diabetes (Ch. 5).

Gonads

The gonads—ovaries in females and testes in males—are the primary reproductive organs in humans. In addition to this function, they also secrete endocrine hormones that are responsible for the secondary sexual characteristics by which we often identify the sexes.

As we mentioned earlier, these sex hormones are steroid compounds derived from cholesterol. The first step in this reaction is the conversion of cholesterol to a precursor or "mother" steroid hormone known as pregnenolone. This reaction is catalyzed in mitochondria by the enzyme CYP11A (also called the cholesterol side-chain cleavage enzyme), which belongs to the family of cytochrome P450s. The gene for *CYP11A* is 20 kb long and lies on chromosome 15. Then, depending on signals received from other protein hormones released by the brain, pregnenolone is converted to the other hormones in the ovaries, testes, and adrenal glands. For example, the follicle-stimulating hormone (FSH) produced by the hypothalamus signals the ovaries to produce estrogens (female-specific hormones), whereas different signals act to stimulate the synthesis of testosterone, the male-specific hormone, by the testes.

Estrogen

Estrogen is not a single hormone but a group of compounds derived from cholesterol and produced by the ovaries and the placenta. There are three principal forms of estrogen found in the human body: estrone, estradiol, and estriol, also known as E1, E2, and E3, respectively. All are produced by enzymatic conversion from precursors called androgens—or the male hormones. Estradiol is produced from testosterone in a reaction catalyzed by the cytochrome P450 enzyme aromatase, also called estrogen synthetase. The aromatase gene, which is on chromosome 15, is 70 kb long. The production of estrone from androstenedione is catalyzed by 17 beta-hydroxysteroid dehydrogenase. Estrone is a weaker hormone than estradiol, and postmenopausal women have more of it than estradiol.

Although estrogens are present in both men and women, they are found in significantly higher quantities in women. They promote the development of female secondary sexual characteristics (e.g., increasing breast size and the onset of menses at puberty). In addition, they are involved in controlling the menstrual cycle in postpubertal women, which is why birth control pills contain estrogens. The condition we refer to as menopause signifies a natural decrease in the production of estrogens by the body. Estrogens act by binding to receptors on these cells and promoting the transcription of specific genes. An example of an estrogen receptor is ESR1, encoded by a gene on chromosome 6.

Testosterone

Testosterone, the male sex hormone, stimulates the physical changes that produce secondary sexual characteristics in boys. Interestingly enough, it is also the precursor for the female hormones as well. At puberty, the pituitary hormones follicle-stimulating hormone (FSH) and luteinising hormone (LH) prompt the testes to produce more testosterone, causing a sharp increase in the levels of this hormone. The conversion of androstenedione to testosterone is catalyzed by 17 beta-hydroxysteroid dehydrogenase, which, in turn, is controlled by LH. The gene for this enzyme is 60 kb long and is located on chromosome 9. Mutations in the gene cause severe complications in the development of male sexual characteristics in the fetus.

HOUSEKEEPING GENES

The building you live in has certain services that are always readily available (unless of course you forgot to pay your bills!). Flip a switch and the lights come on; turn on the tap and you can get yourself a glass of water. Imagine what would happen if these constant services suddenly shut down. Your daily routine would quickly come to a standstill, and your life would be disrupted. So it is with living cells, too. In all of our cells, there are certain genes involved in basic functions that the cells depend on for life. These "housekeeping" genes are constitutively expressed—they are always turned on and are always being transcribed. Such genes are thought to be involved in routine cellular metabolism. They lack strong tissue specificity (i.e., they tend to be expressed in the cells of most tissues in the body). Housekeeping genes encode proteins that carry out essential cellular functions such as DNA replication, RNA transcription, and respiration. They probably represent the majority of the functional human genome. We will give an overview of some examples of intracellular housekeeping genes. Details about different types mentioned earlier are covered in individual sections.

All genes are "expressed" (i.e., transcribed from DNA to mRNA) by the action of special proteins called transcription factors. Transcription factors recognize certain sequences on a gene, called promoters and enhancers (see the entry on promoters in Chapter 2 for details), and regulate gene expression through their interaction with these regions. However, because the housekeeping genes must be expressed constantly and ubiquitously, scientists believe that they must be under less stringent control. Consequently, one would expect certain differences in the structure of the housekeeping genes. It has been found, for example, that the housekeeping genes are generally more compact, with shorter introns, untranslated regions, and coding sequences. Exactly how this affects transcription is not entirely clear.

Another difference in a housekeeping gene is the lack of any TATA or CAAT boxes, AT-rich promoters found on most other genes, which are recognized by RNA polymerase II. Instead, many of these genes are extremely rich in G and C residues and contain features such as a regulatory GC box (GGGCGG) upstream of the coding sequence. These GC-rich features commonly found in the housekeeping genes seem to be a general characteristic of genes that lack tissue specificity. A representative example is a gene on chromosome 20 for *adenosine deaminase* (ADA), a nucleotide-metabolizing enzyme that converts adenosine to inosine. (Because nucleotides are a constant requirement for DNA and RNA synthesis in the cell, their recycling and processing must be regarded as a basic housekeeping function.) An exception is the *beta-actin* gene (on chromosome 7), which expresses a highly conserved protein in all eukaryotic cells—a component of the cytoskeleton and a mediator of internal cell motility. Unlike many other housekeeping genes, the beta-actin gene contains a TATA box.

Glyceraldehyde-3-phosphate dehydrogenase is another example of a housekeeping gene. It catalyzes an important energy-yielding step in carbohydrate metabolism, the reversible oxidative phosphorylation of glyceraldehyde-3-phosphate in the presence of inorganic phosphate and NAD. The gene for this enzyme is found on chromosome 12. It is widely expressed in many tissues, which is not surprising given its important function. Its sequence is also conserved across many species, and in fact its rate of evolutionary change is among the slowest of all known genes.

Perhaps the genes most associated with the label of housekeeping are those for the *heat shock proteins* (hsp). These proteins derive their name from the fact that their first discovered function was in helping to protect cells from damage by interfering with the uncontrolled protein unfolding that occurs under conditions of metabolic stress (such as sudden temperature changes). However, as it turns out, it is not only stressed cells that produce hsp; some of these proteins are actually expressed all the time and perform important housekeeping duties. Some play a key role in the functioning of the immune system, whereas others temporarily stabilize unfolded or partially folded proteins and help proteins form the correct tertiary structure. The latter types of hsp are also known as *chaperonins* because of their function as protein chaperones. They belong to a large family of proteins whose genes are scattered throughout the chromosomes. Two genes belonging to the hsp60 and hsp10 families are linked head to head on chromosome 2, comprising approximately 17 kb and consisting of 12 and 4 exons, respectively, with a single bidirectional promoter region between them. Interestingly, the same genes that encode chaperonins nonspecifically in different tissues also encode crystallins, the transparent structural proteins that make up eye lenses, under specific conditions and locations of expression. This phenomenon where the same gene has dual functions in different conditions is known as gene sharing.

Defects in the housekeeping genes are rare, not because they do not occur but because their function is so critical to life that any mutations would be lethal. Because they are expressed at constant levels, the housekeeping genes are often used in gene expression experiments to calibrate measurements of expression levels of other genes.

See also: DNA Synthesis/Replication and Repair, Respiration, The Senses—Sight/Vision—The Eye Lens—Crystallins, Sugar Metabolism—Glycolysis, Transcription: RNA Polymerases and Transcription Factors.

IMMUNITY

The AIDS epidemic brought immunity and the immune system squarely into the lay vocabulary during the mid-1980s. Words and phrases such as "T cell counts" and "opportunistic infections" became common in day-to-day conversations, and advertisements began to use properties such as "immune-boosting" as selling points to promote herbal teas, vitamins, and dietary supplements.

What is the immune system? What does it do and how does it work? Here we will discuss some of the major components of the immune system, focusing particularly on the genes for these components.

Many biology textbooks refer to the immune system as the body's "third line of defense" against potentially harmful entities such as bacteria, viruses, and toxins. The first line of defense consists of physical barriers against potential pathogens. Examples include the skin and mucus membranes, inhibitory substances in sweat, saliva, tears, and mucus, and the cilia (tiny hairs) along nasal passages to keep bacteria and other things from entering the body. The second line of defense comprises cells, enzymes, and other proteins that act in a nonspecific or generalized manner to protect the body. The immune system, which is the third line of this defense system, deals with pathogens in a specific manner by learning the nature of the enemy before designing and deploying forces against it. This multilevel protection is essential for an organism to survive and preserve its integrity. The demarcation between the three levels just described is a bit misleading because the cells and proteins of the immune system often activate many of the mechanisms of nonspecific defense. We will deal primarily with components of specific immunity, although certain relevant elements of nonspecific protection will also be discussed.

We should clarify the concept of *specificity*, which is a fundamental property of the immune system, to make sense of how our bodies protect themselves. One way of thinking about this concept is in terms of *recognition* and *identity*—an organism must be able to recognize the difference between things (cells, tissues, organs, etc.) that belong to itself and things that do not so that it does not attack itself. The concept of specificity, or the self versus nonself differentiation, is perhaps easier to grasp at the level of whole organisms. For instance, a mosquito resting on an arm is clearly a distinct creature and not part of the human on which it is sitting. The skin, as the body's first line of defense, protects the body from the mosquito. But when the mosquito bites to draw blood, it often deposits something, a toxin or a parasitic organism for example, into the bloodstream. Now the body needs a way to get rid of the foreign substance before it can do harm. This is where the cells and proteins of the immune system step in, but before they can act, they need to realize that the toxin or pathogen is different from the normal constituents of the bloodstream. The immune system recognizes the difference by looking for molecules (usually proteins) called antigens on the surfaces of cells. We already discussed one special set of these antigens, the blood group factors, on the surface of red blood cells. However, there are a number of other proteins—most importantly a group of proteins called the *HLA antigens*—that establish immunological identity and difference. These proteins are discussed in more detail later.

Once a pathogen is recognized as foreign, the immune response comes into play. Depending on the nature of the substance (i.e., soluble or particulate, bacteria or virus), the body uses either a *humoral* or a *cellular* mechanism of immunity. Humoral immunity, typically mounted against freely circulating agents, is due to special proteins called the *antibodies* or immunoglobulins (see the entry on blood), which are produced by a subpopulation of white blood cells (lymphocytes) called the *B cells*. Cellular immunity is mediated by lymphocytes called *T cells*, which protect against foreign cells as well as our own cells when they are infected by foreign agents or damaged in any way, as in cancer.

B and T lymphocytes are indistinguishable, even under the microscope, and can be differentiated only on the basis of their functions and surface composition. They get their names from the organs in the body with which they are associated: B cells were first discovered in birds in an organ called the bursa of Fabricius, and T cells are known to undergo a maturation process in the thymus gland. Both the antibodies and T cells activate a series of nonspecific responses to clear the pathogen from the system.

The blood is not the only site in the body for immunological activity. Immune cells are also present at key sites of the body that are exposed to the environment and are more susceptible to attack by the pathogens. The tissues of the digestive tract and lungs have a population of immunologically active cells, as do the urinogenital tract, nose, and ears. The immune system is physically diffuse and defined strictly in terms of its function, in comparison with other systems in the body, such as the circulatory or digestive systems, which are defined both by anatomy and function. For example, the digestive system is composed of the alimentary canal—the esophagus, stomach, and intestines—to which are connected organs such as the liver and pancreas that provide enzymes for digesting food. The immune system often recruits organs and tissues of other systems (e.g., bone marrow, spleen, and thymus) for its functions.

Any interference with the immune system can have serious consequences for a person's ability to protect himself or herself against threats from the environment. In Chapter 5, we will look at examples of immune system disorders that are linked to specific genes. Here, we will concentrate on the genes that code for the normal functions of the immune system.

See also: Autoimmune Diseases, Cancer—Burkitt Lymphoma, Primary Immune Deficiencies (Ch. 5).

Antibodies/Immunoglobulins

As mentioned earlier, antibodies make up the humoral arm of the immune system. The humoral immune system is most effective against bacteria, bacterial antigens, and certain viruses (before they infect host cells). When these agents enter the body, the B cells "see" them as foreign and potentially dangerous and produce antibodies against them. Depending on the nature of the threat, the antibodies proceed to act in different ways. The following are some examples:

- If the invading organism is a bacterium that produces a toxin, then the antibodies bind to the toxin and neutralize it (i.e., stop it from acting on the body's tissue). A similar reaction is also elicited against certain types of viruses.

- Antibodies bind to the surface of the bacteria and make them susceptible to phagocytosis (engulfment and destruction) by other cells, called macrophages. This activity of the antibodies is called *opsonization*, and the specific subset of antibodies that act this way are called opsonins.

- Certain antibodies bind to bacteria and activate a nonspecific response known as the complement cascade, which ultimately lyses the bacteria cell.

- Antibodies bind to the surface of some bacteria and then cross-link with one another, causing the bacteria to clump together or *agglutinate*, which makes them lyse (die). These types of antibodies are called agglutinins.

A single bacterium or virus can have several different antigens, each of which is capable of eliciting and reacting specifically with a specific antibody. The most immunogenic molecules (i.e., those capable of triggering the strongest immune responses) are sugar-containing proteins. However, many nonglycosylated proteins, as well as polysaccharides, are highly antigenic. Lipids and nucleic acids tend to be weaker antigens.

Chemically, the antibodies are globular proteins (globulins) in the gamma subfamily (see the entry on blood), constructed specially in order to be able to recognize and bind to a variety of antigens. As they are the globulins of the immune system, they are also called immunoglobulins. The simplest antibody molecule is composed of four peptide chains (two heavy and two light chains) held together

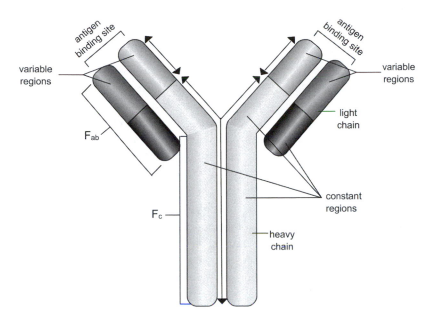

Figure 4.5. Schematic structure of antibody (immunoglobulin) molecule. [Ricochet Productions]

by special disulfide bonds. In humans, there are two types of light chains, called the kappa and lambda chains, and a total of nine possible heavy chains. Each antibody molecule consists of pairs of identical light and heavy chains.

Figure 4.5 shows the manner in which the basic unit of an antibody molecule is constructed. This basic structure was deciphered in the 1960s and 1970s by the work of two scientists on either side of the Atlantic, Rodney Porter (1917–1985) and Gerald Edelman (1929–   ), who won the 1972 Nobel Prize in Physiology or Medicine for this work. As they showed, the four chains of an immunoglobulin can be represented in the shape of a "Y," with the two arms consisting of both heavy and light chains and the stem with only portions of the two heavy chains. The ends of the two arms (Fab in Figure 4.5) form the antigen recognizing and binding part of the immunoglobulins and are called the *variable* regions. Subtle structural differences (i.e., in amino acid sequence) in these variable regions account for the ability of different antibody molecules to specifically recognize different antigens. In fact, there are *hypervariable* regions within the variable regions of the peptides. The portions of the immunoglobulin molecules not involved directly in antigen binding are called the *constant* domains of the molecules. As shown in Figure 4.5, the short chains have one constant domain, whereas each long chain has three such domains. The constant domains of the long chains are attached to one another to form the Fc portion of the molecule (the stem of the Y structure). Structural differences in the Fc region account for differences in such antibody functions as the types of immune reactions they elicit, the different cells to which they can bind, or the tissues in which they are produced.

Based on the type of heavy chains they contain, there are five major classes of immunoglobulins (Ig)—IgG, IgM, IgA, IgE, and IgD. The most common and abundant (75 percent) is IgG, which most closely conforms to the structure described earlier. IgM and IgA are actually polymers consisting of more than one four-chain unit. IgA consists of two such units linked by a short polypeptide called the J chain. It is often found as a secreted antibody in saliva and the Peyer's patches of the intestines. IgM is often found in the serum as a complex of five units, where the J chain links two of the five units. Both IgD and IgE are single units and are produced in very low amounts in comparison with the others. IgE is especially important in immune reactions that attract specific blood cells called eosinophils to the site of immune activity.

Not surprisingly, the genetics of immunity is very complex. The body has no idea which antigens it may encounter during its lifetime but must be prepared for every contingency. Although each of us has the genetic capacity to produce nearly 18 billion different antibody molecules, only a small fraction of our genome is actually devoted to antibody production, no more or less than for any other gene or protein. There are only three major genetic loci involved in antibody production: one each for the light chains (kappa and lambda) on chromosomes 2 and 22 and one for the heavy chain on chromosome 14. Each of these loci has several genes for different segments of the polypeptide chains (i.e., the constant, variable, and hypervariable regions).

The basis of antibody diversity lies in the way these genes are reorganized during gene expression. Through random gene splicing, any combination of the multiple forms of each gene can join, resulting in billions of possible gene combinations! Moreover, specialized enzymes in the B-lymphocyte cause splicing inaccuracies or add additional nucleotides at the different junctions to generate further diversity. The constant regions for the different species of heavy chains lie in tandem, and the final species of antibody produced (i.e., IgA, IgG, etc.) depends on which gene remains after splicing (see Figure 4.6). Normally, the cells produce IgM early in the immune response and later switch to other antibody species by splicing out the IgM constant chain and using one of the others. The 1987 Nobel Prize in Physiology or Medicine went to Susumu Tonegawa (1939–) for working out the basic genetic mechanisms involved in generating antibody diversity.

The (much simplified) mechanism just described tells us how a single cell can adapt its genome to produce an antibody specific for a single antigen, but there is still the problem of attending to the diversity of antigens to which the body is exposed. When an antigen enters the body for the first time, only a few B cells respond to its presence and produce antibodies against it. This first stage is called the *primary immune response*. In the absence of a continued threat, the antibody production drops off, but a small population of B cell clones, with their genes already spliced to form the specific antibody, remain circulating in the bloodstream. When the antigen is introduced into the body again, this population of clones is immediately alerted to multiply. As the population of specific B cells is now larger, the antibody is produced in much larger quantities than before. This is known as the *secondary immune response*. The secondary response is the reason for the success of vaccines against specific diseases. The genetics of this system were worked out by the combined research of many scientists worldwide, and the 1984 Nobel Prize in Physiology or Medicine awarded for this work was shared by a Dane (Niels Jerne, 1911–1994), a German (Georges Köhler, 1946–1995), and an Argentinean (César Milstein, 1927–2002).

T-Cell Receptors

The T-cell receptors (TCRs) are special membrane-bound molecules on the surface of T lymphocytes that recognize and bind to specific antigens. A typical T cell has about 20,000 TCRs on its surface. These molecules may be considered as the equivalent of antibodies for the cellular arm of the

1. DNA Rearrangement
Genes scattered through the chromosomes

2. Transcription and RNA Splicing
Rearranged Gene (many different arrangements possible)

3. Translation and Processing

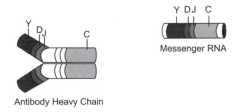

Antibody Heavy Chain

Messenger RNA

Figure 4.6. Immunoglobulin genes and recombination mechanism. [Ricochet Productions]

immune system. But although this parallel is justified, such a description of the TCRs is something of an oversimplification; TCRs are considerably more complex and abundant than the antibodies.

The greater complexity of the TCRs is related to the fact that the T cells have a larger range of immunological functions than the B cells. Whereas the latter primarily produce antibodies against foreign antigens, the T cells perform a variety of regulatory tasks in the immune system in addition to the specific task of antigen recognition and elimination. Regulatory activity includes both enhancing and suppressing various immune responses. Constituting a larger fraction of the lymphocyte population, the T cells can be divided into three functional types:

1. T "helper" cells that activate B cells bound to antigen to produce antibodies,

2. T "suppressor" cells that regulate the overall immune response to a threat, and

3. cytotoxic T cells, which directly attack and kill cells that the immune system deems harmful.

The subclasses of T cells are distinguished by specific membrane-bound molecules or markers. These marker molecules are distinct from the T cell receptors, although the two molecules are functionally linked.

Unlike the antibodies, which simply recognize antigens by virtue of their foreign-

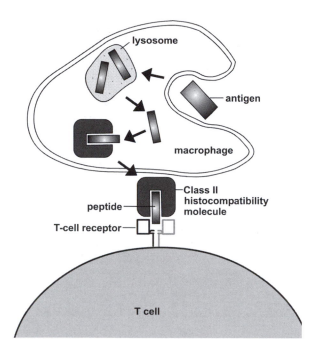

Figure 4.7. Structure of T-cell receptor with antigen presentation in context of MHC. [Ricochet Productions]

ness, T cells can only recognize antigens that are "presented" to them in combination with specific molecules called MHC receptors. These molecules (discussed in detail in the following sections) are present in various cells of the body. Before presentation to the T cells, every foreign antigen (bacterial or viral) is first processed by special antigen-processing cells such as macrophages and other white blood cells. The molecules are broken down into smaller parts, and the antigenic portions of the bacteria or virus combine chemically with the MHC molecules on the cell surface. It is this complex that the T cell "sees" (see Figure 4.7). The antigen is presented to the T cells *within* an MHC molecule. There are actually two molecules on the T cell surface that participate in recognition: The TCR recognizes the antigen, and the MHC portion of the complex is recognized by another cell-surface glycoprotein on the T cells. (Details about these receptors, called CD3, CD4, and CD8, are not discussed here.)

Each TCR consists of two membrane-bound protein chains of approximately 43 kDa called the alpha and beta chains. Different genes, scattered through the genome, code for the different TCR peptides. Examples include *TCRA* and *TCRD* on chromosome 14 and *TCRB* on chromosome 7. Like immunoglobulin chains, the TCR peptides have variable and constant regions. The constant regions are bound to the cell membrane, whereas the variable regions of the two chains form a site that recognizes and binds the antigen presented by an MHC receptor. The antigen-recognizing portion of the TCR chains is assembled in much the same fashion as that of an antibody molecule (i.e., by the splicing of V, D, and J regions).

MHC/HLA

The antibodies and TCRs are the immune molecules that recognize foreign entities, and the MHC molecules are the markers of "self" for the immune system. They prevent the immune system from attacking an individual's own cells and also present foreign antigens to T cells in order to initiate a proper immune response. Biologists first identified MHC molecules as complex cell-surface antigens that allowed them to determine whether or not different tissues were compatible for transplantation. This is the reason they were named as the major histo (tissue) compatibility (MHC) complex. In humans, they are also called the

HLA molecules, for human leukocyte antigens, in parallel with the blood group antigens found on the RBCs.

A typical cell has 50,000–100,000 MHC molecules on its surface. There are three classes of MHC molecules: class I, class II, and class III. Class I and class II molecules play an important role in the communication between cells. They are surface proteins that present antigenic peptides to T lymphocytes. Class III molecules are soluble proteins that do not participate in binding antigenic peptides. They include components of the complement cascade and cytokines such as the tumor necrosis factors.

The first two classes of MHC molecules are specialized to present antigens from different sources. The structure and function of these two classes are somewhat similar but have subtle differences. Class I molecules have one heavy chain (45 kDa) and a light chain called β2-microglobulin (12 kDa). Class II molecules are also made up of two chains (α and β) but these are roughly similar in size (34 and 30 kDa).

Class I MHC proteins are found in virtually all cells of the body. Their main function is to present antigens that are synthesized within the cells themselves (e.g., viral proteins or cancerous proteins to cytotoxic T lymphocytes). The fact that antigens are endogenously synthesized implies that the cell's genome is compromised and the cell should be destroyed so that its alterations cannot spread through the body. As mentioned earlier, CTLs have both T cell receptors (TCRs) and CD8 molecules on their surfaces. The TCR recognizes and binds peptide antigens when they are bound in complexes with MHC class I molecules, whereas the CD8 receptor binds to the MHC class I portion of the complex. Each CTL expresses a unique TCR that only binds a specific MHC–peptide complex. Therefore, if a CTL is capable of binding to an MHC–peptide complex on the cell surface, that cell is most likely presenting a peptide foreign to the host.

How do class I molecules work? First, a protein called ubiquitin binds to the foreign proteins entering the cell and flags them for destruction. (Ubiquitin is a small protein found only in eukaryotes and encoded by a family of genes scattered across the human genome.) The protein–ubiquitin complex is transported into the cell, where the protein is broken down. The antigenic peptides are then chaperoned to the rough endoplasmic reticulum, which is the site in the cell where the MHC I molecule is synthesized. First the alpha chain and then the beta chain of the MHC I molecule binds to the foreign peptide. This MHC I/antigen complex is then transported via the Golgi body to the plasma membrane, where it is presented to a T cell.

In contrast with class I molecules, *class II MHC* proteins are found only on a few cells, notably the B lymphocytes and macrophages. Class II molecules present proteins from external sources, such as bacterial toxins and viral surface proteins that have been engulfed and processed. Regulatory T cells, mainly the helper cells (T_h cells), recognize these complexes and proceed to activate the B cells to produce antibodies. So MHC class II proteins help T cells, B cells, and macrophages communicate with one another.

The first step in antigen processing is the joining together of alpha and beta chains of the MHC II molecule in the antigen-presenting cell (macrophage) helped by chaperone proteins. Once these chains are bound together, the complex binds a third polypeptide, known as the invariant chain. This complex then moves to the Golgi body and is transported to the cellular site containing the engulfed antigenic peptide. The acidic conditions here cause the invariant chain to become disassociated from the complex. It is replaced by the antigenic peptide. The complex is then transported to the plasma membrane, where the class II molecule presents the antigen to T helper cells containing CD4 molecules.

The MHC proteins are encoded by a cluster of genes that lie in close proximity to one another on chromosome 6. In most species, each class of MHC is represented by more than one locus, and in humans these are called *HLA*. The *HLA* complex in its entirety is more than 3,800 kb. The class I loci are known as *HLA-A*, *-B*, and *-C* and the class II loci *HLA-DP*, *-DQ*, and *-DR*. Class I genes cover over 600 kb of the *HLA* complex, and a typical gene consists of eight exons.

All MHC genes are codominant in expression, which means that both paternal and maternal alleles are expressed in each cell. Usually, the entire linked MHC complex is inherited without recombination. The MHC genes are believed to have remained clustered like this for over 600 million years, and a great deal of research has been focused on determining the evolutionary origin of the complex. Scientists have identified a structural resemblance between MHC molecules and immunoglobulins (Ig) and have hypothesized that the Ig genes and MHC genes may be descendants of a common evolutionary ancestor.

The portions of the MHC that encode class I and class II molecules are highly polymorphic, whereas class III molecules, which are not involved in antigen binding, are much less so. There are many genetic variants or alleles at each locus. Many of these alleles are represented at significant frequency (> 1 percent) in the population, and the alleles often differ from one another by several amino acid substitutions, so they are significantly different from one another. For example, in humans there are more than 300 alleles at certain MHC loci. Each allelic product has a unique set of peptides to which it can bind with high affinity. Clearly, therefore, it is advantageous to the species to have many different alleles present. There is selective pressure to maintain this enormous variation—the greater the variation, the more versatile is the immune system, and therefore the more likely the species is to withstand disease. Note, however, that unlike antibodies and TCRs, MHC diversity arises only in the germ line. MHC genes do not have access to somatic cell recombination mechanisms for generating diversity. As a result, the affinity and selectivity of MHC molecules for foreign proteins are lower than those of antibodies (see the preceding discussion on immunoglobulins).

Another interesting property of the MHC antigens and their variation is the apparent link to the incidence of genetic diseases. Certain *HLA* alleles seem to be found more frequently in patients with specific diseases than in the normal population. For example, *HLA-B27* alleles are found in more than 90 percent of patients with ankylosing spondylitis, whereas 10 percent of normal individuals show this phenotype. The functional significance of this sort of association is not clear, but it has important implications for predicting the probability of genetic disease and for genetic testing.

Lymphokines

One of the secondary consequences of lymphocyte stimulation by antigens—the production of TCRs or antibodies is always the first step—is the production of various chemical messengers. These chemicals are known as lymphokines and carry messages to other cells of the immune system to mediate various responses such as the phagocytosis and killing of invading bacteria or the inflammatory responses. In other words, the lymphokines are the primary effectors of the reactions associated with inflammation and healing. Some examples follow.

Interferons

These proteins are produced specifically to initiate defensive reactions against different viruses. Most animals, including humans, produce three main classes of interferons, which were first identified according to where they were produced and which cells they stimulated. It is now known that all of these molecules have a range of activities. The genes—*IFNA1* and *IFNB1*—for the alpha and beta types of interferon reside close together on chromosome 9. These genes are expressed predominantly in the B cells and are also known as the leukocyte interferons. Interferon gamma, encoded by the *IFNG* gene on chromosome 12, is also called the immune interferon and is produced mainly by the T lymphocytes.

Interferons mediate their actions by binding to specific cell-surface receptors and setting off signaling cascades that stimulate the destruction of the viruses. The *IFNAR1* gene on chromosome 21 encodes a receptor that can bind to different interferons, and the *IFNGR1* gene encodes a gamma-interferon-specific receptor.

The discovery of interferons as natural antiviral proteins prompted scientists to investigate their potential for use as antiviral agents. Interferon alpha, for example, has been approved for therapeutic use against hairy-cell leukemia and hepatitis C, besides showing some effectiveness in stalling hepatitis B and some rare blood cancers. More recently, nasal sprays containing alpha interferon were shown to protect against colds caused by rhinoviruses.

Interleukins (ILs)

These proteins activate specific populations of immune cells, including, but not limited to, killer T cells, macrophages, and B cells, thereby mediating the full range of nonspecific responses to pathogens, including phagocytosis, cell lysis, and inflammation. Included in this group are also specific cytokines that induce the growth, differentiation, and proliferation of the lymphocytes (see the entry on growth factors for examples of growth factors for other cell types). About 18 different interleukin families, with one or more members, have been discovered so far, although there may be more. Table 4.2 provides information on the first 12 types.

Table 4.2. The interleukins at a glance

Interleukin	Gene location	Major source in body	Major effect	Receptor genes	Gene location
IL-1 family	Gene cluster on Chm. 2			*IL1R1* and *IL1R2*	Cluster on Chm. 2 separate from IL-1 cluster
IL-1 alpha		Monocytes and macrophages	Induces apoptosis in response to injury		
IL-1 beta		Activated macrophages	Cell proliferation, differentiation, and apoptosis; contributes to inflammatory pain response in central nervous system		
IL-2 (TCGR)	Chm. 4	Activated T cells	Proliferation of activated T cells	*IL2RA*	Chm. 10
				IL2RB	Chm. 22
				IL2RG	Chm. X
				(subunit shared by receptors for IL-2, 4, 7, 9, and 15)	
IL-3 (multi-CSF)	Chm. 5 (clustered with genes for IL-4, 5, and 13 and CSF2)	T lymphocytes	Hematopoietic growth factor; growth of blood cell precursors	*IL3RA*	Chm. X and Chm. Y
				CSF2RB	Chm. 22
				(common subunit for receptors of IL-3 and 5 and CSF)	
IL-4	Chm. 5	Activated T cells and mast cells	B cell proliferation IgE production	*IL4RA*	Chm. 16
IL-5	Chm. 5	T cells and mast cells	B cell and eosinophil differentiation	*IL5RA*	Chm. 3
IL-6 (inter-feron-beta)	Chm. 7	Activated T cells	B cell and hepatocyte differentiation; synergistic effects with IL-1 or TNF-alpha	*IL6R*	Chm. 1
IL-7	Chm. 8	Thymus and bone marrow stromal cells; locally in intestinal epithelial cells	Development of T cell and B cell precursors; co-factor in TCR formation	*IL7R*	Chm. 5
IL-8	Chm. 4	Macrophages	Attraction of neutrophils; major mediator of inflammatory response	*IL8RA* *IL8RB*	Chm. 2 (in same cluster)

Table 4.2. (continued)

Interleukin	Gene location	Major source in body	Major effect	Receptor genes	Gene location
IL-9	Chm. 5	Activated T cells	Promotes growth of T cells and mast cells	*IL9R*	Chm. X and Chm. Y
IL-10	Chm. 1	Primarily monocytes and to a lesser extent B and T cells	Down-regulates helper T cytokines; enhances B cell activity and antibody production	*IL10RA* *IL10RB*	Chm. 11 Chm. 21 (in cluster with IFN receptors)
IL-11	Chm. 19	Stromal cells	Promotes growth of T-cell-dependent antibody-synthesizing B cells; synergistic effects on hematopoiesis	*IL11RA*	Chm. 9
IL-12 Alpha subunit Beta subunit	 Chm. 3	Macrophages, B cells	Acts on T helper and natural killer (NK) cells; promotes T_H1 cells while suppressing T_H2 functions	*IL12RB1* and *IL12RB2* (subunits)	Chm. 19 and Chm. 1

Tumor Necrosis Factors (TNFs)

Like the interleukins, the tumor necrosis factors are cytokines, so named because they were first discovered due to their anticancer properties. These proteins are produced mainly by the macrophages and T cells and have functions that are very similar to those of the interleukins (i.e., they help regulate immune and inflammatory responses as well as blood cell formation). They also enhance the phagocytic function of macrophages and aid in wound healing by stimulating the growth of new blood vessels and tissue. Although best-known for their activity in cases of acute infection, the TNFs seem to be involved in a much wider spectrum of conditions.

There are two types of TNFs, alpha and beta. TNF-alpha, also known as *cachectin*, seems to be the prototype of a family (the "TNF superfamily") of molecules that are involved in immune regulation and inflammation. TNF-alpha is produced mainly by macrophages and affects a variety of cellular functions, such as lipid metabolism, coagulation, and insulin resistance, during an immune or inflammatory response. In fact, its biological effects are also very similar to those of interleukin 1, but it is a structurally distinct protein that binds to different receptors. The *TNFA* gene on chromosome 6 encodes a precursor protein that is cleaved to make the functional molecule.

TNF-beta is also known as *lymphotoxin*. Its gene, called *TNFB* or *LTA*, lies in close proximity to *TNFA* on chromosome 6. The two proteins share about 30 percent amino acid homology and have similar biological activities as well. The main difference is that the beta type is produced predominantly by T lymphocytes. In addition to mediating a large variety of inflammatory, immunostimulatory, and antiviral responses, lymphotoxin is also involved in the formation of secondary lymphoid organs during development and plays a role in apoptosis.

Single-nucleotide polymorphisms (SNPs) in regulatory regions of the *TNF* genes have been shown to be associated with susceptibility to a number of complex disorders. Three SNPs located in the *TNFA* promoter region are all substitutions of adenine for guanine. These polymorphisms have been found to be associated with susceptibility to cerebral malaria, mucocutaneous leishmaniasis, meningococcal disease, lepromatous leprosy, trachoma, and asthma. Scientists have also explored links between TNF and HIV/AIDS, insulin resistance, obesity, rheumatoid arthritis, multiple sclerosis, and osteoporosis, among other disorders.

Like other lymphokines, the TNFs mediate their activities via specific receptors on the surface of their target cells. The receptors activate signal transduction pathways that lead to turning genes on or off. Like the TNFs themselves, the receptors for members of the TNF superfamily also constitute superfamilies of related receptors. An example is the receptor superfamily (*TNFSF*) on chromosome 8. These large families are still growing, with new members being discovered and added. The prospect of controlling the activity of specific TNF family members makes them promising as drug targets.

See also: Growth Factors (Cytokines).

Complement

The production of antibodies just described is only the first step in an immune reaction to a pathogen. Although antibodies may be enough to neutralize certain viruses, the complete removal of the pathogen usually requires something more. The immune complexes, formed by the attachment of the antibodies to the offending antigens, must be removed from the bloodstream or tissues. And pathogens such as bacteria and protozoa, for example, must be killed, as must cells that have somehow become abnormal as a result of infection (e.g., cancer cells). The complement pathway is one of the main mechanisms in place for destroying cellular pathogens.

Complement is not a single protein but actually refers to a complicated network of serum proteins and cell-surface receptors that must be activated in order to properly perform their functions. In this aspect of its activity, complement greatly resembles another protective reaction that we have discussed previously, namely blood clotting. Both processes unfold as a cascade of serial reactions triggered by an injurious event in which the product of each reaction mediates the next step of the cascade. As in the case of blood clotting, the complement cascade may be triggered by different stimuli, but different responses eventually converge on a common pathway. Both reactions also result in the release of various growth factors and cytokines. But whereas the clotting cascade results in the repair of damaged tissues, the end result of complement is destruction.

The components of the complement pathway are glycoproteins numbered *C1–C9*. Proteins C2 through C9 are present in the bloodstream as inactive precursors that require activation by the previous member of the cascade. Each step of the cascade involves activation proteolysis or cleavage of a protein into two fragments, one of which acts as an immune modifier and the other of which (the b fragment) continues the cascade. The end result is the formation of a membrane-attack complex that lyses the cell to which the immune complex is bound (e.g., bacterium or virus-infected cell). There are three routes through which the cascade is stimulated. The *classical* pathway is set off by the formation of immune (antigen-antibody) complexes. Certain carbohydrates are capable of triggering a very similar reaction (barring the participation of one of the subcomponents of the first step, C1), which is known as the *lectin* pathway. Certain bacterial products, such as the lipopolysaccharides or IgA and IgE antibodies, can trigger a third pathway called the *alternative* pathway.

As shown in Figure 4.8, the classical pathway starts with C1, which is a very complex protein composed of 22 different polypeptide chains organized in three subunits designated as C1q, C1r, and C1s. The subunit C1q is a tulip-shaped protein molecule that recognizes and binds to the antigen–antibody complex, specifically the Fc portion (primarily of IgG and IgM antibodies), and sets into motion the serial activation of the enzymes of the cascade. This subunit is composed of three polypeptide chains that are encoded by a cluster of genes on chromosome 1. Deficiencies in C1q have been linked to systemic lupus erythematosus, a chronic inflammatory disease of connective tissue, and glomerulonephritis. The binding of C1q activates C1r, which in turn activates the C1s subunit. These subunits are encoded by adjacent genes on chromosome 12.

The activated C1s enzyme simultaneously activates two complement proteins, C2 and C4, both of which are encoded by genes on chromosome 6. The enzymatic subunits of these proteins then combine together to form C3 convertase, or the C3 activating enzyme. Meanwhile, the C4a subunit released in the previous step has anaphylactic activity and becomes involved in mediating the inflammatory response. The enzymatic portion of the cleaved C3 (encoded on chromosome 19) then combines with the existing convertase enzyme to form a larger complex that functions as the C5 convertase.

C3 is also the point of convergence for the alternative pathway. The classical pathway occurs in response to IgG and IgM, but this pathway of complement activation occurs in response to IgA or IgE, or directly to bacterial lipopolysaccharides (even in the absence of antibodies). These molecules bind and activate C3, with the help of other factors, and form an alternative complex that also functions as the C5 convertase. This additional activity of C3 is perhaps why individuals with deficiencies of this enzyme are more susceptible than normal to bacterial infections.

The enzymatic portion of C5 (C5b) has a dual role as the activator of C6 and

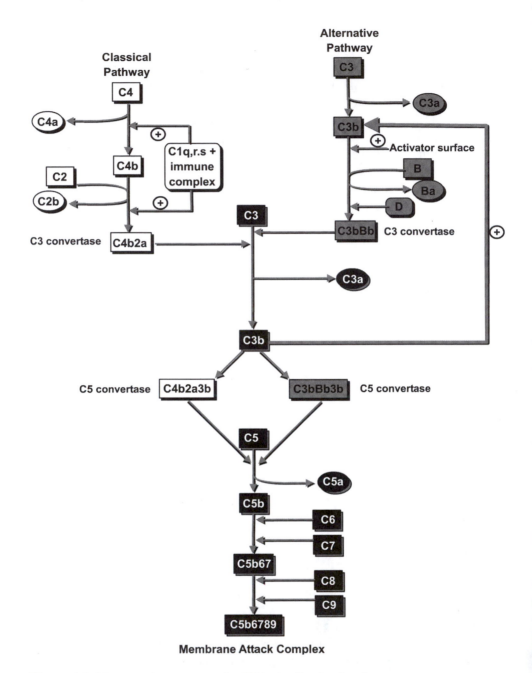

Figure 4.8. The complement cascade. [Ricochet Productions]

also as the anchor to the surface of the cell that is the target of the complement. The membrane attack complex (MAC) is formed by the serial attachment of activated proteins C6 through C9 to membrane-bound C5b. The proteins of these final steps are not encoded contiguously on a single chromosome but are found scattered among different chromosomes: C5 on chromosome 9; C6, C7, and C9 on chromosome 5; and C8 on chromosome 1 (contiguous with the C1q and r sub-units). The polymerization of the complement proteins causes a disruption in the integrity of the cell wall by forming holes, which results in cellular lysis due to the leakage of its contents.

KERATIN

Keratins are tough, insoluble proteins that make up the outer layers of our body, such as the skin, hair, nails, and tooth enamel, as well as the feathers, hooves, and horny tissues of birds and animals. The protein accounts for about 30 percent of the protein content of living cells of the epidermis (skin cells) and up to 85 percent of the dead cells that make up the outermost layers of our skin. The amount of keratin in hair fibers and the nail plate varies anywhere from 65 to 95 percent of the total weight of these structures.

Keratin is not a simple, or even a single, protein but a family of proteins, different members of which combine with one another to make up different structures. There are two main classes of these proteins, the cytokeratins—or cellular keratins—found inside living epidermal cells, and the skin keratins that give substance to the outer skin layers, hair, and nails (essentially dead tissues). The cytokeratins form filaments that run within the cell and weave a kind of basket around the nucleus to lend it support. The skin keratins, as well as those in the nails and hair, are typically present as combinations of acidic and basic (or neutral) proteins organized in layers—the hair shafts, for example, are composed of three concentric layers of protein.

A characteristic feature of the keratins is their high sulfur content due to the presence of many units of the sulfur-containing amino acid cysteine in keratin fibers. The sulfur atoms of cysteines on adjacent protein chains form strong chemical bonds (called disulfide bonds) with one another, which makes the keratin strong, durable, and highly resistant to environmental stress. Various physical properties of our hair and nails depend on the keratins. For example, the greater the number of disulphide bonds between the keratin chains, the curlier hair becomes. The relative strength and brittleness of nails depends on the amount and type of keratins it contains.

The genes for the keratin proteins are found in different parts of the genome, although the majority may be found in two large clusters (of more than 10 genes) on chromosomes 12 and 17. By and large, the genes for the acidic proteins (both cytokeratin and hair keratin components) are located on chromosome 17, whereas the basic and neutral keratin genes are on chromosome 12. An example of genes outside these two clusters is one on chromosome 11 that codes for an especially sulfur-rich keratin that is found in the hair and cuticles. Mutations in different keratin genes have been associated with genetic diseases of the hair, skin, and nails.

See also: Monilethrix (Ch. 5).

LIPID BIOSYNTHESIS

It seems like we can hardly do anything or go anywhere these days without seeing or hearing something about fat! Television programs, books, and food cartons in our grocery stores are all full of advice about how much fat there is in food, what is good and what is not, and what we should do to avoid obesity and live healthier lives. But despite this bad name, and their admitted dangers, fats—or lipids—are actually extremely important and have a variety of functions in living systems. First, they are an important source of energy. Although they consist of the same basic elements as the carbohydrates (sugars and starches) (i.e., carbon, hydrogen, and oxygen), fats are structured differently so that, carbon atom for carbon atom, they release significantly more energy than do carbohydrates. The net result of the oxidation of 1g of fat is 9 kcal of energy, as compared with 4 kcal of energy from the same amount of carbohydrate. Lipids are also essential structural elements in living cells. The phospholipids in particular are the main ingredient in cell membranes and are responsible for maintaining continuity in the structure and functioning as a barrier between the cell and its exterior environment. The covering of cells, both in the brain and peripheral nervous system, also contains different types of complex lipids. Still another substance that is a component of certain animal fats is cholesterol, a very important compound for the synthesis of different steroid hormones such as the cortisones and sex hormones, estrogen and testosterone.

Lipids are large and complex molecules composed mainly of carbon, hydrogen, and oxygen. A characteristic feature that they have in common is that they are almost always very hydrophobic (water repelling) and insoluble in water. Lipids with the different functions mentioned have distinct structural features.

1. Common dietary fats from which we derive energy typically have what is known as a "triglyceride" structure. There is a central core of glycerol

whose three carbon atoms are linked to long hydrocarbon chains called fatty acids. A fatty acid is a COOH (acid) group attached to a chain of carbons, usually 12–24 units long, which are attached to hydrogen molecules (e.g., palmitic acid with 16 carbons):

The fatty acid chain is attached to the central glycerol via an acyl bond with its carboxyl (COOH) carbon; that is,

C   from glycerol
|
O
  \
   ‖—— fatty acid chain
  O

The amount of energy stored by a lipid molecule is proportional to the length of the carbon chains of its fatty acids.

Most of us are probably familiar with the terms "saturated" and "unsaturated" appearing on food labels with reference to the types of fats. These terms refer to the structure of the fatty acid side chains of a triglyceride. Adjacent atoms of the hydrocarbon chain may be held together either by single or double bonds, and the nature of this bond determines the number of hydrogen atoms that can be attached to each carbon. A saturated fatty acid is, as indicated by its name, completely saturated with hydrogen atoms, leaving no room for any double bonds. Palmitic acid, with 16 carbon units, is illustrated in the preceding diagram. Mono- and polyunsaturated fatty acids contain one or more than one carbon–carbon double bond within the hydrocarbon chain, respectively. For example,

$$H$$
$$C_n\diagup C{=}C \diagup C_nCOOH$$
$$H$$

In general, fats with unsaturated side chains are considered healthier than the saturated fats. Examples include olive and canola oils. Corn oil and fats from animal sources, such as butter, tend to be saturated.

The phospholipids that make up the cell membranes share the basic structure of a triglyceride except for the substitution of one or more fatty acids with phosphorylated side groups (i.e., phosphate-containing molecules). The phosphate group forms the point of contact between the lipid side chain and the central glycerol.

O
‖—— fatty acid chain
O
|
HO—P=O
|
O
 \
  C   from glycerol

Because they are charged groups, the phosphates in the membrane lipids give these molecules some polarity, which is responsible for the formation of the characteristic double-layer configuration of various biological membranes.

2. Cholesterol and other structural lipids, such as the glycolipids of the nervous and brain tissues, are not triglyceride fats. They are considerably more complex in structure, with aromatic rings as well as side chains. They are often found in combination with other molecules such as proteins (forming lipoproteins) and sugars (glycolipids and lipopolysaccharides). The structure of cholesterol is:

The body relies on both external sources (i.e., diet) and its own stores of deposited fats for its supply of lipids. The fats we use for the purposes of energy come from both sources, as does cholesterol, but the membrane phospholipids and other structural lipids of the nervous system are synthesized de novo. We are familiar with the fact that we tend to store fats in regions such as the hips, thighs, and belly. This is fat that the body synthesizes and stores itself and mobilizes when it needs energy and none is forthcoming from dietary sources, for example, during hard exercise or conditions of starvation. We now look at the genes and proteins involved as pathways of lipid biosynthesis in the human body before turning to the utilization (metabolism) of these lipids in the following entry.

Fatty Acid Synthesis

Fatty acid synthesis in the human body takes place predominantly in four tissues in the body: the liver; adipose (fat) tissue in the belly, thighs, and upper arms, for example; lactating mammary glands; and the central nervous system. The fatty acids made in the adipose tissue are typically converted into stored fats, whereas those made in the nervous system are used in the synthesis of the structural lipids.

Perhaps the single most important molecule in lipid biosynthesis is acetyl CoA—acetyl coenzyme A to use its full name—a molecule that lies at the heart of most metabolic processes in the human body. Not only is it the required precursor for the synthesis of all major classes of fats (i.e., the triglycerides, membrane phospholipids, and cholesterol), but it is also the key intermediate in the metabolism of sugars (via glycolysis and the Krebs cycle), amino acids, fatty acids (see the entry on lipid metabolism), and ketone bodies. The most common conduit for the acetyl CoA used in fat biosynthesis is the glycolytic pathway (see the section on glycolysis in the entry on sugar metabolism in this chapter), largely because the simple sugars are the quickest, easiest, and most abundant source of this precursor.

Palmitic acid, a saturated 16-carbon fatty acid, is always the first fatty acid that is synthesized. The overall synthesis of palmitic acid (or its palmityl CoA form) requires eight acetyl CoA and seven ATP molecules. It proceeds in the following fashion:

- First, an enzyme called *acetyl CoA carboxylase* converts acetyl CoA to malonyl CoA by the addition of carbon dioxide. It performs this activity with the help of a molecule known as biotin, which acts as the physical carrier for the carbon dioxide unit. Biotin is an essential participant in this conversion. It is not synthesized by humans and is typically obtained from substances such as egg white. The conversion of acetyl to malonyl CoA also requires energy, which is obtained from ATP molecules.

  The enzyme involved in biosynthesis is actually one of two forms of the carboxylase designated as the alpha form and is produced from a gene (*ACACA*) on chromosome 17. The beta form of this enzyme, on chromosome 12, appears to be involved in the regulation rather than synthesis of fatty acids.

- Six molecules of malonyl CoA and one molecule of acetyl CoA then react with one another to form a molecule of palmitic acid. This reaction takes place in a repeated series of reactions that are all synthesized by *fatty acid synthase*, a multifunctional enzyme complex with seven independent catalytic functions. These functions are performed by different parts of the protein, known as its domains.

In humans, the different catalytic domains of fatty acid synthase are encoded into a single, long, 257 kDa protein by a gene on chromosome 17 known as *FASN*. Homologous genes in other animals encode similar multifunctional enzymes. In plants and prokaryotes, however, seven individual enzymes perform the different functions of the synthase in conjunction with a carrier protein for the acyl (acetyl, butryl, etc.) groups.

The palmitic acid so formed undergoes modifications such as elongation and desaturation to form various other fatty acids. The elongation and desaturation reactions may take place in virtually any of the body's tissues. Some of the fatty acids obtained from the degradation of dietary fats may also be modified directly. Elongation may take place either in the endoplasmic reticulum (ER) or the mitochondria. The reaction in the ER utilizes malonyl CoA and resembles the de novo synthesis described earlier but does not involve the synthase complex. In the mitochondria, palmitic and other fatty acids are elongated using acetyl CoA in a process that reverses the beta-oxidation reaction (see the entry on lipid metabolism) used for the degradation of fatty acids. This reaction requires energy.

Desaturation or dehydrogenation of fatty acids is associated with the ER membranes and involves cytochrome b_5, NADP, and enzymes called *desaturases*. These enzymes catalyze the formation of double bonds between specific carbons of fatty acids. Different enzymes target carbon atoms at specific positions of a fatty acid chain. Examples include enzymes encoded by adjacent genes *FADS1*, *FADS2*, and *FADS3* on chromosome 11. Once the different fatty acids are formed, they combine with glycerol to form mono-, di-, and triglyceride fats, which are taken to specific tissues for storage or processed further to synthesize components of the cell membranes and nerve tissues.

Cholesterol Biosynthesis

If the fats are a maligned molecular species in today's world, then cholesterol is the archvillain among the offenders. Most of us are familiar with the dire warnings of heart attacks and blocked veins that are frequently associated with cholesterol. And yet despite the very real dangers posed by an excess of cholesterol, it is an important molecule that forms a part of every cell in the body and serves many vital functions. We actually need cholesterol to maintain healthy cell walls, make hormones such as estrogen and testosterone, synthesize vitamin D, and also produce bile acids, which help to digest dietary fats.

Nearly half the cholesterol found in our bodies is synthesized in our own tissues, while the rest comes from diet. Biosynthesis occurs in several locations—for example, the liver produces about 10 percent and the intestines about 15 percent. As we saw earlier, cholesterol is much more complex than the triglycerides. At the cellular level, it is synthesized in the cytoplasm. Cholesterol synthesis, particularly the early stages, seems to be localized to special compartments in the cytoplasm, called the peroxisomes, as indicated by the fact that patients with peroxisomal disorders such as Zellweger's syndrome have deficiencies in cholesterol biosynthesis. As was the case for the fatty acids, acetyl CoA is the main precursor for cholesterol. Over a dozen enzymes are involved in the multistep process that condenses and converts the acetyl CoA molecules into the ringed structures that make up cholesterol. Individual enzymes will not be detailed here.

LIPID METABOLISM

As we saw in the previous entry, fats play many important physiological functions, and we get them both from internal (i.e., biosynthetic) and external (dietary) sources. How the fats get from food in our digestive tract and how we mobilize fats stored in different tissues is an interesting and very complicated story. In this section, we discuss the genes and proteins used by the human body to obtain and utilize the lipids.

Dietary Fat Absorption

Most of the food we eat needs to be processed in the digestive tract before it can be absorbed into the body and used for various functions. This is especially true of the fats, which are insoluble in the water-based (aqueous) intestinal fluids. Therefore, the first step in their digestion is their breakdown—emulsification—into smaller, more manageable units or droplets. Special salts, called bile salts, which are synthesized in the liver and secreted into the small intestine via the gall bladder, are responsible for this function. The smaller droplets of lipids produced by the action of bile salts are called micelles.

Even after they have broken down into micelles, the lipid molecules are still too large to cross the cell membranes and they have not yet undergone any sort of chemical degradation. Before they can enter the cells of the intestinal lining and be used by the body, they must be digested. Enzymes called lipases, secreted by the pancreas and delivered to the small intestine, digest dietary triglycerides and release free fatty acids and glycerol as well as intermediates such as di- and monoglycerides (i.e., glycerol molecules with two or one attached chains, respectively). The gene *PNLIP* on chromosome 10 codes for a pancreatic lipase enzyme, and mutations in this gene are associated with deficiencies in the ability to digest fats.

Once inside the intestinal cells, the glycerides are reassembled along with proteins to form packages called chylomicrons. This gives the fats a protein coating that helps them dissolve in water. Still too big to enter the bloodstream via the capillary walls, however, the chylomicrons travel from the intestinal cells to the lymph vessels. Eventually the lymphatic system merges with the veins, which is where the fats (in the form of chylomicrons and free fatty acids) enter the bloodstream. They are then delivered to various tissues to meet their fates.

Fatty Acid Degradation

The fatty acids are taken up from the bloodstream by the cells in the liver, where they are broken down into smaller molecules and harvested for their energy. This breakdown happens inside the mitochondria. Once again, the fatty acid molecules are too large to cross the barrier of the mitochondrial membranes and must be helped by carrier proteins. The protein involved in ferrying fatty acids into the mitochondria is known as *carnitine*, which is often found in red meats and dairy products. Humans contain two variants of this protein, encoded by genes on chromosomes 11 and 22 (*CPT1A* and *CPT1B*) and expressed in the liver and muscles, respectively. People with low carnitine levels often have lipid deposition in the muscles and are irritable and generally weak.

Inside the mitochondria, enzymes act to break down fatty acids by an oxidation process called *beta-oxidation*. Before this can take place, the fatty acids must be activated by the addition of a molecule called Coenzyme-A (CoA). A system of enzymes called the *carnitine acyltransferases* catalyzes the transport of the fatty acids from the carnitine carrier to the high-energy CoA group. In humans, the genes for these enzymes—called *CPT1* and *CPT2*—are found on chromosomes 11 and 1, respectively. Beta-oxidation itself is a series of four enzyme-catalyzed steps that occur over and over again (i.e., in a spiral manner) and shrink the fatty acid by two carbon atoms each time (see Figure 4.9). Each cycle also generates a molecule of acetyl CoA.

The first step in beta-oxidation of a fatty acid is the dehydrogenation of the carbon atom closest to the carboxyl group, called the beta-carbon (see Figure 4.8), resulting in the formation of a double bond between the beta carbon and its adjacent carbon (the gamma carbon). It is the involvement of the beta carbon in this and other steps that gives the reaction cycle its name. Enzymes called the *fatty acyl CoA dehydrogenases* catalyze this reaction. There are actually three classes of these enzymes, each specific for fatty acids of different lengths.

The long-chain fatty acyl CoA dehydrogenase (LCAD) acts on carbon chains longer than 12 units. The *ACADL* gene on chromosome 2 codes for this enzyme, deficiencies in which are associated with severe hypoglycemia, skeletal muscle weakness, and enlarged liver and heart, which can be fatal. Medium-chain dehydrogenase (MCAD), which is encoded on chromosome 1, acts on straight fatty acid chains 6–12 carbon atoms long. Genetic deficiencies in this enzyme are believed to be one of the most common inborn errors of metabolism (see the entry in Chapter 5). Fatty acids from 4 to 6 carbon atoms long are acted on by SCAD, which is encoded on chromosome 12. The final step of the beta-oxidation spiral—namely the cleaving of a 4-carbon Co-A molecule into two acetyl CoA molecules—is catalyzed by an enzyme called *acetyl-Coenzyme A acyltransferase*, the gene for which is on chromosome 18 (*ACAA2*).

The dehydrogenation reaction is followed by the hydration of the carbon–carbon double bond, which is catalyzed by an enzyme called *enoyl CoA hydratase*, the gene product of the *ECHS1* gene on chromosome 10. A second dehydrogenation event converts the beta carbon into a keto group. A complex dehydrogenase enzyme that participates in different pathways catalyzes this reaction. The enzyme consists of two subunits, which are encoded on adjacent genes (*HADHA* and *HADHB*) on chromosome 2. Finally, enzymes called *beta-ketothiolases* act on the ketone to split the molecule into an acetyl-CoA and a fatty acid (acyl) chain that is now shorter by two carbons. The beta-ketothiolases constitute a large family of proteins that contain multiple subunits or domains encoded on different genes. An example of a member of this family is the *ABHD5* gene on chromosome 3. Mutations in this gene are associated with a genetic condition characterized by Chanarin–Dorfman syndrome, a triglyceride storage disease with impaired long-chain fatty acid oxidation.

The Fate of the Chylomicrons

One of the main destinations for the chylomicrons is the liver. Here, proteins combine with the partially digested glycerides to form compounds called *lipoproteins* that are then transported via the blood to different tissues of the body, where they are used in the production of cholesterol and other structural lipids. Two main classes of lipoproteins—known as the "low density lipoproteins" (LDLs) and the "very low density lipoproteins" (VLDLs)—participate in this transport. The third major class of lipoproteins, known as the high-density lipoproteins (HDLs), plays

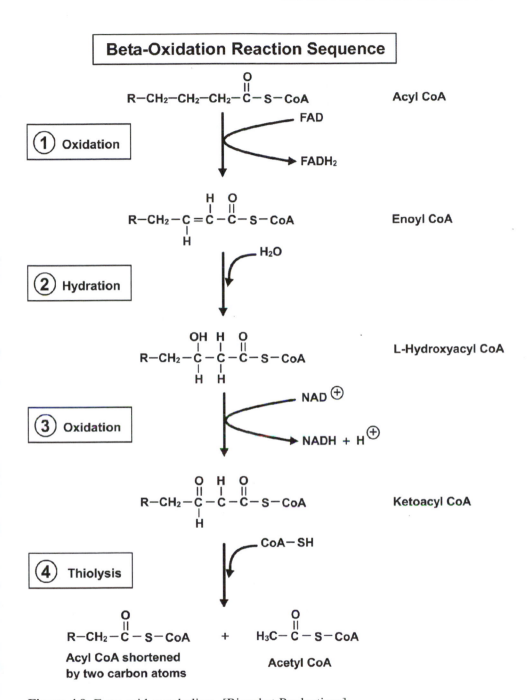

Figure 4.9. Fatty acid metabolism. [Ricochet Productions]

a role in cholesterol transport. The classification of the lipoproteins is based upon their densities and reflects the relative proportions of lipids to proteins in the molecules. These molecules participate in the synthesis and processing of cholesterol (discussed later in this chapter).

Enzymes called *lipoprotein lipases* convert glycerides into lipoproteins. The VLDLs are formed first and are then converted into LDLs. Lipoprotein lipase is an important regulator of lipid and lipoprotein metabolism. Although its primary function is to catalyze the hydrolysis of triglycerides in chylomicrons and VLDLs, it also contributes to the lipid and energy metabolism of different tissues in varying ways. Although the synthesis, manner of secretion, and mechanism of endothelial binding of lipoprotein lipase appear similar in all tissues, the factors that control gene expression and posttranslational events related to processing vary from tissue to tissue. The best-characterized gene for this enzyme in humans is the *LPL* gene on chromosome 8.

Mobilization of Fat Stores

So far, we have been discussing the fate of the lipids we get from the foods we eat. But not all the fat our bodies use comes directly from our diet. In fact, a significant portion of our energy comes from mobilizing the fat we store in various tissues. These fat reserves are synthesized de novo, as described in the entry on lipid biosynthesis, and stored until needed. The adipose tissues in the regions of the belly, thighs, hips, upper arms, and other areas are common sites for the storage of these fats.

A variety of stimuli can trigger the mobilization of the fats stored in our tissues. Our bodies usually turn to this source of energy only in times of extreme need when none is provided from the digestive tract. A significant regulator of the metabolism of stored fats is insulin (see the entry on hormones and the endocrine system for details). When a person is well-fed, insulin released from the pancreas prevents the inappropriate mobilization of stored fat. Instead, any excess fats and carbohydrates in the diet and bloodstream get stored in adipose tissue. A complex cascade of reactions is needed to release metabolic energy from fats. Hormones such as glucagon and epinephrine trigger the cascade. These hormones bind cell-surface receptors on adipocytes. The receptors are coupled to the activation of adenylate cyclase. The resultant increase in cAMP leads to activation of a protein kinase, which in turn phosphorylates and activates hormone-sensitive lipase.

The mobilization of stored fats once again calls for the action of an enzyme called *lipoprotein lipase*. The triacylglycerol components of VLDLs and chylomicrons are hydrolyzed to free fatty acids and glycerol in adipose tissue and skeletal muscles. The free fatty acids diffuse from adipose cells, combine with albumin in the blood, and are transported to other tissues, where they diffuse into cells and get used for energy. The glycerol enters the bloodstream and eventually goes back to the liver and kidneys for disposal.

Cholesterol Metabolism

Cholesterol, like the triglyceride fats, is obtained both from the diet and via biosynthesis. Dietary cholesterol, which accounts for a little more than half our total supply, comes mainly from animal sources because plants do not seem to need or manufacture it. (Information about cholesterol synthesis in our tissues is covered under the entry on lipid biosynthesis.) Once in the gut, cholesterol from our food gets packaged into chylomicrons, which pass directly from the intestinal tract into the bloodstream to the liver. In the liver, these chylomicrons are broken down and the cholesterol converted to LDLs, which are then secreted back into the blood for transport to other tissues. The cells of these tissues have a surface receptor protein called the *LDL receptor* (LDLR), whose job it is to recognize and bind cholesterol-carrying lipoproteins in the bloodstream so that the cells can use cholesterol. In healthy people, LDLR helps to maintain cholesterol in balance. The protein seems to be a "mosaic" made up of exons shared with several different proteins. Its gene is large, more than 45 kb in length and having 18 exons, and is found on chromosome 19. Mutations in the gene can disrupt the function of the receptor. For example, mutations may cause incorrect folding of the protein, which may interfere with the receptor's ability to bind calcium. If LDLR malfunctions, cholesterol levels in the blood can be very high. Familial hypercholesterolemia due to LDLR defects affects about one in 500 people.

Evidently, the dangers of cholesterol are associated with its excess. So how does the body deal with the leftovers once the body's tissues have used all the cholesterol they need? The only organ in the body that can break down sterols (not just cholesterol but other related compounds as well) is the liver. Here the sterols are converted into bile acids, which are secreted into the gut via the gall bladder and eliminated from the body with the feces. The most abundant bile acids produced are chenodeoxycholic acid and cholic acid, which account for 45 percent and 31 percent of the total end products, respectively. The conversion of cholesterol to these compounds occurs via a long metabolic pathway. The first step, which also determines the rate at which the rest of the degradation proceeds, is the addition of hydroxyl groups to cholesterol by the enzyme *7 alpha-hydroxylase*. This enzyme is a member of the cytochrome P450 family encoded by a gene on chromosome 8. Details of the intermediary steps are beyond the scope of the present volume.

Because most cells do not have the enzymes needed to break down sterols into bile acids, the excess cholesterol in various tissues must be transported to the liver. This happens via a process called *reverse cholesterol transport*. High-density lipoproteins (HDLs) that are made by the liver remove cholesterol from the arteries and transport it back to the liver for excretion. The HDLs are the smallest and densest lipoproteins—they have a high proportion of protein (which increases their density relative to other lipoprotein particles, hence the name).

Excess cholesterol moves from cell membranes into the HDL particles. An enzyme called *lecithin-cholesterol acyltransferase* (LCAT) catalyzes the formation of cholesterol esters from lipoproteins. The *LCAT* gene is located on chromosome 16. The cholesteryl esters in HDL particles are then transferred by *cholesteryl ester transfer protein* (CETP) from HDLs to VLDLs and LDLs, which are then transported to the liver for breakdown. CETP is an extremely hydrophobic protein that catalyzes the transfer of insoluble cholesterol esters in the "reverse transport of cholesterol." This transfer of insoluble cholesteryl esters among lipoprotein particles is very important for maintaining normal levels of cholesterol. The 22 kb gene is located on chromosome 16. A number of *CETP* polymorphisms are associated with atherosclerosis, the formation of plaques in arteries.

MITOCHONDRIAL GENES

The mitochondrial chromosome, or plasmid, in humans is a circular piece of double-stranded DNA containing about 17,000 base pairs. It is autonomous (i.e., independent of the nuclear chromosomes in its replication and the expression of its genes). About 93 percent of this DNA is contained in the sequences of the 37 genes that are encoded in this genome. This is in sharp contrast with the 3 percent coding capacity of the chromosomes. Relative to the nuclear genes, the mitochondrial genes are very closely spaced and have no introns, making them more akin to the genes of prokaryotes such as bacteria than to the genes of the eukaryotic organisms of which they are a part. Another interesting feature of the mitochondrial plasmid is the uneven distribution of nucleotide bases. One of the two strands is much richer in guanines (Gs), which makes it much heavier that its complement, which correspondingly contains the significantly lighter cytosines (Cs). The heavier strand encodes roughly three times as many genes as the light strand (28 versus 9).

The 37 genes in the mitochondrial plasmid include 24 genes that code for RNA molecules—2 rRNA species (see the entries on ribosomes in Chapter 2 and this chapter for details) and 22 tRNAs—and 13 genes for polypeptides or proteins. These proteins are components of the very large and complex *oxidative phosphorylation* system (containing over 90 subunits) that is responsible for the essential respiratory activities performed by the mitochondria. These activities are described in detail elsewhere in this chapter. The following specific respiratory proteins are encoded in the mitochondrial plasmid:

- 7 of the 43 subunits of the NADH dehydrogenase (also called complex I) enzyme of the mitochondrial respiratory chain.

These proteins are designated as MTND1–MTND6 and MTND4L. Complex I is responsible for accepting electrons from NADH and transferring them to the next member of the electron transport chain (coenzyme Q10), thereby generating energy to pump protons out across the mitochondrial inner membrane. The different polypeptides of this complex make up three different protein fractions. The polypeptides encoded by the mitochondrial genes 1, 3, and 4L are part of the hydrophobic fragment, which is also the probable location for the subunits 2, 4, and 5. MTND6 differs from all of the others in that it is located in the iron-protein fragment of complex I. It is also the only protein from this set that is encoded by the light, cytosine-rich strand of the mitochondrial plasmid (between positions 14149 and 14673). Defects in this protein can cause Leber's hereditary optic neuropathy, a maternally inherited condition wherein the patients suffer from acute bilateral blindness due to retinal degeneration.

Cytochrome b (MTCYB)

Cytochrome b is a hydrophobic protein with two heme (iron-binding) groups. It forms the central functional portion of complex III of the respiratory chain along with two core proteins that are synthesized in the cytoplasm. Cytochrome b is the sole component of this 10-subunit complex that is encoded within the mitochondrial genome itself. This complex functions as the second enzyme in the electron transport system of the mitochondria and catalyzes the transfer of electrons from the coenzyme Q10 to cytochrome c. Cytochrome b forms the electron-binding region of complex III. The coding sequence lies on the heavy strand of the mitochondrial plasmid from nucleotide positions 14747 and 15887. It is very highly conserved in evolution.

- 3 polypeptides of the cytochrome c oxidase complex (complex IV).

Complex IV is the third and final enzyme of the mitochondrial respiratory chain and catalyzes the transfer of electrons from cytochrome c to oxygen, forming water and releasing energy. The three subunits of this complex that are encoded by the mitochondria are designated as MTCO1, MTCO2, and MTCO3. The CO1 and CO2 subunits form the scaffolding for the enzyme's metallic ions (iron and copper). Less is known about the exact function of the CO3 subunit, but it is one of the most conserved parts of the enzyme.

- 2 out of 14 polypeptides that make up the ATP synthase complex, or complex V.

The energy released by the transfer of electrons across the respiratory chain is stored in the form of ATP molecules. This synthesis is catalyzed by the synthase complex, in which the subunits *ATPase 6* and *ATPase 8* are encoded in the mitochondria. Extending respectively from nucleotides 8527–9207 and 8366–8572 of the heavy chain, these two genes actually overlap with one another.

Defects in any of these genes lead to genetic disorders that are maternally inherited and do not follow classical Mendelian inheritance. Specific examples are discussed in Chapter 5.

See also: Mitochondria (Ch. 2); the section on Mitochondrial DNA in Chapter 3.

MUSCLE

The muscular system needs no introduction—all of us are aware of how much we depend on different muscles in our body for our movements, such as walking or running, working out, and lifting heavy objects. But are you aware of how many times you use your muscles without even consciously thinking about it? Virtually all of the body's movements, whether voluntary actions such as those just mentioned or automatic activities such as breathing and blinking (researchers have estimated that the eye muscles often move more than 100,000 times a day, making them the busiest muscle in the body), the movement of food through the stomach and intestines, and the rhythmic contraction of the heart, depend on muscular tissue.

All told, each human being has over 630 muscles, comprising 40–50 percent of the total body weight, which are used for various types of activities, both voluntary and involuntary (automatic). The secret of a muscle's ability to move lies in its composition and structure. A closer look at the muscular tissue reveals that it is made up of special fibers capable of contracting (and subsequently relaxing back to their original shape like an elastic band). The coordinated contractions of hundreds of these fibers enable a muscle to perform its particular locomotive function. Actually, the translation of the contraction of individual muscle fibers to motion takes place through a system of connective tissues and muscle sheaths that give the muscles continuity and also link muscles to the skeleton.

A fundamental aspect of muscular motion is that it works only in one direction. In other words, a muscle can only pull and cannot push. This may seem counterintuitive when we think about the range of movements of which we are capable, but in fact this flexibility is due to the fact that most often muscles work in pairs, with each partner pulling in the opposite direction.

Depending on the types of movements they govern, the muscles of the human body may be classified into three basic types:

- The *skeletal* muscles, so named because they are found connected to different bones of the skeleton, comprise about 40 percent of the total muscle mass. These are the muscles that we use in our conscious, or *voluntary*, movements such as walking, running, lifting, and so on. Examples include the muscles of the back and limbs. Each muscle is composed of bundles of cylindrical muscle fibers with several nuclei. Skeletal muscle fibers are arranged in alternating dark and light bundles, which is why they are also known as the "striated" muscles.

- Most of the *involuntary* movements in the body are under the control of the *smooth* muscles. These muscles perform a variety of tasks (e.g., peristalsis—the movement of food through the digestive tract—the contractions of the bladder and uterus, and breathing, the latter through the action of the diaphragm). In contrast with the skeletal muscles, the tissues in the smooth muscles are organized in sheets rather than striated bundles.

- The *cardiac* muscles are responsible for the heart's ability to pump blood. These muscles share characteristics of both types of muscles; they resemble the skeletal muscles in that they are striated, but their action is involuntary, like that of the smooth muscles. Where the skeletal and smooth muscles form discrete, compact muscles that are separate from other muscles, the cardiac muscle has a branched structure, which makes virtually the entire organ somewhat contiguous.

Here we will discuss the proteins that enable muscles to contract and direct movement. In addition, we will also discuss some other muscle proteins that perform some important functions in these tissues.

Actin

Actin is the protein of the cytoskeletal system that allows movement of cells or cellular processes or projections, such as cilia. It is especially abundant in muscle cells but in fact accounts for at least 10 percent of the total mass of all of the other cells in the body as well. Actin is highly conserved in animals, implying that it has not changed significantly through the course of evolution.

Actin performs various functions in different cells. For example, in the epithelial cells that line the intestines, long, finger-like projections into the intestinal lumen are supported by bundles of actin filaments that increase the surface area of the intestinal lining. This improves the absorption of nutrients from the digestive tract that are then transported through the bloodstream. Red blood cells, which are responsible for transporting oxygen through the blood, have a two-dimensional network of short filaments of actin connected by linker proteins that support the cells and help them maintain their flattened, biconcave disc shape. This network is flexible and can be distorted so that a red blood cell can squeeze through capillaries one-quarter its size. And, of course, as mentioned earlier, actin plays a fundamental role in the physiology of muscles, enabling them to contract and perform their different activities.

Structurally, actin is a globular protein with an ATP-binding site at the center of the molecule. Single units of actin (monomers) can very quickly assemble themselves into long polymeric fibers called microfilaments. The microfilaments have a range of roles: As part of the cell's cytoskeleton, they form the scaffold to which other proteins bind; they interact with myosin to allow contraction of the muscles; and during cell division they cause the cell to constrict in the middle and divide into two. The microfilaments have the special ability to switch between a rigid fiber state and a more mobile, almost liquid form called the sol state, which enables the activities just described to proceed without damaging the cell. For instance, when cells change their shape (e.g., during contraction or relaxation), the relatively rigid microfilaments, which form a gel-like mesh just beneath the cell membrane, change to the sol state, enabling the layer of actin to move. A conversion from the liquid sol to the filamentous gel then keeps the cell in its new shape.

There are actually six known actins in mammals, classified into three types—alpha, beta, and gamma—according to their function and location. The alpha actins are specific to muscle cells and are responsible for contractile movement. These include actins of the skeletal muscles (encoded on chromosome 1), cardiac muscles (chromosome 15), and smooth muscles of the aorta (chromosome 10). The beta and gamma actins coexist in most cells as components of the cytoskeleton and as mediators of internal cell motility. Beta actin maps to chromosome 7, and the gene for a gamma actin in the smooth muscles of the intestines has been mapped to chromosome 2. Defects in cardiac actin are implicated in dilated and hypertrophic cardiomyopathies, and mutations in the skeletal muscle alpha actin gene are associated with two different muscle diseases: congenital myopathy, in which the affected person is born with an excess of thin filaments, and nemaline myopathy. Both diseases are characterized by structural abnormalities of muscle fibers and variable degrees of muscle weakness.

The movement of the muscles is probably best understood by what is known as the *sliding filament model*, which although best described in the skeletal and cardiac muscles is applicable to all muscle types. The basic contractile unit in a muscle, called a *sarcomere*, consists of alternating bundles of actin (thin filaments) and myosin (thick filaments). Muscle contraction results from chemical changes that actin and myosin undergo, which cause them to alternately link and unlink. The bipolar filaments of myosin (see the next section) use ATP to pull on the opposing sets of actin filaments from each end to shorten the sarcomeres. The energy for this reaction is supplied by adenosine triphosphate (ATP). The force generated between the thin-filament protein actin and the thick-filament protein myosin causes the thick and thin muscle filaments to slide past each other during the contraction. These muscle contractions power beating of the heart, movements of skeletal muscles, and involuntary motions such as the contraction of the muscles of the gut during peristalsis.

Actin filament lengths in sarcomeres vary between fast- and slow-twitch skeletal muscles, a mixture of which are present in human muscles. The two fiber types generally produce the same amount of force per contraction, but fast-twitch fibers produce that force at a higher rate (i.e., they fire more rapidly). Therefore, fast-twitch fibers are useful to a sprinter when there is a limited amount of time to generate maximal force. Slow-twitch fibers, on the other hand, fire less rapidly but can go for a long time before they tire. They contain more mitochondria and myoglobin, which increases their efficiency in burning fuel to generate ATP without lactic acid buildup. As a result, they can withstand repeated contractions over a long time, such as those required for endurance activities such as a marathon. Different muscle groups in the body have different combinations of fast- and slow-twitch muscles, and the actual combination depends both on the type and level of activity.

Myosin

As described earlier, myosin is the major protein responsible for muscle contraction. It makes up the thick filaments that lie parallel to the microfilaments of actin in the sarcomeres of the muscle fibers. Both myosin and actin also function in the motility of diverse nonmuscle cells, as discussed for actin. For instance, the interaction between the two proteins helps to change cell shape and permit some movements.

During contraction, the myosin thick filaments attach themselves to the actin thin filaments by forming chemical bonds known as cross bridges. Initially, the cross bridge is extended with adenosine diphosphate (ADP) and inorganic phosphate (Pi) attached to the myosin. As soon as the cross bridge is formed, the myosin head bends, thereby creating force and sliding the actin filament past the myosin. This process is called the power stroke. During the power stroke, myosin releases the ADP and a phosphate group (Pi). Once ADP and Pi are released, a molecule of adenosine triphosphate (ATP) binds to the myosin. When the ATP binds, the myosin releases the actin molecule. When the actin is released, the ATP molecule gets split into ADP and Pi by the myosin. The energy from the ATP resets the myosin head to its original position so that the process can be repeated. The actions of different myosin molecules are not synchronized—at any given moment, some myosins are attaching to the actin filament while others are creating force and still others are releasing the actin filament.

There are two structures in the grooves of each thin filament that enable the thin filaments to slide along the thick ones: a long, rodlike protein called tropomyosin and a shorter beadlike protein complex called troponin. Troponin and tropomyosin are the molecular switches that control the interaction of actin and myosin during contraction.

Several members of the myosin gene family have been identified by genomic analysis—most recent accounts put the number at 40! The functions for most of them are still being determined. It appears that myosin I and myosin II are the most abundant in cells. Myosin II powers muscle contraction and cell movement, and myosins I and V are involved in membrane interactions such as the transport of membrane vesicles. All of the proteins in the myosin gene family are composed of multiple components—one or two heavy chains and several light chains—which are seldom (if ever) encoded on the same chromosomes. The heavy chain makes up the bulk of the myosin head, which contains the ATP (or ADP) binding site and the actin-binding region. The most conserved region in various myosin proteins is the globular head domain, which interacts with actin to effect motion. All of the myosin proteins have a tail domain that is unique to each type of myosin and determines its location and function in the cell. Defects in different genes have been associated with a number of diverse genetic conditions.

Dystrophin

Dystrophin is an important cytoskeletal protein found at the inner surface of muscle fibers, especially enriched in the regions where the muscles are connected to the nerves and tendons. The protein plays a key structural role as part of a large complex (along with membrane glycoproteins) in muscle fiber membranes that links the intracellular cytoskeleton (actin) with the extracellular matrix, thereby maintaining the strength of the muscle fibers. Dystrophin is also involved in stabilizing the cell membranes during the cycles of muscle contraction and relaxation, so that the intracellular contents do not leak out. A basal level of this protein is also found along the membranes of cells in other tissues besides the muscles.

Located on the X chromosome, the dystrophin gene, at 2.4 Mb, is the largest known gene in the human genome. It consists of no less than 79 exons, which nevertheless constitute a mere 0.6 percent of the total gene, interspersed amid very large tracts of introns. This arrangement of the gene allows for the mRNA to be spliced in many different ways, leading to the production of a large number of transcripts, which are then translated into slightly different forms (called isoforms) of a protein containing nearly 4,000 amino acids. The dystrophin gene also has at least eight independent promoters that direct the production of specific isoforms of the protein in specific cells and tissues such as the skeletal and cardiac muscles, retinal muscles, and cells of the nervous system such as Purkinje (brain) and Schwann (nerve fibers) cells. Deletion mutations in the dystrophin gene are associated both with mild and severe forms of a muscle-wasting disease known as muscular dystrophy.

See also: Muscular Dystrophy—Duchenne/Becker Type (Ch. 5).

Myoglobin

We have already seen how vital oxygen is for the proper functioning and, indeed, survival of warm-blooded organisms such as ourselves. Perhaps the most important physiological role of oxygen is in the production of energy (or ATP). This energy is deployed for various metabolic and biosynthetic purposes. It is very important for the tissues most involved in energy transformations to have a storehouse of oxygen to which they can turn when their normal supply from the blood is interrupted or stalled for some reason. In both the cardiac and striated (skeletal) types of muscles, this function is performed by the protein myoglobin, which not only stores oxygen but also facilitates its intracellular diffusion to the mitochondria, where the energy is produced.

Encoded by a single gene on chromosome 22, myoglobin, like hemoglobin, is a globular metalloprotein whose oxygen-binding capacity resides in its iron-containing "heme" group. Unlike hemoglobin, which is a large tetrameric protein, myoglobin is a monomer (i.e., it has a single polypeptide chain) that is folded into a globular structure containing its heme cage. This smaller size is necessary for its role in the intracellular trafficking of oxygen. Myoglobin is most abundant in those muscle fibers that primarily depend on oxidative phosphorylation for their energy. These fibers are known as "red muscles," in contrast with "white muscles" that rely on glycolysis for energy and hence lack the red oxy-myoglobin. It is this protein that gives dark meat its characteristic red color. Interestingly, whales and dolphins also have a lot of myoglobin because they need to store oxygen in their muscles when they dive underwater.

Important as it is, myoglobin is, surprisingly enough, not essential for survival, according to a 1998 discovery made by scientists at the University of Texas. The researchers showed this by developing a knockout mouse lacking the myoglobin gene. The only difference between the knockout mutants and the normal mice was the lack of myoglobin in the muscles. This had no apparent effect, as the mutants were as active as control mice. What could explain this? The researchers think that although myoglobin plays a pivotal role in oxygen storage and transport, there may be other genes that compensate for its activity when the protein is deficient or missing. Clearly, the puzzle of oxygen transport and storage in muscles is still far from complete. Meanwhile, within the environs of the laboratory, myoglobin has been of great importance in the elucidation of protein structure. In 1962, the British crystallographer John C. Kendrew (1917–1997) won a share of the Nobel Prize for Chemistry for his use of the technique of X-ray diffraction to construct a three-dimensional model of crystalline sperm whale myoglobin.

Brief mention should also be made in this section of a protein called *neuroglobin*, which is encoded by a globin gene mapped to chromosome 14. This gene was actually discovered by conducting a database search of the human genome for globin-like sequences. Subsequent molecular and biochemical studies of the protein encoded by this gene revealed it to be a monomer, expressed predominantly in the brain tissue and with oxygen-binding capacities similar to myoglobin, indicating a parallel role for it in the brain. Some researchers have also noted that regions of the brain such as the hippocampus that have lower levels of neuroglobin have lower resistance to ischemia (the blockage or cutting off of the blood supply) and are also more frequently affected by neurofibrillary tangles in Alzheimer's disease.

See also: Blood—Hemoglobin.

ONCOGENES AND PROTO-ONCOGENES

There are many genes in a cell that code for proteins involved in maintaining the cell cycle of cell division, growth, differentiation, and programmed cell death (apoptosis). As we have discussed previously, it is important for multicellular organisms such as ourselves to coordinate the activities of the cells in various tissues and organs. The main effectors of this communication and coordination are the proteins of the signal transduction pathways. These proteins perform the function of relaying messages—from one cell to another, from the environment to cells, or within cells (e.g., from the surface to the nucleus)—that instruct the cells to grow or divide or die.

Mutations in any of the proteins involved in the signal transduction pathways can wreak havoc in the body. One of the most common outcomes is the loss of control over cell growth, which leads to the development of cancer. The genes that code for the proteins are called oncogenes (i.e., cancer-causing genes). An oncogene is defined as a signaling gene that has suffered some genetic change and therefore produces a protein that malfunctions and changes the cell cycle in some way, thus leading to cancer. In other words, an oncogene is a gene that induces the transformation of a normal cell to a cancerous form. The genes of many signal transduction proteins were first discovered in their role as oncogenes, even before their functions in normal cells were elucidated.

A proto-oncogene is a potential oncogene, or a precursor, one that has not yet suffered a genetic mutation. The proto-oncogenes are involved in elaborate cell-signaling systems that carry signals from the cell surface to the nucleus. The protein products of proto-oncogenes transmit these signals. There are many subgroups of proto-oncogenes. Some important groups include growth factors, kinases, G-protein receptors, and nuclear proto-oncogenes. Kinases, which are present on the cell surface, have the ability to add phosphate groups to amino acids such as tyrosine (an important signaling tool). G-protein receptors, of which Ras is a member (see the next section), are present in the cytoplasm. Signaling proteins such as Myc, Myb, Fos, and ErbA direct the transcription of DNA into RNA and play a critical role in deciding which proteins the cell must synthesize depending on the signal it receives.

Specific mutations lead to the transformation of a proto-oncogene to an oncogene. For example, deletions in the *EGFR* proto-oncogene, on chromosome 7, result in continuous and uncontrolled signal transduction by the epidermal growth factor receptor it encodes. The cell's machinery is always turned "on," and the cell becomes cancerous. Another oncogene, called *Neu* or *Her2*, is a mutated form of a receptor that is overexpressed in certain breast cancer cases. A monoclonal antibody against Her2/Neu, called Herceptin, has been used with some success in treating certain breast cancer patients.

The history of oncogenes began in 1911 when a chicken breeder brought a hen with a tumor (sarcoma) to the attention of virologist Peyton Rous (1879–1970). His investigation of the phenomenon led Rous to the conclusion that the tumor was caused by a virus (which was later named the Rous sarcoma virus—RSV). At the time, however, the scientific community was reluctant to believe that an infectious agent such as a virus could cause cancer. Rous's findings were ignored for nearly half a century until the discovery of retroviruses, a special class of viruses that could insert their genome into the DNA of host cells. Many retroviruses, especially those of hosts such as birds and rodents, were found to induce tumors (i.e., they were oncogenic). During the 1960s and 1970s, a lot of research was done on these retroviruses because scientists thought they would find similar viruses that cause tumors in humans. Many scientists thought that the oncogenic viruses caused all cancers and that a cancer grew by the spread of the virus from cell to cell. This turned out not to be true, and it is now clear that viruses are not the cause of most human cancers. But interestingly there is a germ of truth to the idea, at least with respect to a subset of human and other cancers. Michael Bishop (1936–   ) and Harold Varmus (1939–   ) used a nucleic acid probe complementary to RSV to probe the genetic material of various organisms. What they found was that the chromosomal DNA of several different species contained sequences that were very similar (but *not* identical) to the viral sequences. Thus oncogenes, or rather proto-oncogenes, exist in the cells of many organisms. Rous (in 1966) and Bishop and Varmus (in 1989) received the Nobel Prize for Physiology or Medicine for their discoveries.

We have already mentioned some examples of oncogene–oncogene pairs in human cells. We now discuss some examples in detail to give a better picture of how these genes can cause cancer.

See also: Cell Cycle, Signal Transduction (Ch. 2).

Ras Oncogenes

The *Ras* oncogenes belong to a family of *r*etrovirus-*as*sociated (hence their name) DNA sequences that normally encode signal proteins called G-protein receptors. They were originally isolated from Harvey (H-ras) and Kirsten (K-ras) mouse sarcoma viruses. These genes are conserved across several animal species, including humans, birds, mice, and some invertebrates. Human homologs to these genes are found on chromosomes 11 and 12, respectively. The closely related N-ras gene has been detected in human neuroblastoma and sarcoma cell lines (chromosome 1). All genes of the family have a similar layout of exons and introns, and they all encode a "p21" protein. Mutations in H-ras, K-ras, and N-ras convert these genes into active oncogenes.

The protein expressed by the Ras proto-oncogene is involved in activating the transcription of genes, namely the conversion of DNA into mRNA. It receives a signal from a receptor protein on the cell surface to bind a particular molecule called guanosine triphosphate (GTP) that causes a subtle change in the shape of the protein, which then triggers a cascade of protein kinases. The pathway signals nuclear proteins to start transcribing genes important for cell growth. Ras can also signal the cell to stop growing by releasing the bound GTP.

When the Ras gene is mutated (i.e., it becomes an oncogene), its protein product loses the ability to break up and release GTP. As a result, the signaling pathway is always on, and transcription and cell growth continue uncontrolled. The Ras oncogene has been identified in cancers of many different organs, including the pancreas, colon, lung, thyroid, bladder, ovaries, breast, skin, liver, and kidney, and certain leukemias. Roughly half of all colon cancers and 90 percent of pancreatic carcinomas are associated with the *Ras* oncogene.

Rous Sarcoma Oncogenes

As we saw earlier, RSV holds a permanent place in the history of cancer virology. The proto-oncogene (*Src*) in avian (bird) species and homologous genes in the human genome encode tyrosine kinase components of signal transduction. These enzymes relay messages from cell-surface receptors to various sites within the cell, telling the cell to grow and divide. Normally, the activity of the kinases is very tightly regulated by an internal "switch" that enables the protein to turn on and off in response to other events within the cell. When the kinase is turned off, for instance, the cell knows to stop proliferating. In the Rous sarcoma virus, this gene contains a point mutation that changes a crucial amino acid in the enzyme and causes the switch to be permanently "on." The cells infected by this virus (or, rather, containing this viral sequence) cease to respond to normal signs that they should stop growing and develop into tumors.

As we know, the human genome codes for a large number of tyrosine kinases that act specifically in different cells and tissues of the body. Of these, there are two genes in close proximity to each other on chromosome 20 that are very similar to RSV. The first, known simply as the Rous sarcoma proto-oncogene, seems to play a role in the regulation of cell growth in the embryo. Point mutations in this gene are associated with certain forms of colon cancer. A second gene, called the hemopoietic cell kinase (*HCK*), encodes a kinase that is expressed in developing blood cells (i.e., hemopoietic cells). This kinase is not associated with any known cancers.

PROTEIN SYNTHESIS: THE TRANSLATION APPARATUS

The translation of the genetic code into proteins is the essential step in accessing the instructions for life encoded into the genome of all living organisms. Even viruses, which have no translation apparatus of their own, use the machinery of their host cells to make viral proteins, which are then assembled to form new virus particles. This section is an overview of the genes and their products that participate in this fundamental process.

Protein synthesis occurs in two phases: an initial transcription, in which the gene on the DNA template is read and synthesized into a messenger RNA, followed by translation, the conversion of the code of nucleotide triplets into sequences of amino acids (i.e., the peptides). Details of transcription, as well as certain individual components of the translation apparatus (such as the ribosomes), are discussed in independent entries of this chapter. Here we will review translation in a stepwise manner and supply details of important genes that were missed in other sections.

Translation—the phase where the nucleotide sequence is actually read and rewritten into a sequence of amino acids—takes place in the cytoplasm. It involves three main structural components: the newly synthesized and processed mRNA (the nucleotide template), ribosomes (which provide the enzymes and the physical scaffolding for the process), and the tRNAs, which supply the amino acids, the building blocks of the proteins.

The process begins with the attachment of a newly synthesized mRNA molecule recently extruded from the nucleus to a ribosome. A special initiator tRNA charged with methionine binds to this mRNA–ribosome complex at a designated site known as the P-site. Protein synthesis then proceeds in a repeated sequence of a three-step cycle, which elongates the protein (peptide) one amino acid at a time (see Figure 4.10):

1. First, tRNA charged with the appropriate amino acid binds to the mRNA, which is held at a designated site on the ribosome known as the A-site.

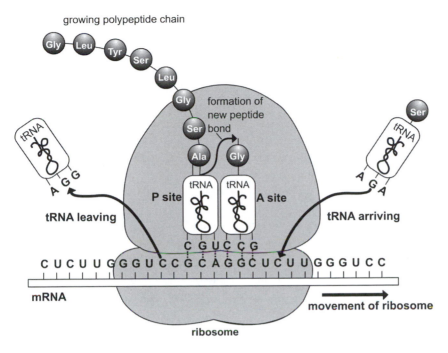

growing polypeptide chain

P site

A site

tRNA leaving

tRNA arriving

formation of new peptide bond

tRNA

mRNA

movement of ribosome

ribosome

C U C U U G G G U C C G C A G G C U C U U G G G U C C

Figure 4.10. Protein synthesis. [Ricochet Productions]

2. Next, ribosomal enzymes (for details, see the entries on ribosomes in this chapter and Chapter 2) catalyze the formation of a peptide bond between this amino acid and the amino acid chain at the P-site.

3. Finally, the ribosome ejects the now empty tRNA from the A-site and moves a distance of three nucleotides (i.e., one codon) along the mRNA, so that the next charged tRNA can bind and begin the next cycle.

The cycle of chain elongation comes to a stop when the ribosome encounters the stop codon on the mRNA. It then releases the peptide, which undergoes further modifications to form the final, functional protein.

A tRNA is a small molecule, 73–93 nucleotides in length, which is specifically configured to discharge its function in as efficient a manner as possible. Each tRNA is capable of binding only one type of amino acid, the identity of which is determined by the sequence of a triplet of nucleotides known as an anticodon, which is complementary to a triplet codon on the mRNA. Given that there is a total of 64 possible codons, we might reasonably expect that the human genome contains at least as many tRNA coding sequences or genes. In reality, the human genome contains anywhere from 1,000 to 2,000 tRNA genes scattered throughout its chromosomes, which make approximately 60 to 90 distinct tRNA molecules. The most significant feature that distinguishes these tRNAs from one another is the anticodon sequence. Only 61 of the 64 anticodons are represented in these molecules, as there are no tRNAs for the three stop codons. There is no clear pattern for the distribution of these genes and the genes for redundant tRNAs (i.e., tRNAs with different codons that are specific for the same amino acid may be found close together in a single cluster or on different chromosomes altogether). For example, there is a cluster of four genes on chromosome 14—*TRP1*, *TRP2*, *TRL1*, and *TRT*—that encode tRNAs for proline, leucine, and threonine, respectively, and another cluster on chromosome 17 has a second gene for leucine tRNA (*TRL2*) in addition to tRNA genes for glutamine and lysine (*TRQ1* and *TRK1*). Transfer RNAs for methionine deserve special mention because all translation begins with this amino acid, regardless of its presence or absence in the final protein product. Researchers have isolated 12 species that they have identified specifically as initiator methionine tRNAs and more genes for tRNAs that supply methionine to a growing peptide chain. Examples of the former include the adjacent genes *TRMI1* and *TRMI2* on chromosome 6. A third gene, called *TRM1*, is also located on chromosome 6 but at a different locus from the initiator genes, in a cluster along with genes for alanine and arginine tRNAs (*TRAN* and *TRR3*).

Once it has performed this function, the empty tRNA is released into the cytoplasm, where it can be charged with a new amino acid residue and sent back to work on the protein assembly line. The charging of the tRNAs takes place with the help of enzymes called the *amino-acyl tRNA synthetases*. There is a specific synthetase for each amino acid (e.g., methionyl tRNA synthetase (MARS) for methionine, histidyl tRNA (HARS) for histidine, etc.). The genes for these synthetases reside on different chromosomes (*MARS* is on chromosome 12 and *HARS* is on chromosome 5), and their location bears no relationship with the location of the tRNA clusters for the cognate amino acids.

Protein Folding and Posttranslational Modifications

Unlike DNA and RNA, whose nucleotide sequences are the most important determinants of function, proteins (and ribozymes as well) depend as much on their shape or three-dimensional conformation as they do on the primary sequence for their proper functioning. A protein gets its final shape by folding itself soon after translation. The amino acid sequence plays a large role in folding, allowing certain residues to form bonds among themselves and keeping others apart by virtue of properties such as size, charge, and distance from one another. In addition, special proteins known as the *chaperonins* (see the entry on housekeeping genes) help newly synthesized proteins with their folding by stabilizing unfolded and intermediate forms so that they do not fold into the wrong shape. An example of a chaperonin is a 60 kDa heat-shock protein encoded by the *HSPD1* gene on chromosome 2, which is essential for the folding and assembly of newly imported proteins in the mitochondria. Another gene for a tubulin-specific chaperonin on chromosome 1 (*TBCE*) encodes a protein that is specifically involved in the correct folding of beta-tubulin proteins.

In addition to folding, the newly translated proteins may need to undergo further chemical changes before they are ready to function. These changes, which may take place before or after the protein is folded and transported to its correct location in the cell, can involve the addition of chemical groups, such as acetate, phosphate, carbohydrates, or lipids. In some cases, sections of the protein are removed (i.e., they undergo cleavage), whereas in others, two or more peptide chains must come together to form a functional protein complex. The modifications may alter the protein's physical and chemical properties, folding, conformation distribution, stability, activity, and, consequently, function. Here we will briefly describe a few of these modifications.

Phosphorylation

The addition of a phosphate group to specific amino acid residues of a protein plays a critical role in the regulation of many cellular processes, including cell cycle, growth, apoptosis, and signal transduction pathways. Most phosphorylations work to regulate the biological activity of a protein and are transient, which means that phosphates are added and later removed. Phosphate groups are usually added to the amino acid residues serine, threonine, and tyrosine. Enzymes that phosphorylate proteins are known as *kinases*. The kinases are encoded by members of a very large family of genes. Scientists have cataloged hundreds of protein kinases across the human genome (the "kinome"). These genes appear to be scattered in groups throughout all of the chromosomes. (See the entries in Chapter 2 and this chapter on signal transduction and its pathways for examples of specific kinases and their genes.)

Glycosylation

Glycosylation refers to the addition of carbohydrate chains, usually to the residues asparigine, hydroxylysine, serine, and threonine of a protein. Many cell-surface and secreted proteins undergo glycosylation, which often serves to enhance the antigenicity of a protein. Glycosylation takes place in the lumen of the endoplasmic reticulum (ER) or in the Golgi bodies of cells. The carbohydrate then attached can be O-linked or N-linked, depending on the type of amino acid residue to which it is linked. Specific enzymes known as the *glycosyltransferases* catalyze the transfer of a specific sugar molecule to the proteins.

The human A, B, and O blood-group antigens of the red blood cells are good examples of proteins with specific glycosyl groups. All people have the enzymes needed to glycosylate the O antigen. The A antigen is similar to O, except that it contains an N-acetylgalactosamine attached to the outer galactose residue, whereas the B antigen has an extra galactose residue attached to the outer galactose. So people with type A blood also have the *GalNAc transferase*, which adds the extra N-acetylgalactosamine, whereas those with type B blood have the *Gal transferase*, which adds the extra galactose. Interestingly, the sequences of *GalNAc* (*A antigen*) and *Gal* (*B antigen*) *transferase* genes differ in just three codons, yet these differences are enough to cause fatal accidents during blood transfusion. Genes for both proteins are allelic variants found at the same locus on chromosome 9.

Proteolytic Cleavage

Many proteins undergo some proteolytic cleavage following translation. The simplest form of this process, which occurs in most cases, is the removal of the initiation methionine. Often, many enzymes are synthesized as inactive precursors that are activated under proper physiological conditions by limited proteolysis. Pancreatic enzymes and enzymes involved in blood clotting are examples. These proteins have signal peptides that are cleaved by specific peptidases. Another example is insulin. A signal peptidase in the endoplasmic reticulum within pancreatic beta cells catalyzes the initial cleavage of pre-proinsulin. In this process, the 24 amino acid signal peptide (consisting mostly of hydrophobic residues) is removed, and the protein folds into proinsulin. Proinsulin is then transferred to the Golgi apparatus. There it is cleaved by another peptidase to yield active insulin, which is composed of two peptide chains linked together through disulfide bonds. These peptidases are trypsin-like proteases, a large family of enzymes encoded by genes that are found across the entire genome.

RESPIRATION

Earlier in this chapter (see Food and Digestion), we discussed the way in which our bodies get and process food, one of the essential things we need for survival and carrying out various activities. But as important as food is to us, air—specifically oxygen—is of even greater significance. We rely on an involuntary system to get it from our surroundings and deliver it to our tissues and cells. This process is known as respiration, something we do for our entire lives from the instant we are born until we die—quite literally, "draw our last breath" means to die.

Respiration actually takes place at many levels. We are all familiar with breathing—the process by which we draw in the air with the use of our lungs. This is the first level of respiration. Linked to this mechanical process of breathing is a physiological–biochemical event whereby the oxygen from the air is actually absorbed into the body. Hemoglobin in the blood binds to the oxygen that is drawn into the cavity of the lungs, where it exchanges it for carbon dioxide and transports it to all other organs and tissues of the body (see the entry on blood for details). The large internal surface area of lung tissue, which is well-endowed with capillaries bearing blood, facilitates this oxygen–CO_2 exchange. The oxygen-rich blood is then taken to the heart via the pulmonary vein, where the heart pumps it out to the rest of the body. Once the oxygenated blood reaches the tissues where it is needed, it is released by hemoglobin and diffuses across the cell membranes into the cells.

The next level of respiration takes place within the cells. This is where the oxygen is actually used to generate the energy needed by the body. This process may be viewed as a sort of intracellular combustion where the oxygen "burns" carbon-based fuels such as sugars and fats and produces energy in the form of ATP molecules. The oxidation of a single molecule of a sugar such as glucose generates an average of 36 molecules of ATP and that of a fat (triglyceride) molecule about 5–10 times more, depending on the lengths of its fatty acid side chains. But whereas the net chemical reaction described is quite simple, intracellular respiration is actually much more complicated than simple combustion. It involves many different molecules and cellular components and takes place in several stages so that energy (ATP) is released gradually and utilized more efficiently. In the remainder of this section, we discuss the processes involved in intracellular respiration, with special attention to the genes involved.

The early stages of the breakdown of sugars, namely glycolysis, and fatty acids are discussed in detail in other sections of this chapter. (See the entries on lipid metabolism and sugar metabolism.) Both of these processes produce molecules of acetyl CoA, which is the entry point of the carbon molecules into the main oxygen-utilizing ATP-generating pathways of the cells. For example, the breakdown of one glucose molecule produces a total of 30 ATP molecules, of which only four are generated during glycolysis. The remaining 26 ATPs come from the subsequent processing of the acetyl CoA molecules via two closely linked biochemical processes that take place within the mitochondria. Details of these processes follow.

The Krebs Cycle

The Krebs cycle is a stepwise cycle of oxidation whereby acetyl CoA gives up its energy to the cell and is eventually converted to carbon dioxide, which is removed from the cell into the bloodstream and then returned to the lungs (via hemoglobin). Named for the British biochemist Sir Hans Krebs (1900–1981), who worked out the details of the cycle in the 1930s, the Krebs cycle is also known as the citric acid cycle or the tricarboxylic acid (TCA) cycle after some of the principal intermediates in the cycle.

In brief, the cycle proceeds as follows. Acetyl CoA reacts with a four-carbon molecule called oxaloacetic acid to produce citric acid (with six carbons, three of which are acid—COOH—groups), which serves as a substrate for seven distinct enzyme-catalyzed reactions that occur in sequence and proceed with the formation of seven intermediate compounds. The last in the series, malate, is converted back to oxaloacetic acid, which is then set to start another cycle. Each turn of the citric acid cycle produces two molecules of carbon dioxide and eight molecules of NADH as by-products. The NADH molecules then go through oxidative phosphorylation, as we will describe, to produce energy (see Figure 4.11).

As mentioned earlier, the link between glycolysis and the Krebs cycle is the conversion of pyruvate into acetyl coenzyme A and carbon dioxide. The *pyruvate dehydrogenase* complex catalyzes this conversion. This complex contains multiple copies of three enzymes: pyruvate dehydrogenase (E1), dihydrolipoamide acetyltransferase (E2), and lipoamide dehydrogenase (E3), which are encoded by unlinked genes on different chromosomes. The gene for the E1 alpha subunit gene, for example, is located on the X chromosome. Scientists have found more than

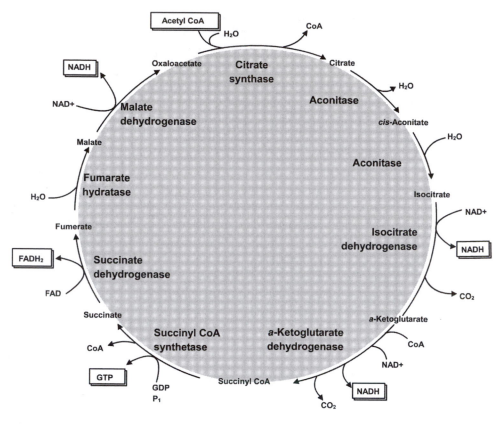

Figure 4.11. The Krebs cycle. [Ricochet Productions]

90 mutations of the E1 alpha enzyme subunit that impair either protein stability or catalytic ability. The gene for the E1 beta enzyme subunit of the PDC has been mapped to chromosome 3, and the E3 gene is found on chromosome 7. Pyruvate dehydrogenase complex deficiency (PDCD) is one of the most common neurodegenerative disorders associated with abnormal mitochondrial metabolism, which deprives the body of energy.

1. Following its formation, acetyl CoA transfers its two-carbon acetyl group to oxaloacetic acid

to form citric acid.

(The carbon atom that forms the linking point between the oxaloacetate and acetyl CoA is marked with an asterisk in both structures.) This reaction is catalyzed by the enzyme *citrate synthase*, which is found in the mitochondrial matrix. The gene for this enzyme is on chromosome 12 and, interestingly, contains no introns.

2. The next step in the cycle is the two-step conversion of citrate into isocitrate catalyzed by *aconitase*. The gene for this enzyme is on chromosome 22 and contains 18 translated exons distributed over approximately 35 kb of DNA. Aconitase is an iron–sulfur protein whose active site contains a cluster of four irons and four sulfurs that are bound by three cysteine residues. Unlike the metal-ion center of other iron–sulfur proteins, which function mostly as electron carriers in oxidative phosphorylation, the Fe–S cluster of aconitase reacts directly with an enzyme substrate.

3. The conversion of isocitrate into alpha-ketoglutarate with the release of CO_2 and NADH is catalyzed by *isocitrate dehydrogenase*. The gene for this enzyme is found on chromosome 15. This reaction is the main rate-limiting step in the citric acid cycle.

4. Alpha-ketoglutarate is converted to succinyl coenzyme A by the *alpha-ketoglutarate dehydrogenase* complex. This is the second carbon dioxide–releasing step of the Krebs cycle—in this case it is the terminal carbon of the oxaloacetate molecule that is released. The enzyme complex is composed of three proteins, each encoded on a different and well-characterized gene. The first, 2-oxoglutarate dehydrogenase, is encoded by a gene on chromosome 7. The gene for dihydrolipoamide succinyltransferase (DLST) has been mapped to chromosome 14. The third enzyme, dihydrolipoamide dehydrogenase (DLD), is also encoded on chromosome 7, albeit in a distant locus from the first subunit. Mutations in the *DLD* gene have been found to be associated with a form of maple syrup urine

disease, a disorder where the breakdown of the branched-chain amino acids such as leucine, isoleucine, and valine is impaired. The urine of affected individuals has the odor of maple syrup.

5. In the next step of the cycle, succinyl coenzyme A is converted into succinate through the action of the enzyme *succinyl coenzyme A synthetase*. Succinyl CoA synthetase (SCS) is composed of an alpha subunit and a beta subunit that determines whether the enzyme binds GTP or ATP. The alpha subunit is coded by a gene on chromosome 6, and the beta subunit is coded by a gene found on chromosome 13.

6. Succinate is dehydrogenated to form fumarate under the action of *succinate dehydrogenase*, a membrane-bound enzyme consisting of four subunits. The gene for subunit A, a flavoprotein, has been mapped to chromosome 5. Mutations in this gene are associated with Leigh syndrome, a progressive neurodegenerative disorder. Subunit B (an iron–sulfur protein) is encoded by a 40 kb gene on chromosome 1. Mutations in this gene cause susceptibility to familial pheochromocytoma, a rare disease in which tumors cause the adrenal medulla to overproduce epinephrine and norepinephrine hormones. Subunits C and D are integral membrane proteins that anchor the enzyme complex to the matrix side of the mitochondrial membrane. The genes for these two subunits are located on chromosomes 1 and 11, respectively.

7. The next reaction adds water to fumarate to make malate, with *fumarase* acting as the catalyst. The fumarase gene locus is on chromosome 1. Fumaricaciduria is a very rare disorder caused by mutations in the fumarase gene, which results in a rare metabolic disorder causing severe neurological disease and often death in early childhood.

8. The TCA cycle starts over with the conversion of malate to oxaloacetate and the production of NADH, and this reaction is catalyzed by *malate dehydrogenase*. The gene for this enzyme has been mapped to chromosome 7.

Oxidative Phosphorylation

Oxidative phosphorylation is the final process involved in releasing energy from the carbohydrates within the cell. It entails the transfer of electrons from the NADH produced during the Krebs cycle, through a series of electron carriers embedded in the inner mitochondrial membrane, to O_2. NADH donates hydride ions (one proton and two electrons) to the first of this series in the electron transport

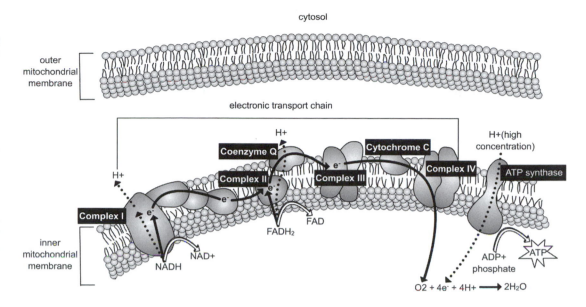

Figure 4.12. Electron transport chain in mitochondrial membrane. [Ricochet Productions]

chain. As the hydride ion is passed from one enzyme to another in the chain, energy is made available to power the formation of ATP from adenosine diphosphate (ADP) and inorganic phosphate. At the end of the electron transport chain, the hydride ion combines with oxygen and a proton to form a water molecule (see Figure 4.12).

The mitochondrial electron transport system consists of five main enzymatic components that participate serially to transfer the electrons and produce energy and water. Each of the enzymes is a complex aggregate of cytochromes (i.e., proteins with iron-containing heme groups) and various coenzymes. Although these proteins are present almost exclusively in the mitochondria, which we know to be autonomous in their replication, only 13 out of more than 90 known components are actually encoded in the mitochondrial DNA. The rest of the genes are scattered among the nuclear chromosomes.

1. *NADH:ubiquinone oxidoreductase*—Also called NADH coenzyme Q oxidoreductase or simply complex I, this enzyme system catalyzes the transfer of electrons from NADH (produced during the Krebs cycle) into the electron transport system. During this process, complex I also translocates protons, which helps to provide the electrochemical potential needed to produce ATP. It is a large complex, consisting of 43 separate protein subunits that make up three distinct chemical fractions: a flavoprotein fraction, an iron–sulfur protein fraction, and a hydrophobic protein fraction. Examples of genes for these subunits include the following:

• A gene on chromosome 11 for a 52 kDa enzyme called NADH dehydrogenase flavoprotein I. As indicated by its name, this protein is part of the flavoprotein fraction. It is believed to form the binding site for NADH and FMN (flavin group).

- *NDUFS1* on chromosome 2. This gene encodes the largest subunit of complex I, a 75 kDa component of the iron–sulfur fragment of the enzyme. It may form part of the active site where NADH is oxidized. The immediate acceptor of electrons for this enzyme is believed to be ubiquinone.

2. *Succinate dehydrogenase (complex II)*—This enzyme complex specifically oxidizes succinate to fumarate, providing a link between Krebs cycle intermediates and oxidative phosphorylation. Complex II contains four subunits and catalyzes the conversion of ubiquinone to ubiquinol. The genes for subunits of this system were described earlier (see step 6 of the Krebs cycle). Of the five enzyme complexes described in this section, this is the only one with no subunits encoded by the mitochondrial genome.

3. *Cytochrome c reductase (complex III)*—The third enzyme complex in the mitochondrial chain catalyzes the transfer of electrons from ubiquinol to cytochrome c, a small peripheral membrane protein found in the space between the inner and outer membranes of the mitochondria. This transfer is coupled with the translocation of protons across the mitochondrial inner membrane. Two of the subunits are ubiquinol–cytochrome c reductase core proteins I and II, encoded by the genes *UQCRC1* and *UQCRC2* on chromosomes 16 and 3, respectively.

4. *Cytochrome c oxidase (complex IV)*—The terminal enzyme in the electron transfer chain, this oxidase complex reduces oxygen to water and utilizes the excess energy to translocate protons across the mitochondrial membrane. It consists of several protein subunits. Interestingly, subunit VIII of cytochrome c oxidase maps to approximately the same chromosomal region as NADH:ubiquinone oxidoreductase.

5. *ATP synthase*—ATP synthase in the mitochondrial membrane catalyzes the synthesis of ATP from ADP and phosphate driven by a flux of protons across the membrane generated by electron transfer. One half of the 1997 Nobel Prize in Chemistry was given to P. D. Boyer (1918–   ) and J. E. Walker (1941–   ) for their work on the enzymatic mechanism of ATP synthesis.

The ATP synthase complex consists of 14 polypeptides that make up two major parts: a mostly water-soluble portion called F1, and a largely membrane-bound F0 portion. All but two are encoded by nuclear genes, which have remained highly conserved through evolution. The F1 segment provides the catalytic activity for the interconversion of ADP and ATP, whereas the F0 segment contains a proton channel that couples the proton gradient generated by electron transport to the F1 domain.

The F1 domain consists of five kinds of protein subunits—alpha, beta, gamma, delta, and epsilon. A gene on chromosome 12 labeled as *ATP5B* encodes the beta subunit, and the *ATP5C1* gene on chromosome 10 encodes the gamma subunit. Genes for different components of the F0 protein are found on chromosomes 2, 12, and 17. A soluble component of the F0 protein, called coupling factor 6 (F6), is required for the interaction between the F1 and F0 segments. The gene for this enzyme, called *ATP5J*, lies on chromosome 19.

See also: Lipid Metabolism, Mitochondrial Genes, Sugar Metabolism.

RIBOSOMES

Protein synthesis is such a fundamental property of all living organisms that its mechanisms are well-conserved throughout evolution from the bacteria to the most complex, multicellular beings. Even viruses, which have no protein-synthesizing machinery of their own, commandeer the components of their host cells to synthesize their (viral) proteins using the same mechanisms as in the rest of the living world. One of the primary pieces of this protein-synthesis equipment, one might even say the factory site, is the ribosome (see the entry in Chapter 2). Mammalian cells such as ours consist of two species of ribosomes: the larger kind that are typical of eukaryotes are scattered throughout the cytoplasm, and a smaller type, more similar to the prokaryotic ribosomes, are found sequestered within organelles such as the mitochondria. The differences in these structures are mainly structural, and the essential mechanisms of action of both eukaryotic and prokaryotic ribosomes are very similar. (By convention, the ribosomes, their subunits, and ribosomal RNA species are identified by their sedimentation coefficients (S) and this is how we refer to them in the rest of this discussion.)

See also: Protein, Ribonucleic Acid (RNA)—rRNA, Transcription, Translation (Ch. 2).

Eukaryotic Ribosomes

The cytoplasmic ribosomes in human cells consist of about 80 structurally distinct proteins and four types of RNA organized into two unevenly sized subunits: a larger 60S subunit and a smaller 40S subunit. The genes for the various ribosomal proteins (rps) are widely dispersed throughout the genome—at least 20 of the 22 autosomes, the X, and even the Y chromosome have genes for functional ribosomal proteins. Chromosome 19, with 12 rp genes, has the highest number in one location. With so many genes so widely and randomly scattered in the genome, it is no wonder that the identification and characterization of the functional genes for these proteins has been a very long, and indeed still ongoing, undertaking for genome researchers. Their task is complicated by the presence of more than 2,000 pseudogenes that are also distributed throughout the genome. Naturally, a discussion of all the rps is well beyond the scope of this volume, but we consider some representative examples.

The nomenclature of the ribosomal proteins is according to the subunit in which they are present—proteins of the 60S subunit are designated as RPL (for large), whereas the 40S proteins are called RPS (small). There are approximately 49 RPLs and 33 RPSs. An interesting example of a 60S gene is *RPL23a* on chromosome 17. This protein, which has been studied more extensively in yeast, binds to a specific site on the 26S ribosomal RNA molecule. In humans, it appears to be the access route for interferon to inhibit the growth of the cell, presumably by preventing protein synthesis. The first four introns (i.e., the noncoding portions of the gene that are spliced out prior to translation) in this gene code for sequences of small nucleolar RNAs (snRNA). These RNAs play important roles in the modification of the precursors to the ribosomal RNA molecules.

Another example of a large-subunit protein is RPL29 encoded on chromosome 3. In addition to its ribosomal functions of RNA-binding and protein synthesis, this protein is expressed on the cell surface, where it binds specifically to heparin. An example of a small-subunit component is RPS19, whose gene is on chromosome 19. Like the L29 protein, it also has an extraribosomal function. S19 appears to be involved in the differentiation and proliferation of red blood corpuscles, as evidenced by the fact that mutations in its gene cause Diamond–Blackfan anemia, a medical condition characterized by absent or decreased RBC precursors.

With just one or two exceptions, every ribosomal protein has just one gene, although each of these genes has multiple pseudogenes that are also processed. One exception is RPS4. This protein has two forms (isoforms) that are, interestingly enough, encoded on the X and Y chromosomes and designated as RPS4X and RPS4Y, respectively. Although these forms are not identical in structure, they have the same RNA-binding capabilities and functions. One gene can partially compensate for the loss of function if the other gene is mutated or absent.

rRNA

There are four RNA molecules in the cytoplasmic ribosomes in the cytoplasm of eukaryotic cells. The 40S subunit contains a single molecule (SSU rRNA) of about 1,900 bases in length with a sedimentation coefficient of 18S. The 60S subunit has one very large RNA molecule (28S with 4,700 nucleotides) and two smaller species—5S and 5.8S—with 120 and 160 bases, respectively. Although once believed to play mere supporting roles in protein synthesis, we now know that these rRNAs are actually ribozymes that carry out the catalytic functions of the ribosomes. For instance, the large-subunit (LSU) RNA is a ribozyme that catalyzes the formation of the chemical bond between the amino acid on a tRNA molecule (see the entry on translation in Chapter 2) and the growing peptide chain.

The human genome has anywhere from 150 to 300 copies of the genes that code for ribosomal RNA. This DNA, known as RNR (for ribosomal RNA), is present on the satellite regions of chromosomes 13, 14, 15, 21, and 22 (RNR1–RNR5). Gene transcription, which is initiated by the binding of RNA polymerase I to the promoter site, results in the production of a single large 45S precursor RNA. This precursor is cut into the four individual rRNA species (by certain small nucleolar RNAs) that then interact with appropriate proteins to form the ribosomal subunits.

The number of rRNA genes on a given chromosome is highly variable in the population. Although they are not identical, the genes on the different RNRs show a high degree of similarity to one another, which indicates that they did not evolve independently on each chromosome after some distant ancestral separation. Currently, scientists support a model wherein concerted evolution of these genes brought about recombination among the RNR regions of the different chromosomes.

Mitochondrial Ribosomes

The ribosomes in the mitochondria (also called the mitoribosomes) of our cells are distinct from those in the rest of the cytoplasm. They are more closely related in structure and RNA sequence to the prokaryotic ribosomes, but they are not identical to the latter either. With a sedimentation coefficient of 55S, the mitochondrial ribosomes are considerably smaller than either of the others. They are involved specifically in the translation of the 13 proteins encoded on the mitochondrial genome.

Like the other ribosome species, the 55S ribosomes have two subunits, each consisting of rRNA and protein. The larger, 39S subunit has about 48 proteins and a single 16S rRNA molecule containing 1,560 bases, and the smaller subunit (28S) has 29 proteins and a 12S rRNA molecule with only 950 bases. Interestingly enough, the mitoribosomal proteins, although homologous to bacterial ribosomal proteins, are encoded by nuclear and not mitochondrial DNA. These genes are widely distributed all over the genome. The proteins are synthesized in the cytoplasm and then imported into the mitochondria, where they combine with the rRNAs to form the ribosomes. The system of nomenclature is the same as in the case of the cytoplasmic ribosomes: MRPL for the large-subunit proteins and MRPS for the smaller subunit. A representative example is MRPS12. This protein is expressed by a gene on chromosome 19 and appears to control the ribosome's decoding fidelity as well as its susceptibility to aminoglycoside antibiotics. Scientists think that this gene, together with the gene for mitochondrial seryl-tRNA synthetase (located upstream and adjacent to *MRPS12*), may be a candidate for autosomal deafness.

Only the genes for mitochondrial rRNAs reside on the mitochondrial plasmid itself: The 16S rRNA is encoded by nucleotides 1671–3229 and the 12S species by nucleotides 648–1601. These RNAs as well as the tRNAs encoded in the mitochondrial genome participate exclusively in mitochondrial protein synthesis.

See also: Mitochondria, Ribonucleic Acid (RNA)—Transfer RNA (tRNA) (Ch. 2); Protein Synthesis: The Translation Apparatus.

THE SENSES

The ability to perceive and respond uniquely to different stimuli is one of the characteristics that distinguishes living beings from nonliving objects. The stimuli we receive come in myriad forms—heat, light, sound, and chemicals are just a few. Some of these are so familiar that we take them for granted; can you even imagine life without the sights, sounds, smells, and tastes of various things in the world around us? Other stimuli may be more subtle, such as the signals emanated and felt by the release of pheromones, but they exist nevertheless. In this section, we look at the genes, proteins, and structures with which we perceive and make sense of the world around us.

A quick caveat before we begin: The perception of stimuli takes place at many levels, particularly in complex multicellular organisms. Here we look primarily at the first level of perception, the reception of the raw data, as it were. The mechanisms by which we process this information (i.e., transmit the messages to the brain or respond to the stimuli) are not discussed in this section. Let it suffice to say that mechanisms and internal systems such as the signal transduction pathways and ion channels at the cellular and molecular levels and the endocrine, circulatory, and nervous systems at the level of the organs and tissues play a role in these processes. Most of these systems have been covered in other sections of this chapter.

Sight/Vision

Conduct a little experiment. Pick any object and ask a friend to describe it to you. Chances are that he or she begins with visual descriptions (i.e., the color, shape, and size of the object) before mentioning other attributes such as its texture, hardness, taste (should it be edible), or smell. This is because our vision—the ability to "see" shapes, sizes, and colors, distinguish between light and dark, and even be reading this book—is probably the sense that a majority of humans most take for granted, unless of course they are blind.

The organs we use for this most basic of perceptions are, as we all know, the eyes. Like little cameras, our eyes capture images of various things in our surroundings and send these pictures to the brain, which then processes the information so that we can "see" them and recognize or wonder at them. And, just like cameras, the eyes are complex structures with many different intricate parts that work together to make the images. A complete discussion of all these parts and their functions is beyond the scope of this volume, but the principal actors involved in making images are discussed.

The Eye Lens—Crystallins

Lenses are an integral part of any light-focusing system—consider cameras, magnifying glasses, and microscopes, for example—and the eyes are no exception. Optical lenses are transparent spherical or convex objects whose function is to focus the light rays from different sources to form images of these objects. But whereas the lenses of various man-made optical devices are made of materials such as glass or plastic, eye lenses are made of living cells. A thin layer of epithelial cells lines the outer rim of the lens. The body is composed of bundles of transparent fiberlike cells that lose their nuclei and become filled with special transparent proteins called the *crystallins*, which function as the transparent, "glassy" substance that focuses light. About 80–90 percent of the lens is made up of crystallin.

The crystallins found in the human eye belong to a large family of proteins that includes a number of metabolic enzymes as well as stress proteins (the heat shock proteins—see the entry on housekeeping genes) whose genes are widely distributed throughout the genome. In fact, due to a curious phenomenon known as "gene sharing" or "gene recruitment," the proteins that function as the transparent crystallins in the eye lens are actually encoded by the same genes that encode the enzymes and stress proteins in other tissues. By changing the pattern and location of gene expression, as well as the concentration of proteins, gene sharing enables a single protein to play different roles in different tissues.

Examples of these versatile crystalline genes include the *CRYAA* and *CRYAB* genes on chromosomes 21 and 11, which encode proteins that function as subunits of crystallin in the lens and as heat-shock proteins in other tissues. These proteins also have kinase activity. The *ENO1* gene on chromosome 1 produces an enolase enzyme that functions in sugar metabolism in the cytoplasm of most cells. However, it also produces tau-crystallin in the lens tissue. Mutations in these as well as other crystalline genes are often associated with the development of cataracts and other problems of vision relating to the ability to focus.

Just as the light entering a camera is guarded by a system of shutters that adjusts the aperture of the lens and the time period for which it is exposed to light in order to control the images, so is the eye lens. Very briefly, the pigmented iris forms a barrier that prevents the indiscriminate reception of light, and it is only through a narrow inlet known as the pupil that light actually makes its way to the lens. An intricate system of small muscles controls the size of the pupil, thereby regulating the amount of light it receives.

See also: Housekeeping Genes—Heat Shock Proteins.

Retina/Photoreceptors

Focusing the light from various objects in our surroundings to form images is the first step in visual perception. The next essential step is the reception of this image on a screen. In a camera, this screen is provided by the film, which has special photosensitive pigments that change when exposed to light and records the images. In the eyes, the *retina* forms this screen or film, capturing the image focused by the lens and transmitting it via the optic nerve to the brain, which then processes this information. The retina is composed of a single layer of light-sensitive cells coating the inner layer of the eyeball. Special pigment-bearing receptors located in the cell membranes of the retinal cells absorb the light from the images and initiate signal transduction pathways, which form the images that are relayed to the brain. In this section, we take a brief look at some examples of the genes that control the pigments involved in vision.

Two major properties of light—brightness and color—contribute toward the formation of images. The eye uses different populations of retinal cells to receive information regarding these two properties, which it then processes and coordinates in order to perceive not only these attributes but also other information such as depth or distance. The cells that receive information about the relative brightness or darkness of objects are called *rods*. More abundant around the periphery of the retina, the rods are the primary mediators of vision under low levels of light (in dim or dark conditions). The activity of rod cells is the reason we are able to perceive shapes of objects, without necessarily knowing about their colors, in dim or near-dark conditions. As we all know from experience, we perceive colors better when the light is bright. This is because the cells that are responsible for color perception—known as the *cones*—are more active when light levels are high. Cones are present in higher density toward the center of the retina.

Have you ever heard the adage about carrots being good for the eyes? Well, there's some truth to it. This is because carrots contain large quantities of carotenoid pigments (or vitamin A), which play an important role in dim light absorption. A derivative of vitamin A known as retinal is the actual site of light absorption in the rod cells. Retinal is held in place by a protein called *rhodopsin* that is present in the membranes of the rod cells. Rhodopsin is a G-protein-coupled receptor (see the entry on signal transduction in this chapter) encoded by the *RHO* gene on chromosome 3 that initiates the signal transduction pathway ultimately responsible for dim light (495 nm wavelength) perception by the eyes. When retinal absorbs light energy, it undergoes changes in its structure and can no longer fit inside its place on rhodopsin. The resulting dissociation causes conformational changes in the intracellular domains of the protein, which triggers the signaling cascade. Mutations in the *RHO* gene may cause retinitis pigmentosa (see the entry in Chapter 5) or night blindness. In addition, mutations in genes involved in the synthesis and metabolism of vitamin A also cause vision defects.

Color vision is somewhat more complex because it involves more than a single type of pigment or light receptor. Human vision is trichromatic, which means that the human eye can distinguish among three different colors. The reason for this ability is that we have three different populations of cone cells, each of which contains receptors for different wavelengths of light. The three colors (or more accurately ranges of color) detected by our cones are blue, red, and green. These colors are known as the primary colors. The ability to see other colors and nuances in shade is a result of comparing signals from different cones. Beneath the pigmented epithelium of the retina lies a layer of cells called the opponent cells. Their function is to compute the signals received by different types of cones and send this information to the brain so that differences are perceived. Genes for the receptors—known as the *opsins*—of the different cones are found on different chromosomes. Genes for the opsins that absorb red and green light lie adjacent to one another on the X chromosome, whereas the gene for the blue cone opsin is present on chromosome 7. Mutations in one or more of these genes lead to defects in color vision, such as red/green color blindness, and in very rare cases to a total loss of ability to discern color.

In addition to the images we see, our eyes also receive a certain amount of excess light. This excess must be dealt with if images are to form properly (think of overexposed photographs). So, in addition to the rods and cones, the retina has cells whose main function is to absorb the unused light and divert its energy. An example of a protein that may be involved in this function is the product of the *RPE65* gene on chromosome 1. This gene is expressed specifically in cells of the retinal epithelium, where it produces a receptor for retinal-binding proteins. Mutations in this gene have been implicated in certain vision defects, including retinitis pigmentosa.

Hearing

The ability to hear is arguably at least as important for our interaction with our surroundings as the sense of sight. Like vision, hearing is something most of us take for granted and rely on in ways we may not even be aware of.

Although both light and sound are forms of energy that we perceive in our environment, there are some fundamental differences in their nature that affect the way in which we perceive them. Light is a form of electromagnetic energy. Sound, on the other hand, is a form of mechanical energy. It is transmitted in the form of waves, which are made by the vibrations of particles in air, water, and even solid surfaces. Sounds audible to humans fall within a frequency range of 16 to 20,000 Hertz (wave cycles per second), although the range varies from one individual to another and may narrow as a person gets older. Many animals have better hearing than humans—dogs, for example, can hear sounds at frequencies that are higher than what we can hear.

The ears are the organs responsible for receiving sound vibrations and transmitting their messages to the brain. They are elaborate and complex organs with many different parts whose function is to transmit information about sound, such as loudness and pitch, from our environment to the brain. The recognition of language or music is a higher-order function that is the result of how the brain processes sound and is not related to the function of the sound receptors.

The pinnae, or external ears, funnel or concentrate the sound waves into the auditory canal of the outer ear. Sound waves pass through this canal and strike the eardrum, causing it to vibrate and transmit the vibrations across the tiny linked bones of the middle ear (called the hammer, anvil, and stirrup), which also act as amplifiers for the sound. The stirrup then transmits the vibrations through a flexible membrane known as the oval window to a compartment of the inner ear known as the cochlea. The cochlea is a snail-like coiled tube of about 3.5 cm that is divided into two compartments and filled with a special potassium-rich fluid called the endolymph. As the vibrations pass through the endolymph, they are transmitted to sound-sensitive hair cells that line the inner chamber of the cochlea. The hair cells form the site in the body where the message received from the vibrations is transformed into a nervous impulse that is then transmitted to the brain via the auditory nerve. The vibrations of the endolymph are transmitted via special extensions of the hair cells known as the *stereocilia*. As the hair cells receive vibrations via their stereocilia, their potassium channels are opened, which leads to the influx of these ions into the cells. This triggers the nervous

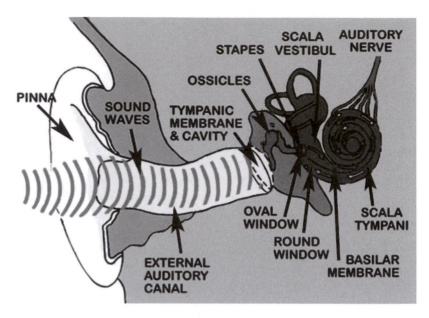

Sound transmission in the human ear. [© Educational Images/Custom Medical Stock Photo]

impulse that is transmitted to the auditory nerve (see the accompanying photograph).

Much of the early pathway of sound is based on the movement of vibrations across specially formed structures such as the membranous eardrum and the tiny bones of the middle ear. The proper formation of these structures is not related to specific proteins but to regulatory genes that are expressed in a tissue-specific manner during embryonic development and make sure that the various parts of the ears develop properly. Examples of such developmental genes include the *DFN* genes that are found on the X chromosome. Labeled as *DFN-1, -2, -3*, and so on, these genes have been found to be associated with various forms of congenital X-linked deafness when mutated. *DFN1* encodes a homolog of an intracellular trafficking protein in yeast that transports various metabolites from the cytoplasm to the mitochondrial membrane. The association of this gene with deafness indicates that it might perform a similar function, specifically in the cells of the developing ear. *DFN3*, more commonly labeled as *POU3F4*, makes a transcription factor, which is involved in the development of the stirrup bones of the middle ear.

Scientists have also identified genes that are expressed specifically in the hair cells of the inner ear and seem to regulate hearing and deafness. Of particular importance are the genes involved in organizing the stereocilia. Examples include a gene on chromosome 11 for a type of myosin (muscle protein) known as Myo7A that is expressed specifically in the inner ear and the *DIAPH1* gene on chromosome 5. The latter is a homolog of a *Drosophila* gene called *diaphanous*, whose gene product is involved in actin polymerization. Scientists think that the *DIAPH1* gene product plays a similar role in the hair cells. Mutations of the gene have been linked to an autosomal dominant form of deafness. *CDH23* on chromosome 10 makes a calcium-dependent adhesion protein called cadherin 23 that coordinates hair cells. Still another gene, called *TMIE*, on chromosome 3 encodes a transmembrane protein seen exclusively in the inner ear and lies in close proximity to a locus that is associated with a recessive form of deafness.

Touch

Touch refers to our ability to perceive sensations of temperature (heat and cold), pressure, friction, and pain. Nerve endings distributed throughout the body, but especially abundant at the fingertips, tongue, and sexual organs, are the receptors of these sensory perceptions. Hairs on the skin magnify the ability of the skin to experience these sensations. The nerves involved in these sensations are called the sensory nerves and are distinct from the motor nerves, which control the movement of the muscles.

The transmission of various sensations appears to depend on the activation of different ion-channel-like proteins in the sensory nerves. The *TRPV3* (transient receptor potential) gene on chromosome 17, for example, encodes a cation channel protein that is activated between 22 and 40°C and seems to plays a role in the sensation of warmth. A second gene, called *TRPV1*, that lies in close proximity to *TRPV3* encodes a receptor for a chemical called capsaicin, the ingredient in chili peppers that triggers the burning sensation in the mouth when we eat them.

Smell (Olfaction) and Taste

The senses of sight, hearing, and even touch are perceptions of different forms of energy (i.e., light, sound, and temperature (heat) and pressure, respectively). In contrast, the senses of smell and taste respond to chemical stimuli. The reception of these stimuli is largely a matter of initiating specific signal transduction pathways in the cells receiving the chemical signals.

The nose is the organ responsible for receiving sensations of smell. Its cavity is lined with mucus membranes that receive signals from various volatile chemicals via receptors and transmit this information to the brain via the olfactory nerves. Smell or olfactory receptors appear to be sensitive to seven distinct types of signals. These include both smells that we identify with distinct substances such as camphor, mint, and musk, as well as more general sensations such as acrid (burning) and putrid (rotting) odors. Olfactory receptors are members of the family of G-protein-coupled signal transduction proteins and constitute a large family of proteins whose genes are scattered widely throughout the genome. Examples include olfactory receptor (*OR*) genes such as *OR1D2* on chromosome 17 and *OR2H3* on chromosome 6. See the entry on food and digestion for examples of genes and receptors involved in taste sensations.

See also: Signal Transduction.

SEX DETERMINATION

Considering the many differences between men and women, it is pretty amazing that the main thing that determines gender is the presence or absence of one half of a single chromosome pair out of the 23—namely, the Y chromosome. If a fetus has a Y chromosome and an X chromosome it will become a boy, and if it has two X chromosomes it will become a girl. What is it about this miniature chromosome that makes all the difference? How does the Y chromosome impart the property of "maleness" to a fetus?

The route to sex determination is not always as clear and simple as the difference between X and Y. Scientists have found that the answers lie at the level of development and gene regulation. True, the Y chromosome is involved, but it acts by the switching on or off of particular genes at the right time during development and not merely by encoding for the phenotypic differences of males versus females.

The search for the sex determination genes has closed in on a region of the Y chromosome called *SRY* (*S*ex-determining *R*egion on the *Y* chromosome). Identified in 1990, this gene is the master switch that triggers the events that convert the embryo into a male. It acts as a transcriptional activator binding to other genes on the Y chromosome, initiating male sex determination and regulating male development. *SRY* encodes a member of the "High Mobility Group" (HMG) protein family whose members are characterized by an 80 amino acid DNA-binding motif called the HMG domain. This family of proteins includes many transcription factors. The SRY protein's target sequence has been identified in the promoter region of genes controlling sexual differentiation such as P450 aromatase, an enzyme that converts testosterone to estradiol. Among other genes that it activates are *Sox9* and *Dax1* (a member of the nuclear hormone receptor superfamily and located, rather surprisingly, on the X chromosome). Mutations in *Sox9*, on the Y chromosome, lead to "XY sex reversal" in humans. *Sox9* is up-regulated in developing testes shortly after the onset of *SRY* expression but down-regulated in developing ovaries. Like *SRY*, these downstream genes play important roles in sex determination.

The human fetus initially has the basics for both the male and female accessory sex organs, the Wolffian and Müllerian ducts, respectively. *SRY* begins its action when the embryo is about 6 weeks old, taking the first step toward the formation of the testes. Once the testes are formed, they secrete two factors that masculinize the fetus. The first factor, secreted by testicular cells called the Leydig cells, is the hormone *testosterone*, which induces the Wolffian ducts to differentiate into the epididymis, vas deferens, and seminal vesicles. Testosterone is metabolized into dihydrotestosterone, which in turn induces the development of male external genitalia. The second factor, secreted by Sertoli cells in the testes, is called the *anti-Müllerian hormone* (AMH) and induces the Müllerian ducts to regress. AMH is a dimeric glycoprotein, and its gene has been mapped to chromosome 19. In the female fetus, the absence of these two factors causes the Wolffian ducts to degenerate and the Müllerian ducts to develop into female reproductive organs.

So it turns out that our default program is female, and we would all be female if it were not for *SRY* and associated male development genes. Studies of people with Klinefelter or Turner syndromes (XXY males and XO females—i.e., females with only one X chromosome—respectively; see Chapter 5 for details) also confirm that, in general, the presence of a Y chromosome is necessary for male sexual characteristics to develop. The number of X chromosomes present does not seem to play a significant role in sex determination (i.e., if a normal Y chromosome is present, it will overrule any number of X chromosomes and induce maleness). This is because *SRY* is typically present only on a complete Y chromosome.

There are some rare exceptions, however. For example, a fetus with two X chromosomes might develop male instead of female characteristics. In these cases, *SRY* seems to be present and active, even though the Y chromosome is clearly absent. Somehow, through a process of recombination, the *SRY* region gets transposed to the X chromosome in the sperm. Fertilization then produces an XX embryo but one that develops into a male fetus because *SRY* directs it to develop testes. These XX males tend to be infertile because although they express male hormones and appear male, they nevertheless lack the remaining Y chromosome genes that are essential for proper development of the male reproductive system. Alternatively, some XY individuals may appear to be normal females at birth. This happens when there are mutations in male development genes that interfere with the normal development of male characteristics. At puberty, it becomes apparent that these individuals are not normal females because they do not develop secondary sexual characteristics, do not menstruate, and their ovaries do not develop eggs. This is known as "*XY reversal*." At the 1996 Atlanta Olympics, a test based on a molecular probe for *SRY* was used to ensure that woman athletes did not have this masculinizing gene. However, many XY females seem not to have *SRY* mutations, which led researchers to investigate mutations in other genes, such as *DAX1*.

There are several other conditions in which sex differentiation is disrupted where the culprits appear to be mutations in developmental genes downstream from *SRY*. For example, in 5-alpha reductase deficiency, conversion of testosterone to dihydrotestosterone (DHT) is disrupted. Although individuals with this deficiency possess testes, the external genitalia remain small and underdeveloped. Dihydrotestosterone fits perfectly into the androgen receptor on developing cells and is the most potent androgen. It is responsible for the development of external male genital anatomy in the fetus. The gene for the isoform of 5-alpha reductase that is involved in this disorder is found on chromosome 2. More than 20 different mutations of this gene have been reported.

See also: Hormones and the Endocrine System; Klinefelter Syndrome, Turner Syndrome (Ch. 5).

SIGNAL TRANSDUCTION

Signal transduction is the process by which cells and tissues in a large, multicellular organism communicate with one another and with their external environment. An organism has several pathways of signal transduction to transmit signals among its different tissues and organs. There is one signal transduction pathway, for instance, that tells a cell in a developing organ that it is time to stop growing because it is crowding its neighbor and a different one that enables a B cell to proliferate and produce specific antibodies in an immune response. The body's ability to respond to different sensory stimuli—tastes and odors, temperature, light, and pressure—is also due to the transmission of these signals into the appropriate cells. In fact, it is hard to imagine the functioning of most of our organs and organ systems without the coordinating work of signal transduction.

A good strategy for understanding the basic mechanism is to think of the signal or message as the baton or torch in a relay race that is passed along from runner to runner until the finish line. The changing over of the signal or baton is the trigger that tells one runner to start running and the other one to stop. In other words, a molecule is turned "on" when it receives the signal and must be turned "off" after it passes the signal on to the next messenger in the series. An example of a typical mechanism for switching signaling molecules on and off in cells is the addition and removal of high-energy phosphate groups (i.e., phosphorylation and dephosphorylation), respectively.

Figure 4.13 shows the key steps involved in signal transduction. The first step is the reception of an external signal by the cell, usually through a specific receptor for the signal. Examples of these signals include hormones, growth factors, molecules on the surfaces of other cells (e.g., MHC markers), tastes and odors, light, temperature, and pressure. Note that this external signal is not the physical signal that is transmitted into the cell. Instead, each signal (or a special family of signals) excites a specific series of players into action. The ultimate response of the cell depends on which team members were chosen.

Once the receptor receives its signal, it amplifies this signal to ensure that it is "heard" inside the cell. The amplified signal is then transmitted to the appropriate target within the cell, where it "tells" the cell how to respond. The final step is quite often the activation of one or more genes by a transcription factor. Depending on the nature of the signal, the transcription factor might activate a single specific gene or a group of genes. This in turn determines the way in which a cell responds to the signal (i.e., by the production of a single specific protein or in a more generalized manner by stepping up its growth, for example, or initiating DNA replication and hence cell division). Other responses include the production of metabolic energy, cell locomotion, and other changes in cell structure, such as the opening of membrane channels.

In the following section, we discuss some representative examples of members involved in key steps in signal transduction pathways.

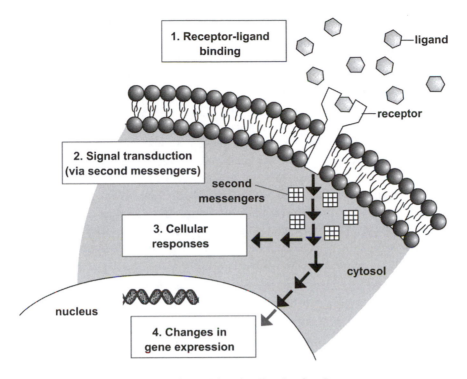

Figure 4.13. Signal transduction. [Ricochet Productions]

Receptors

As mentioned, the receptor molecule is the first point of contact between a cell and the signal it receives. Based on their mode of action, there are three main classes of signal-transducing receptors.

Receptors in the Plasma Membrane That Have Intrinsic Enzymatic Activity

These receptors are found embedded in the cell membrane and typically consist of three distinct portions or domains: an extracellular part that binds to the signal, a transmembrane domain that spans the membrane, and a portion that is exposed to the internal surface of the cell. This internal portion comprises the enzymatic portion of the protein. When the external signal (e.g., a hormone) binds to the portion of the receptor molecule that is exposed on the membrane, it changes the shape of the entire molecule. This causes the activation (turning "on") of the enzymatic part of the molecule. The activated enzyme then acts on the substrates inside the cell and passes the signal on to its next destination. Different receptors have different types of enzymatic activities. Examples of receptor-linked enzymes include protein kinases, tyrosine phosphatases, and guanylate cyclases. Each receptor enzyme has specific intracellular substrates, and this specificity is the basis for choosing the particular pathway along which a signal is transmitted.

1. *Protein kinases* are enzymes that catalyze the transfer of high-energy phosphate groups to proteins using ATP (or sometimes GTP) as the source. They can also catalyze reactions in the reverse direction—namely the removal of a phosphate from a protein to ADP or GDP. The kinases are classified into subcategories according to the specific amino acid residues that they phosphorylate on a protein, so *tyrosine kinases*, phosphorylate tyrosine residues, and *serine kinases* transfer phosphates to serine. The number of receptors with protein kinase activity is very large indeed—the receptor tyrosine kinases (RTKs) alone are classified into at least 14 families based upon structural features in their extracellular portions. Examples include receptors for insulin (the *INSR* gene on chromosome 19) and for different fibroblast growth factors on chromosomes 8 (FGFR1), 10 (FGFR2), and 4 (FGFR3), for example. The receptor for a growth factor called activin encoded by a gene on chromosome 2 has intracellular serine/threonine kinase activity.

2. *Tyrosine phosphatases* are enzymes that can add or remove phosphate groups from tyrosine residues of proteins without the help of ATP using inorganic phosphates as the source. A gene called *PTPRC* on chromosome 1 encodes one such receptor molecule that is expressed specifically on blood cell lineages and appears to be an essential component in B antigen receptor signaling pathways.

3. *Guanylate cyclases* catalyze the removal of a diphosphate group from GTP molecules to form cyclic GMP, which then acts as a means of amplifying the message communicated to the cell via the receptor. Receptors for certain molecules, called the natriuretic peptides (which play a role in maintaining the function of the heart muscles), have this type of enzyme activity. Genes for these peptide receptors are present on chromosomes 1, 9, and 17.

Receptors That Are Coupled or Linked Inside the Cell to GTP-Binding Proteins

These receptors are also called the G-protein-coupled receptors (GPCRs). They are membrane-bound receptors like those just discussed but unlike them have no catalytic activity of their own. Instead, the part of the GPCRs inside the cell is linked to G-proteins that serve as the switches that turn the signals off and on. Well over 1,000 different GPCRs have been identified and cloned, although the specific signal or ligand whose messages they transmit has not yet been identified for many of them. There are three different categories of these receptors, depending on the specific type of signal pathways in which they participate:

1. GPCRs that modulate adenylate cyclase activity: These GPCRs activate an enzyme that produces cyclic AMP (cAMP), which acts as the amplifier of the external signal inside the cell. Receptors of this class include the beta-adrenergic, glucagon, and odor molecule receptors. In the case

of the beta-adrenergic (e.g., on chromosomes 11, 19, and 22) and glucagon receptors (on chromosome 17), cAMP leads to an increase in the activity of protein kinase A. In the case of odorant molecule receptors, it causes the activation of ion channels.

2. GPCRs that activate the enzyme phospholipase C: The phospholipases are enzymes involved in the metabolism of phospholipids, which are important components of various membranes. The activation of phospholipase C by various GPCRs leads to the hydrolysis of phospholipids, which generates free molecules of fatty acid building blocks such as diacylglycerol and inositoltrisphosphate within the cell. These molecules then serve as "second messengers" that continue the transmission of the signals. Examples of GPCRs that activate phospholipase C include receptors for angiotensin (*AGTR1*) on chromosome 3, bradykinin (*BDKRB1*) on chromosome 14, and vasopressin on chromosome 12.

3. Photoreceptors: This class is coupled to a G-protein called transducin that activates a phosphodiesterase enzyme, which in turn causes a decrease in the level of cGMP in the cell. The drop in cGMP then results in the closing of certain ion channels. Retinal cells contain a membrane receptor called *phosducin*, encoded by a gene on chromosome 1, which is coupled to transducin and acts as described.

Intracellular Receptors

These receptors are inside the cells and usually receive signals in the guise of small molecules that can enter the cells without binding to any receptor molecule, for example via diffusion or through pores in the membrane. The receptor forms a complex with the molecule (called a ligand) and migrates to the nucleus, where it can directly affect gene transcription without help from other enzymes. In other words, these receptors transduce their signals single-handedly. Steroid hormones such as the sex hormones transmit their signals via such receptors (e.g., receptors for estrogen (*ESR1* on chromosome 6) and androgen receptors on the X chromosome).

Second Messengers and Signal Amplifiers

The message initiated by the reception of the signal by the receptor must be relayed within the cell to its ultimate destination, the genes. Although there are a few receptors that perform this task themselves (e.g., the steroid receptors), most signals are transmitted, as we said, via a series of messengers. Messengers that act as the go-betweens from the receptor to the nucleus are often called the "second messengers." In reality, the "second" messenger may actually be a series or chain of messengers. The amplification of the external signals is an important feature of many, but not all, signal cascades. As we saw in the previous section, several receptors respond to the stimulus of a signal by producing many molecules of cAMP or cGMP. Each one of these small molecules then acts to activate the next member of the signal-transducing cascade, which means that the information about the single event of a ligand binding to a receptor is multiplied or magnified by several factors. As an example, a single molecule of cAMP activates a protein kinase (protein kinase A) that has several catalytic subunits. Each one of these subunits can catalyze the phosphorylation of many proteins.

G-Proteins

G-proteins (or GTP-binding proteins, to use their full name) are enzymatic proteins that are coupled or attached to the cytoplasmic portion of the GPCRs. In fact, the enzymatic activities that we described as the consequences of GPCRs actually reside within these proteins. Regardless of the diversity of results, the basic structural organization of the G-proteins as well as the mechanism of their activation is the same. It consists of three subunits: an active (enzymatic) alpha subunit and two passive bits called beta and gamma subunits.

When an external signal (e.g., a hormone such as adrenaline) binds to the GPCR on the cell surface, it causes the intracellular portion of the molecule to change its shape slightly. (Think about a pair of pliers or scissors with the hinge as the part that is embedded in the membrane. When someone grasps the handles on the outside and pulls them together, the blades on the inside move apart). When this happens, the alpha subunit of the G-protein drops the GDP it was holding onto and binds GTP instead. Charged with this high-energy molecule, it splits off from the beta and gamma subunits and finds a target enzyme to activate. For example, it may attach to a channel protein and activate it to open its pore to allow the flow of ions into the cell. Once it has activated its target, the GTP is hydrolyzed to GDP, which inactivates the alpha subunit, and it then returns "home" to the beta and gamma subunits so that the whole process can start again.

As mentioned, the G-protein-linked second-messenger pathway often acts as an amplifier. As soon as the activated alpha subunit has diffused away from the receptor, another identical but inactive subunit can take its place and get activated in turn. As long as the external messenger remains bound to the receptor, a steady stream of activated alpha subunits may be produced. A second amplification step can occur at the level of the target enzyme if the target enzyme remains activated long enough to manufacture more than one molecule of the second messenger.

How does this work in real life? Let us consider the example of the toxin liberated by *Vibrio cholerae*, the bacterium that causes cholera. When this bacterium enters the human GI tract, it releases its toxin, which binds to G-protein-coupled receptors on the surface of cells lining the small intestine. The binding of cholera toxin turns on the alpha subunit of a G-protein (Gs). The activated Gs in turn activate the enzyme adenylate cyclase, which catalyzes the synthesis of cAMP, an important chemical messenger. The continuous high levels of cAMP cause a massive loss of salts from the cells of the intestine. To maintain the balance, water moves out of the cells into the intestine. The result is the diarrhea typical of cholera, which can be fatal if the salts and water are not quickly replaced.

Kinases

G-proteins do not always directly activate the final target protein or enzyme. In fact, they often play the role of the go-between in the cascade of events that is initiated by the binding of a signal to its receptor. In many instances, they pass the message on to intracellular protein kinases, which then activate the transcription factors that cause the desired final effect.

The kinases, as we described earlier, are enzymes that mediate the transfer of high-energy phosphate groups between their substrates and nucleotide triphosphate (ATP or GTP) molecules. The substrates for protein kinases are proteins. Most protein kinases are specialized for a single kind of amino acid residue, usually at a specific position of the protein, and are further categorized into tyrosine kinases and serine–threonine kinases, according to the amino acids that they phosphorylate. There are, however, some protein kinases that show dual activity and phosphorylate two different kinds of amino acids. The ability to transfer phosphate groups enables these proteins to activate different proteins, and hence they play important roles in signal transduction, both at the membrane (as we saw already) and within the cell.

An example of an important intracellular protein kinase is protein kinase A, or cAMP-dependent protein kinase. This protein is part of the signal transduction pathway that promotes the break down of fat in the cell during its growth. This 80 kDa enzyme is activated by amplifier molecules such as diacylglycerol and phospholipid and is thought to undergo a conformational change upon binding to the membrane. Protein kinase C phosphorylates a variety of target proteins that control growth and cellular differentiation, such as MAP kinase, transcription factor inhibitor IXB, the vitamin D3 receptor VDR, and the epidermal growth factor (EGF) receptor. At least 12 members of the protein kinase C family have been identified in mammals due to their high sequence homology. Although the genomics of this family of genes has not been worked out completely, some important loci have been identified for protein kinase C, including those on chromosomes 1, 2, 3, 17, and 19.

Another group of kinases, called the mitogen-activated protein (MAP) kinases, were identified by virtue of their activation in response to growth factor stimulation of cells in culture. These kinases are also called ERKs, for extracellular-signal regulated kinases. Mitogen-activated protein kinase activation was first observed in response to activation of the EGF, PDGF, and insulin receptors and other cellular stimuli such as T cell activation. They go on to transmit their signals to transcriptional regulators such as serum response factor (SRF) and the proto-oncogenes *Fos*, *Myc*, and *Jun*. A representative example is the protein product of the gene called *MAP2K3* on chromosome 17. This protein is a dual-action kinase (mitogen-activated protein kinase) that can be activated when insulin binds to its receptor (INSR). Activated MAP2K3 phosphorylates yet another kinase, called MAP14 (the gene for which is on chromosome 6), which has several substrates, among them the transcription factor that activates the glucose transporters.

G-proteins and kinases are only two examples of intracellular proteins that participate in signal transduction. A protein kinase such as those described often activates more proteins such as Ras (see the entry on oncogenes and proto-oncogenes) or still other kinases to continue the cascade. Each messenger or protein is therefore a potential point of divergence of the path along which a signal travels in the cell.

Transcriptional Activation

The final recipient of a signal is most often (but not always) a gene or cluster of genes that are activated to begin producing their gene product. When the signal is a growth factor, for instance, the activated transcription factors stimulate the expression of the genes involved in cell growth and development. The binding of insulin or an antigen to their respective receptors results in the transcription of genes for glucose transporters and immunoglobulins, respectively. Often a single signal may trigger more than one transcription factor and hence the expression of more than one gene or gene cluster. Insulin, for example, also stimulates a cell to higher metabolic activity—in other words, it stimulates both the uptake of sugar (via glucose transporters) and its usage by the cell. In addition, certain signaling pathways involve the activation of new receptors or membrane proteins such as ion channels rather than the nucleus (e.g., odorants).

As we can see from this description, signal transduction constitutes an extremely complicated network of both inter- and intracellular communications, as well as communications between cells and their environment. Different pathways may converge (the same kinase or transcription factor might be activated by different messengers, for example) and still others might diverge when a single kinase activates more than one substrate. Because different signal proteins are involved in many of the fundamental aspects of cellular life, defects in the proteins have the potential to do great harm. Many cancers, for instance, are the result of mutations in the genes of one or another signal protein (see the entry on oncogenes and proto-oncogenes in this chapter and the entry on cancer in Chapter 5). Many endocrine disorders, such as diabetes, are the result of mutations in receptors that interfere with the ability of the signal to deliver its messages to the correct cells (see Table 4.3).

The growing recognition of the fundamental importance of signaling pathways is also evident in the number of recent Nobel Prizes in Physiology and Medicine awarded to scientists whose contributions have directly or indirectly illuminated aspects of signal transduction. The first "signal-related" prize went to Stanley Cohen (1922–   ) and Rita Levi-Montalcini (1909–   ) in 1986 for their discoveries of nerve growth factors, which, as we know, initiate certain signaling pathways. Other Nobels have been more explicitly for signal transduction, such as the 1994 award to Alfred Gilman (1941–   ) and Martin Rodbell (1925–1998) for "their discovery

Table 4.3. Signal transduction: examples of signal initiation and final effects

Signal/ ligand	Nature of signal	Receptor	Receptor type	Intermediary messengers (amplifiers, second messengers, etc.)	Final effect
Insulin	Hormone	Insulin receptor	Receptor tyrosine kinase	Several intermediary messengers—final messenger is a molecule called TC10	TC10 activates a glucose transporter, which fuses with the cell membrane to increase glucose uptake by cells
Fibroblast growth factor 2 (FGF-2)	Growth factor	FGFR	Receptor tyrosine kinase	SOS → Ras → ERK	Growth and proliferation of fibroblasts
Atrial natriuretic peptide	Peptide (produced by muscle cells in the heart)	NPR1	Guanylate cyclase	Cyclic GMP → Protein kinases (PKG) → activate transcription factors and other effectors	1. Vasorelaxation 2. Inhibition of aldosterone secretion 3. Inhibition of rennin secretion by kidneys
Transform-ing growth factor	Growth factor	TGF-beta receptor	Serine/ threonine kinase	Multiple messengers	1. Proliferation of cells 2. Migration of cells 3. Activation of *c-myc* oncogene
Adrenalin	Hormone	beta-2-adrenergic receptor	G-protein-coupled receptor	Adenylate cyclase → cyclic AMP → kinases → transcription factors, etc.	Muscle cells convert glycogen to glucose and are thus energetic and ready for fight-or-flight response
Estrogen	Steroid hormone	ESR1	Intracellular ligand-activated transcription factor	No intermediary messengers	Transcription of specific genes that allow the expression and development of female-specific characteristics

of G-proteins and the role of these proteins in signal transduction in cells," the prize in 1998 to Robert Furchgott (1916–   ), Louis Ignarro (1941–   ), and Ferid Murad (1936–   ) for their discoveries "concerning nitric oxide as a signaling molecule in the cardiovascular system," and in 2000 to Arvid Carlsson (1923–   ), Paul Greengard (1925–   ), and Eric R. Kandel (1929–   ) for their work on signal transduction in the nervous system. The discovery of the importance of phosphorylation in different biological processes (we saw that it was a basic means of relaying messages between different signal transducers) also garnered the Nobel Prize for biochemists Edmond Fischer (1920–   ) and Edwin Krebs (1918–   ) in 1992.

SKIN, EYE, AND HAIR COLOR

Look around you and it is almost impossible to miss the fact that human beings come in many colors, ranging from very pale pinks to extremely dark, nearly black tones, as well as a variety of shades of brown in between. Indeed, the differences are sometime so stark that humans have often ascribed the color differences to some deeper or more fundamental differences among the people and even classified people into different races according to color. But color, like beauty in the saying, is only skin-deep. We will consider the genetic and biochemical bases for color differences.

Skin Color

The primary determinant of color—not just of skin but also of the hair and eyes—is a pigment called *melanin* that is produced by special epithelial cells of the skin called the melanocytes. Other substances, such as hemoglobin and carotenes, may also contribute to variations in skin tones or to sudden changes in color such as blushing or flushing, but those changes are incidental to the main functions of those pigments (i.e., oxygen transport and vision) and make a lesser contribution to overall complexion. On the other hand, the principal function of melanin is to interact with light (radiant energy) and protect the body from its harmful effects (e.g., from UV light). As it interacts with light, melanin undergoes chemical modifications that change its color, which in turn changes the color of skin.

Regardless of skin color, we have about 1,500 melanocytes per square millimeter of skin that lie between the keratin-producing cells of the epidermis. They synthesize melanin and package it into vacuoles called melanosomes that are passed along to the keratinocytes via long branched extensions of the cytoplasm. A single melanocyte supplies pigment to about 40 keratinocytes, and the ultimate color of the skin depends on the way in which the pigment is distributed in the latter. In general, there is a gradient of melanosome distribution across the spectrum—people with darker skin have larger, more numerous, and more deeply pigmented (i.e., darker) melanosomes, whereas lighter-skinned individuals have smaller, fewer, and lighter-colored vacuoles. This affects the way in which light is absorbed and reflected by the skin cells.

How deeply a melanosome (and hence a cell) is pigmented depends on the total amount of pigment it contains as well as the relative proportions of two types of melanin produced: a lighter, red to yellow type called pheomelanin and a brown to black eumelanin. Both pigments are polymers whose building blocks are molecules that are modified versions of the amino acid tyrosine. Pheomelanin is rich in cysteine (a sulfur-rich amino acid) and contains more iron as well. It is the more abundant type of melanin in people with fair skin, freckles, and light hair and eyes. Eumelanin is present in higher proportion in people who have darker complexions, eyes, and hair.

The preceding explanation might suggest that the genetics of color is simply a matter of the dominance of the dark eumelanin pigment over the lighter pheomelanin. This intuitive connection may even seem to be borne out by the observation that children of mixed parentage tend to be darker than their lighter-skinned parent. But the genetics of color is considerably more complex, and, in fact, if we were to look at most families of mixed heritage, we would see a considerable range in the skin color of the children. Also, as we know, every fair person does not have blond hair and blue eyes, and not all dark-skinned people have dark eyes and hair either. People show a remarkable variety of phenotypes in hair, skin, and eye color—we know of blondes with brown eyes and fair-skinned people with blue eyes and dark hair; indeed, the combinations are endless.

The first thing to understand is that color is a complex or polygenic trait governed by many different genes. For instance, there are at least six distinct genetic loci on chromosomes 1, 2, and 4 that play a role in the types and amounts of melanin produced. These genes do not express themselves in a simple dominant versus recessive fashion but show incomplete dominance. This means that the phenotype—color in this case—is often a blend of the parental types. This accounts for gradients of color among people within a family. Another complicating factor is differential expression (i.e., the expression of the same gene at different levels in different tissues). In a dark-haired, light-skinned person, for example, the gene governing "dark" complexions may be actively expressing itself in the melanocytes surrounding the hair follicles and be silent in the skin.

As if that were not enough, it is necessary to realize that the melanins are not the direct products of specific genes but polymers consisting of modified tyrosine units. This means that their synthesis proceeds in a chain of reactions catalyzed by different enzymes, each of which in turn is encoded by a different gene. Examples of genes for such melanogenic enzymes include the *TYR* gene on chromosome 11 that codes a protein called tyrosinase, *TYRP1* on chromosome 9, *DCT* on chromosome 13, and *Pmel17*, also called *SILV*, on chromosome 12. Mutations in any of these genes can break the chain and hence prevent the formation of melanin, which leads to disorders of pigmentation such as albinism (see the entry in Chapter 5). The *OCA2* gene on chromosome 15 encodes a membrane protein that is similar to a protein in mice that is involved in transporting small molecules such as tyrosine across the cell membranes. It appears to play a role in regulating the rate of melanin synthesis by controlling the amount of precursor available in the cell. Support for such a function comes from observations that mutations in this gene are linked to certain forms of albinism.

Yet another level of complexity in the determination of skin color is introduced by the fact that in addition to the melanin-producing genes, there are other genes that control their expression by the melanocytes. Perhaps the best-studied example is the *MC1R* gene on chromosome 16, which encodes a membrane receptor in these cells for the melanocortin hormone. Under normal conditions, the binding of melanocortin (which is activated by energy from light) to its receptor sets off a signal transduction pathway that tells the cell to switch from pheomelanin to eumelanin synthesis. Color is directly related to the level of activity of the receptor. In populations where fair skin and red hair are common (e.g., among people of Irish and other Northern European descents), the tendency to have carrot-colored hair, freckles, and to burn quickly in the sun is often associated with loss-of-function mutations in the *MC1R* gene. However, this correlation cannot be extended to other populations, and this receptor is believed to contribute little to variations in the color of other ethnic populations. A second gene, *MATP* on chromosome 5, also encodes a membrane-associated protein that mediates melanin synthesis. The *ASIP* gene on chromosome 20 encodes a signaling molecule that also plays a regulatory role in melanin synthesis, albeit in a pattern different from that of *MC1R* or *MATP*.

Finally, the environment has also played a role in applying selective pressures that have affected skin color during the course of the evolution of the human race. You may have noticed that people descended from ancestors who lived in sunny (equatorial and tropical) regions of the world tend to be dark, whereas lighter-skinned people come from the less sunny parts of the world. The desirability of these respective traits with respect to environment is explained as follows. In very sunny regions, people need protection from the harmful effects of ultraviolet rays, which can wreak havoc in DNA, both by inducing mutations and by destroying the B vitamin folate, which is essential for DNA synthesis. High levels of melanin in the cells that are exposed to the sun (skin, hair, and nails) serve as protection because it absorbs and diverts the radiant energy. Scientists believe that the ancestors of modern-day humans were protected from the sun by the hair covering their bodies. During the course of evolution, the amount of hair lessened and the underlying skin, which was initially not pigmented, became exposed. Humans in the sunny regions of the world may then have developed their melanin-rich dark skin as a response to sun exposure. Meanwhile, in the less sunny regions of the world, the evolving humans faced a different problem, namely the deprivation of a different nutrient, vitamin D, which is essential for calcium absorption and the proper development of bones. The synthesis of vitamin D in the body requires light, and so, because these populations did not require the diversion of energy provided by high levels of melanin, they remained light-skinned. This need for vitamin D is also the explanation that scientists offer for the relatively lighter complexion of women as compared with men in a given population.

An apparent contradiction to this argument is the evolution of the relatively dark skin of Inuit populations despite their sun-deprived environment. Scientists speculate that it was the unique combination of both climate and diet that led to the evolution of darker skin. Although severely sun-deprived for long stretches of time, the intensity of sunlight in the polar and subpolar regions due to the brightness of the snow is very high, and there is a need for the protection afforded by melanin against ultraviolet light. The concomitant deficit in vitamin D was compensated for by the diet, which has traditionally been based on animals and particularly fish.

Eye Color

Earlier in this chapter, we discussed the role of the ocular (eye) tissue in vision. The eyes focus light from different objects in our environment and form images on the retina, which then transmits the information to the brain for further processing. This means that the tissues of the eyes need to be transparent in order to let the light through. The problem is that in order to be transparent to light, the tissues also become more vulnerable to other forms of radiant energy. Consequently, they must be protected. As in the case of skin and hair, this photoprotective function falls to melanin. In the eyes, this pigment is present in the iris, which is the layer of cells coating the eyeball. The interaction between light and the melanin in the transparent tissues of the iris, as well as the inner retinal layer, results in the reflection of colors (such as blue and green) that are not seen in the opaque tissues of the skin and hair.

Besides the melanogenic and regulatory genes we talked about earlier, there are genes that specifically influence eye color. These include the *BEY* and *OCA2* genes on chromosome 15 and the *GEY* gene on chromosome 19. The *OCA2* gene encodes a transport protein for tyrosine called the P-protein, which is expressed specifically in the ocular tissues. Evidently, it regulates the rate of melanin biosynthesis by controlling the precursor to the pigment. Mutations in this gene are linked to certain forms of oculocutaneous albinism (see the entry in Chapter 5).

The functions of the *BEY* and *GEY* genes are not known, and the mechanism by which they cause different phenotypes is not well-understood. The genes are named for their predominant association with certain colors (i.e., brown and green, respectively). But the relationship is not exact because each of these genes is known to have more than one allele, each of which is associated with a different-colored phenotype. For instance, the *BEY* gene has a dominant form that is believed to correlate with brown eye color and a recessive allele that corresponds with blue. The *GEY* gene very likely has even more than two alleles, the most dominant of which is green and the least dominant blue. The specific combinations of different alleles in a person play a role in determining the final phenotype.

Hair Color

Hair color results from the transport of the melanins to the hair follicles. There are at least three known genes that specifically influence this trait in addition to those mentioned previously. Two of these genes, *HCL1* on chromosome 19 and *HCL3* on chromosome 15, are associated with brown hair color and lie in close proximity to the eye-color genes *GEY* and *BEY*, respectively. This accounts for the linkage of certain hair and eye colors, although the relationships are not com-

pletely predictable because of multiple alleles of the *HCL* genes. In addition, there is a third gene, *HCL2* on chromosome 4, that seems to be responsible for red hair. We still do not know enough about the proteins encoded by any of these genes.
See also: Albinism (Ch. 5).

SUGAR METABOLISM

By far the most direct and abundant source of energy or fuel for the human body is sugar. As we have seen previously, there are other sources of energy. Some, such as the fats and certain proteins, even yield more energy per gram than sugar. But faced with various alternatives, the body will always choose sugar first, turning to fats and other sources only under conditions of sugar deprivation. This is true not only of humans and other animals but indeed of virtually all living creatures that depend on organic materials for their energy source, beginning with the prokaryotes. Here we consider the biochemistry and genetics of the way in which we utilize sugars.

Humans, like other animals, are heterotrophic, which means that they are incapable of synthesizing organic molecules such as sugar from the raw materials of carbon dioxide and water. Organic substances such as carbohydrates are among the essentials—along with air (oxygen) and water—that our bodies need in order to survive. The main source for our sugars are the foods that we eat. All of us are familiar with the myriad forms in which we consume carbohydrates—breads, grains, fruits, and vegetables, not to mention the more obvious "sweet" items such as honey and table sugar, are all examples of this valuable commodity. In fact, sugars, in the form of carbohydrates, typically constitute the largest fraction of any diet.

No matter how complex the source, the eventual fate of the carbohydrates we eat is their degradation—cellulose and a few other indigestible plant materials are the only exceptions. As we explained previously (see the entry on food and digestion), a number of different enzymes in the gut act on foods to produce simple sugar monomers such as glucose, fructose, and galactose. These simple molecules are then taken up by the cells of the intestinal lining and further transported to different tissues and organs, where they are used in the metabolic processes we will describe for the production of energy.

To briefly discuss sugar chemistry, the basic constituents of a sugar molecule are indicated in the generic chemical name for these compounds (i.e., the carbohydrates, or water-carrying carbons). In other words, a sugar monomer is simply a group of hydrated carbon atoms linked together in a specific manner. The most common sugar monomers in biological systems are the pentoses (with five carbon atoms per unit) and hexoses (with six carbons). The sugars that we find in RNA and DNA (i.e., ribose and deoxyribose) are examples of pentoses. Most of our energy is gained from the hexoses, whose basic chemical formula is represented as $C_6H_{12}O_6$. The most common and usable hexose is glucose, whose structure is represented in the accompanying diagram. The molecule is typically in the form of a ring, which is created by the formation of a chemical bond between the first carbon atom and the hydroxyl group of the fifth carbon. For simplicity's sake, only the hydrogen atoms linked to heteroatoms such as oxygen are shown. (The unlabeled positions on the ring represent the carbons at these positions.)

Fructose, another common sugar monomer, found in many fruits, has exactly the same number of carbon, hydrogen, and oxygen molecules as glucose but forms a somewhat different ring structure. Here the chemical bond is forged between the fifth and second carbons of the hexose.

The difference between galactose (a monomer found in milk sugars) and glucose is even subtler and involves a simple rotation of the hydroxyl group about the fourth carbon so that it is now in the plane above the ring rather than below it.

Sugar Storage and Mobilization: Glycogen

More than 99 percent of the carbohydrates consumed by the body are used for the production of energy. However, the body does not always need all this energy immediately. So it converts this excess—present in the form of simple sugars— to a polysaccharide called glycogen, which is stored in the muscles until energy is needed. The process by which glycogen is synthesized is called *glycogenesis*. *Glycogenolysis* refers to the chemical pathways used to break down glycogen and mobilize the sugar for use. A number of different hormones, including insulin, glucagon, and epinephrine, regulate the rate at which various tissues synthesize or utilize glycogen (see the entries on homeostasis and hormones and the endocrine system for details).

A small amount of glycogen is synthesized in nearly all tissues, but it is the liver and muscles that make and store substantial amounts. On average, glycogen accounts for about 4 percent of the total weight of the liver and about 0.5–1 percent of the total weight of skeletal muscles. The function of the glycogen in muscles is to act as an energy reserve for the tissue itself, whereas the glycogen stored in the liver supplies glucose to various other organs at night or other periods when the body is at rest and not actively obtaining or absorbing glucose. Structurally, glycogen is a complex branched polysaccharide made of units of glucose. Monomer units are linked together by chemical bonds, called carboxylase bonds, in which carbon atoms of adjacent sugars are linked via oxygen. The position of the carbon atoms is not consistent, which is the reason for the branched nature of this final product.

As we already mentioned, glycogen is synthesized as a response to the excess of both sugar *and* energy. The first step in glycogenesis is the activation of glucose with a high-energy phosphate group. The catalyst for this reaction is an enzyme called *hexokinase*, which transfers a phosphate group from ATP to form glucose-6 phosphate. Hexokinase exists in several isoforms or isoenzymes, which are produced in different cells and tissues. The predominant form of the enzyme found in our skeletal muscles—which are among the body's most active sites of glucose metabolism and energy production in general—is called hexokinase II and is made by a gene called *HK2* mapped to chromosome 2. The protein is found localized to the outer membrane of the mitochondria. The enzyme glucokinase, encoded by the *GCK* gene on chromosome 7, is an alternative form of this enzyme that is expressed primarily in the liver and the beta cells of the pancreas, where it acts as a receptor to monitor blood glucose concentrations. Although scientists have detected mutations and polymorphisms in this gene, they have found that these mutations most likely do not contribute to skeletal muscle insulin resistance in non-insulin-dependent diabetes mellitus (NIDDM).

Glucose is not the only sugar whose excess prompts glycogenesis. Two other common monomers, fructose and galactose, are also easily absorbed and used in these processes. The enzymes that activate these molecules are similar to the hexokinases in their mode of action but differ with respect to substrate specificity. The *KHK* gene on chromosome 2, for example, makes an enzyme that phosphorylates fructose, and the *GALK* gene on chromosome 17 makes the equivalent enzyme for activating galactose. Deficiencies in the latter are linked to problems with congenital and presenile cataracts.

Regardless of the sugar that is phosphorylated to begin with, it is glucose-6-phosphate that is eventually used in making glycogen. Other sugar phosphates are converted to this version with the help of various enzymes. *Glucose-6-phosphate isomerase* (GPI), for example, catalyzes the interconversion of fructose and glucose phosphates. Basically this is a simple matter of rearranging the existing atoms in the molecules into slightly different configurations. GPI is a two-way enzyme that catalyzes the conversion in whichever direction is required. It is encoded on a large gene located on chromosome 19 that spans more than 40 kb and consists of 18 exons ranging in size from 44 to 153 bp. A number of polymorphisms of this gene are implicated in forms of anemia. The interconversion between the galactose and glucose phosphates is a little more complicated. An enzyme called *galactose-1-phosphate uridylyltransferase* (GALT) catalyzes the interconversion with the help of an additional molecule called uridine monophosphate. The *GALT* gene has been mapped to chromosome 9, and the enzyme is so crucial for galactose metabolism that its absence can prove fatal for newborns unless lactose is completely removed from the diet.

Glycogen is made by the polymerization of glucose phosphates, one molecule at a time. An enzyme called *glycogen synthase* catalyzes the addition of the monomer units to a pre-existing glycogen polymer or a shorter primer molecule that is already in the tissues. Two forms of this enzyme, produced predominantly in the muscles and liver, are produced by the genes *GYS1* on chromosome 19 and *GYS2* on chromosome 12, respectively.

Glycogenolysis, as its name indicates, is the lysis or breakdown of glycogen. Although the body calls on this process when it needs energy, the chemical reaction involved does not release any energy on its own. Instead, it frees up glucose that is then harvested for its energy in the processes of glycolysis and the Krebs cycle. The breakdown of glycogen mainly requires the enzyme *glycogen phosphorylase*, which both catalyzes and regulates the breakdown of glycogen to glucose-1-phosphate. The gene *PYGM* on chromosome 11 makes this enzyme in muscles. The *PYGB* gene on chromosome 20 encodes a less common version of this enzyme that is produced predominantly in the brain. The phosphate molecule is then moved from the C1 to the C6 position, readying the glucose for entry into the glycolytic pathway. Enzymes known as the *phosphoglucomutases* catalyze such a transfer of a phosphate group within the same molecule. The human genome has at least three known genes for enzymes that move the phosphate group between the first and sixth carbons of glucose phosphates: *PGM1* on chromosome 1, *PGM2* on chromosome 4, and *PGM5* on chromosome 9.

Glycolysis

Universal among the organotrophs, glycolysis is the first chemical pathway through which the body breaks down glucose. It is a multistep process that involves many enzymes, which we will describe. Glycolysis actually uses some energy in its early reactions but eventually results in the net production of a small amount of energy—two molecules of ATP together with two molecules of NADH per molecule of glucose. Most of the energy in glucose is released in later processes. Glycolysis does not require any oxygen and hence is an anaerobic process, which makes it especially important under conditions of low oxygen availability. The end product of this chain of reactions is a three-carbon molecule called pyruvic acid.

1. The first steps in glycolysis are the hexokinase-mediated activation of glucose and the isomerization of the glucose phosphate to fructose-6-phosphate. The enzymes and gene involved in these reactions were discussed earlier.

2. *Phosphofructokinase* (PFKL) adds on a second phosphate group to fructose-6-phosphate to form fructose-1,6-diphosphate. The *PFKL* gene is on chromosome 21 and is at least 28 kb long and divided into 22 exons. This reaction is the second one to use energy from ATP.

3. *Fructose-1,6-bisphosphate aldolase* catalyzes the next step, which is essentially the cutting of the glucose molecule into two similar pieces, each with a single phosphate attached. The products of this reversible reaction are glyceraldehyde-3-phosphate and dihydroxyacetone phosphate.

There are three aldolase isoenzymes (A, B, and C), encoded by three different genes, which are differentially expressed during development. The 7.5 kb *ALDOA* gene on chromosome 16 is expressed in the developing embryo. It is also abundant in adult muscle, where it can comprise up to 5 percent of the total cellular protein content, and is also seen in the brain and other nervous tissues. Deficiencies in this enzyme have been associated with myopathy and hemolytic anemia. Aldolase A is highly repressed in the cells of the adult kidneys, liver, and intestines, which get their aldolase from the expression of the *ALDOB* gene on chromosome 9. Defects in this gene cause hereditary fructose intolerance. The third gene, *ALDOC*, is found on chromosome 17 and bears a high degree of homology to the A isoform. The enzyme is found mainly in the brain and other nervous tissues.

4. The next step in the glycolytic chain is the conversion of glyceraldehyde-3-phosphate to 1,3-diphosphoglycerate. This is an oxidative phosphorylation reaction, catalyzed by *glyceraldehyde-3-phosphate dehydrogenase* (GAPD), which uses inorganic phosphate and kicks glycolysis into its energy-yielding phase. The enzyme is composed of four identical polypeptide subunits encoded by the *GAPD* gene on chromosome 12.

5. *Phosphoglycerate kinase* helps to convert 1,3-diphosphoglycerate to 3-phos-

phoglycerate. The gene for this enzyme, *PGK2*, maps to the X chromosome. It has several polymorphisms that sometimes have clinical consequences, such as hemolytic anemia. In addition to its role in glycolysis, PGK2 also seems to have an additional function as a primer recognition protein for DNA polymerase.

6. *Phosphoglycerate mutase* (PGAM) catalyzes the interconversion of 3- and 2-phosphoglycerate. Scientists have found at least two isoforms of this enzyme encoded by different genes: *PGAM1* is found on chromosome 10 and is expressed primarily in brain tissue, and *PGAM2* on chromosome 7 is expressed in muscle tissues. Certain missense mutations in the latter are a cause of myopathy, which is characterized by exercise intolerance, cramps, and myoglobinuria (muscle destruction).

7. The next enzyme in the sequence is an *enolase* that catalyzes the conversion of 2-phosphoglycerate into phosphoenolpyruvate by removing a water molecule. There are three forms of the enzyme, each of which is made of two identical subunits. The 17.7 kb *ENO1* gene on chromosome 1 encodes the version known as enolase alpha, which is found in most tissues. An alternative, monomeric form of the polypeptide from the same gene functions as a structural protein, crystalline, in the eye tissue. Enolase beta, which is found mostly in skeletal muscle tissue, is encoded by the *ENO3* gene on chromosome 17. Enolase gamma is specific for neurons, and its gene *ENO2*, is found on chromosome 12.

8. The final enzyme in the glycolytic pathway is *pyruvate kinase*, which removes the phosphate groups from phosphoenolpyruvate to generate pyruvic acid (or pyruvate).

There are two *PK* genes in the human genome, each of which produces two of the four isoenzymes of the enzyme. The pyruvate kinase gene (*PKM2*) on chromosome 15 encodes the M1 and M2 isoforms and seems to be expressed mostly in muscle cells. The *PKLR* gene, located on chromosome 1, appears to be specific for the liver and red blood corpuscles. Scientists have detected over 125 different mutations of the *PKLR* gene that cause a rare inherited disease known as pyruvate kinase deficiency (PKD). In this condition, red blood cells break down too easily, resulting in low levels of these cells (hemolytic anemia). Certain populations, such as the Amish people, have a somewhat higher incidence of PKD, although the condition afflicts people of all ethnicities.

Pyruvic acid serves as the intersection for many different biochemical pathways, depending on where and when glycolysis occurs. Yeasts, for example, convert it to alcohol in beer, bread, and wine. Our muscles often convert excess pyruvate into lactic acid, which acts as an irritant to our nerve endings and is the basis for muscle pain, especially when we exert ourselves under anaerobic conditions. Finally, pyruvate is also the source of acetyl CoA, which is one of the central metabolites in the human body and is used both in the biosynthesis and breakdown of different molecules. Acetyl CoA is the point of entry into the Krebs cycle, which is the high-energy-yielding phase of carbohydrate metabolism.

See also: Respiration.

Gluconeogenesis

Even though humans and other organotrophs cannot make glucose or other sugars from the basic building blocks of carbon dioxide and water, they do have mechanisms for synthesizing sugars from certain noncarbohydrate sources. Gluconeogenesis is the name given to such processes, which in the human body take place in the liver and kidneys. Without going into too much detail, the process is similar to, although not exactly, a reversal of glycolysis (i.e., a stepwise conversion of pyruvate/pyruvic acid to glucose). Some of the steps are in fact catalyzed by the same enzymes that participate in the breakdown but act in the reverse direction.

The raw materials used for gluconeogenesis fall into two main categories. The first set includes simple compounds such as certain amino acids and propionic acid that are absorbed directly into the blood from the gut. They are recruited into gluconeogenesis by enzymes that help convert them into pyruvate. The second category comprises molecules that are not derived from external sources but are intermediates in different metabolic pathways—including glucose degradation—and get diverted (or recycled) to glucose synthesis in times of need. Examples include lactic acid, which is formed in the muscles when glycolysis takes place in the prolonged absence of oxygen, oxaloacetic acid (a Krebs cycle intermediate), and glycerol (from the breakdown of triglyceride lipids). Both the liver and kidneys are equipped with a number of enzymes to perform these conversions.

TRANSCRIPTION: RNA POLYMERASES AND TRANSCRIPTION FACTORS

As we have remarked so many times in this book, the information in our genes (DNA) is of virtually no use unless we have the means to access it. Otherwise, it is like having a book in a language that no one can understand. The first step in accessing the information in the genome is the transcription of genes into RNA molecules. This is accomplished by an enzyme called RNA polymerase with the help of a host of other proteins known as the transcription factors.

A brief recap of the main steps in transcription (see the entry in Chapter 2 for details) may be useful to understand the complexity of the enzymes and factors involved in the process (see Figure 4.14).

- The most important step in this process is the *initiation*, whereby the polymerase binds to a specific region of the gene known as the promoter and begins the synthesis of an RNA molecule that is complementary to the gene sequence. This step also involves the separation of the two strands of the DNA at the location of the gene so that the appropriate strand is read and processed at the right location. Many different transcription factors play a role in this important first step.

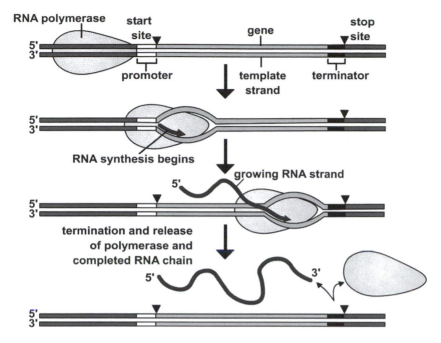

Figure 4.14. Transcription. [Ricochet Productions]

- Next comes chain *elongation*, the stepwise addition of nucleotides to the 3' end of the growing RNA chain. During this process, certain transcription factors help to keep a short stretch of the DNA template in a single-stranded state so that it may be read and transcribed.

- The final step in transcription is *termination*, in which the enzyme recognizes the transcription termination sequences and is released from the DNA.

- All of the primary transcripts produced must undergo posttranscriptional processing steps to produce functional RNA molecules for export out of the nucleus.

RNA Polymerase II

The enzyme responsible for transcribing messenger RNAs from genes in eukaryotic cells is called RNA polymerase II. It got this name because it was the second such enzyme to be discovered and characterized (after RNA polymerase I, which transcribes ribosomal RNA genes in bacteria). Human RP II consists of 12 subunits whose genes are present on different chromosomes. The gene for the largest subunit, POLR2A, is located on chromosome 17. The carboxyl terminal domain of this protein is essential for the polymerizing activity of the enzyme. In addition, this peptide forms part of the enzyme that is responsible for binding the DNA template. *POLR2B* on chromosome 4 produces the second-largest subunit, which, in combination with two other smaller subunits, forms a structure within the enzyme that maintains contact between the DNA template and the newly forming RNA transcript. The activity of one of the smaller subunits—POLR2E, encoded on chromosome 19—suggests that it may be involved in the interaction between RNA polymerase and certain transcriptional activators. Another gene on the same chromosome, *POLR2I*, also codes part of the DNA-binding domain of the enzyme.

Transcription Factors

Basically, any protein that is involved in transcription without actually being a part of the polymerase enzyme complex is called a transcription factor. Different types of transcription factors exert their effects in a variety of ways: by associating directly with the polymerase enzyme, by binding other factors, or by binding to DNA sequences. The end result of the activity of any of these types of subunits is the regulation of transcription. Whereas some of the factors have a general effect on the process, others modulate the transcription of specific genes. The following are some examples of genes for different transcription factors.

The Basal or Core Promoter

In order to properly execute its function of transcribing a gene, RNA polymerase must be able to start at precisely the correct location. This precise placement of the enzyme is achieved with the help of a promoter sequence called the TATA box, which sits a few base pairs upstream of the actual gene. A protein called *transcription factor IID* (TFIID) is the first protein that recognizes the promoter sequence. It binds to the TATA box to position the polymerase properly and also serves as the scaffold for assembly of the remainder of the transcription complex. The TFIID complex consists of a protein called TATA-binding protein (TBP) associated with several other peptides called the TAFs (for TBP-associated factors).

TBP, which is encoded on chromosome 6, consists of a distinctive string of N-terminal glutamines that control the rate of transcription initiation by modulating the DNA-binding activity of the transcription complex. The region of the *TBP* gene that encodes the polyglutamine tract is susceptible to mutations that expand the number of CAG repeats in this region, a genetic defect that is often associated with spinocerebellar ataxia 17, a neurodegenerative disorder. Highly conserved in evolution, the TAFs play numerous roles in transcription initiation, including promoter sequence recognition, coactivation of the enzymes, and modification of other transcription factors. The *TAF1* gene on the X chromosome, for example, encodes the largest subunit of the TFIID, which controls the rate of transcription through its interaction with other subunits. *TAF2* on chromosome 8 encodes another subunit, which interacts with the DNA sequences downstream of the initiation site and helps determine the response of the transcription complex to different activators. A protein called transcription factor IIB, encoded by the

GTF2B gene on chromosome 1, serves as a bridge between TFIID and the polymerase II enzyme.

Upstream Factors

The basal promoter complex just described is universal to all transcription systems. There are, however, many other elements—often found upstream of the actual genes—whose sequences and associated binding factors differ from gene to gene and from cell to cell. A good analogy to this system of multiple promoter sequences is the working of safe deposit boxes in a bank. In order to open any single box, one requires two keys. The first key comes from a bank employee and can activate the unlocking mechanism of any box in the bank. However, it alone is not capable of opening any of the boxes. This key is the equivalent of the basal promoter. The second key, which acts like the upstream promoter, is the one you own for your safe deposit box. This key can only fit the lock of the box to which it is assigned and must be used in order to open the box. However, it cannot unlock the box (or activate the gene) unless the banker uses his key first.

Although different genes and cells may share various transcription factors, what turns on a particular gene in a given tissue at any specific time is the unique combination of upstream sequences and transcription factors that are chosen. Some hormones and their receptors represent a class of transcription factors. The DNA sequences—known as hormone response elements—to which they bind and initiate the expression of specific genes are promoter sequences. Embryonic development also requires the coordinated activity of various transcription factors at different specified times.

Remotely Acting Transcription Factors

In addition to the promoter sequences, there are other DNA sequences, called *enhancers*, that control gene expression. These sequences may be located at different positions in relation to the gene—they may be present several hundreds of base pairs upstream or downstream and, on occasion, even within the gene. The transcription factors that interact with these sites are called *enhancer-binding proteins*. It is believed that these factors control gene expression by changing the configuration of the DNA molecules (e.g., by drawing DNA into loops that bring the polymerases closer to the initiation site). An example of a remote enhancer that is known to regulate the expression of several genes is CCAAT. The *CEBPB* gene on chromosome 20 encodes a typical CCAAT-enhancer-binding protein that is important in the regulation of genes involved in immune and inflammatory responses and also acts as a stimulator of collagen type I genes. The protein acts by forming dimers, which then bind to the DNA and change its configuration. It has a characteristic structure, consisting of an activation domain, a DNA-binding basic region, and a leucine-rich domain that forms dimers rapidly (for which reason it is also known as a leucine zipper).

Silencers are similar to the enhancers in their remote locations but act to repress or inhibit rather than stimulate transcription. Still other sequences, called *insulators*, keep the activation and repression of genes by different transcription factors in check.

Posttranscriptional Processing

The process of transcription is not quite complete, even after the polymerase reaches the end of the gene sequence. The nascent mRNA (or "pre-mRNA") still needs to be modified before it can be released from the nucleus in a biologically active form that can be taken up by the protein-translational machinery of the cytoplasm. These modifications, known collectively as posttranscriptional modifications, entail three major steps.

Splicing

As we know, many eukaryotic genes are present in a punctuated form, with coding regions known as the exons interspersed with noncoding, sometimes regulatory, sequences called the introns. RNA polymerase is unable to distinguish between the introns and exons, and intronic sequences must be removed from the primary transcript in order to produce a functional mRNA. The process by which an intron is removed and the flanking exons are joined together is known as RNA splicing.

There are several different mechanisms of splicing among the organisms of the living world. In humans, specialized RNA–protein complexes called *small nuclear ribonucleoprotein particles* (snRNPs) bind to the primary transcripts at the exon–intron junctions and catalyze the splicing of the introns. A comparison of the DNA sequences of many different genes has shown certain similarities in sequence at the intron–exon junctions. The sequence of the intron usually begins with "GT," and the exon usually ends with "AG." These sites seem to be very important for snRNP binding and intron removal. The RNAs found in snRNPs are identified as U1, U2, U4, U5, and U6. The genes encoding these snRNAs are highly conserved in many different organisms. In humans, there is a family of genes for each of these species. The U1 family is found on chromosome 1, and U2 and U3 are on chromosome 17. People with autoimmune diseases such as Systemic Lupus Erythematosus (SLE) have antibody molecules that attack many different nuclear components, including the snRNPs. An example of a disease resulting from splicing error is β-thalassemia, where a mutation in the intron leads to incorrect splicing and a faulty β-globin protein.

Capping

The 5' ends of all eukaryotic mRNAs are "capped" with a 7-methylguanosine residue that is attached to the terminal nucleotide by a unique 5' to 5' diphosphate bond. This modified guanine "cap" covers the emerging mRNA and protects it from being degraded by cellular enzymes. The cap also flags the mRNA for recognition by the specific proteins of the translational machinery. *RNGTT* is a gene on chromosome 6 that encodes a bifunctional capping enzyme that catalyzes both the transfer of a GMP molecule from GTP to the nascent mRNA and the methylation of the guanine N7 position of the newly formed cap.

Polyadenylation

After transcription is complete, the transcript is cut at a particular site (which may be hundreds of nucleotides before its end), and a stretch of adenine nucleotides called the poly(A) tail is attached to the exposed 3' end. Polyadenylation is a complex process that involves an enzyme called polyA polymerase (PAP), which acts in conjunction with other factors such as a cleavage-stimulating factor (CstF) and a cleavage/polyadenylation specificity factor (CPSF). An example of a gene that produces a protein with PAP-like activity is the *ADPRTL* gene on chromosome 3. Scientists are still investigating the genetics of other factors. The formation of the poly (A) tail completes the mRNA molecule, which is now ready to be exported to the cytosol as the remaining portion of the original transcript is degraded and the RPII falls off the DNA.

TUMOR SUPPRESSOR GENES

As their name suggests, the tumor suppressor genes play a role in inhibiting or preventing the formation of tumors. Actually, they are genes whose normal function is to put the brakes on cell division during the course of the cell cycle. When such genes are absent or inactivated due to mutations, the cell loses control and begins to proliferate unchecked, which in turn can trigger the development of a cancer. The tumor suppressors thus offer protection from cancer.

The cancer-related mutations of the tumor suppressor genes are recessive in nature. This means that both the maternal and paternal copies of the gene in an individual must be defective in order for a cancer to develop because the presence of even one normal, functional copy will produce enough protein to maintain the normal cell cycle. Individuals who are born with a single defective copy, however, are at greater than normal risk for developing cancer because any somatic mutation to the normal copy of the gene can interfere with its function and release the brakes on the normal cell cycle.

The human genome has several tumor suppressor genes widely distributed throughout the chromosomes, and their proteins function in all different parts of the cell, including the nucleus, cytoplasm, and cell surface. Examples of tumor suppressor genes include the p53 tumor suppressor, the retinoblastoma gene, the genes *BRCA1* and *BRCA2* associated with breast cancer (see the section on breast cancer in the entry on cancer in Chapter 5), and the *APC* gene, implicated in the development of polyps in the colon. It is worth repeating here that although many of these genes are named for the cancers where they were first discovered, they do not cause the cancer. In fact, as we have seen, it is their *absence* or loss of function that results in disease. Only a few representative examples will be discussed here and the others mentioned in the context of specific cancers.

p53

Described by some scientists as the "grand-daddy of all human cancer genes," the p53 tumor suppressor gene is perhaps the most frequently mutated gene in human cancers and is found in 50–55 percent of all tumors, including those of the colon, lung, breast, brain, blood (leukemias), and bones (osteosarcomas). The normal version of this gene, found on chromosome 17, codes for a cytoplasmic protein that regulates cell growth and induces cells to die (apoptosis) at the appropriate time in the cell cycle.

There are many reasons why this gene is frequently associated with different cancers. First, the fundamental function of p53 in cell-cycle regulation means that it is expressed in many different organs and tissues, all of which can be sites for tumor formation. Second, the p53 gene is susceptible to somatic mutations during a person's lifetime. So even an individual who receives normal copies of the p53 genes from both parents may develop cancer because of mutations arising in the p53 gene during his or her lifetime. Generally, people who inherit a single copy of normal p53 are at a somewhat higher risk for acquiring cancer than those with two copies of the normal gene. Finally, it is important to realize that p53 is not a solitary actor but part of a network of proteins and other factors that act in a coordinated fashion to maintain the cell cycle. Consequently, mutations in the genes for other members of this network can also have an effect on the cell cycle.

p53 mutations are also the cause of an inherited cancer syndrome called Li–Fraumeni syndrome, a disease in which members of different generations of a single family show a heightened tendency to develop cancers (particularly soft-tissue sarcomas, breast cancer, leukemias, and adrenal gland tumors) at a young age. Li–Fraumeni syndrome was the first evidence of an actual inherited predisposition to cancer.

Retinoblastoma (*Rb*) Gene

The *Rb* gene is mentioned here because it was the study of certain malignant retinoblastomas—tumors in the retina of the eyes—that first led to the identification of tumor suppressor genes in 1990. Although p53 and its association with cancer were already known by then, scientists did not know how this gene caused cancer and at one time even thought it was an oncogene. The normal retinoblastoma gene, designated as *Rb*, codes for a nuclear protein (pRb) that functions as one of the major cell-cycle inhibitors. It is located on the long arm of chromosome 17. Defects in the gene are associated with retinoblastomas as well as osteosarcomas (tumors in the bone).

See also: Oncogene (Ch. 2); Cancer—Retinoblastoma (Ch. 5).

Chapter 5

Genetic Diseases

Having looked at representative examples of genes that perform the functions to help us conduct various activities of life, we now turn to examples of the conditions that occur when there is some problem with these genes. As emphasized in earlier chapters, genes do not cause diseases, despite a tendency to identify many genes and genetic loci on chromosomes by the diseases with which they are associated. Rather, it is aberrations in the genes, mutations that cause the gene product to be absent or faulty, or, in some cases, confer new activities on the gene products, which result in disease. In this chapter, we look at examples of disease conditions caused by different genetic defects such as chromosomal translocations, nucleotide or gene deletions, base substitutions and insertions, and others (see the entry on mutation in Chapter 2 for details).

Diseases are organized alphabetically and, whenever possible identified by their most common name. In each entry, we identify the genes that are affected and explain how the disruption of their normal function leads to the disease. Environmental factors affecting these diseases are also discussed in as comprehensive a manner as possible. Most of the entries in this chapter should appear in the tables presented in Chapter 3. Readers are encouraged to use this chapter interactively with the previous one to get a better understanding of the complex and intricate connections between our genes and various disease conditions.

ALBINISM

Albinism is the name given to a group of related conditions characterized by a deficiency of the pigment melanin in the skin, hair, and eyes (the irises). Depending on the extent of the deficiency, the loss of melanin may be partial or complete. The most typical sign associated with albinism is an extreme pallor of these tissues—i.e., very pale blonde, nearly white hair, eyebrows, and lashes and very pale blue or pink eyes. In certain cases, this absence of pigment may occur in patches. The deficiency in melanin leaves albinos extremely vulnerable to sunlight. They get sunburn very easily and are also at a higher risk for skin cancer, especially melanomas. More complete forms of albinism are often accompanied by problems with the eyes such as rapid, nearly uncontrolled eye movements, trouble with visual tracking, extreme sensitivity to light (photophobia), and decreased visual acuity, ranging from myopia (short-sightedness) to functional blindness.

As we know from the previous chapter (see the entry on skin, eye, and hair color), color is a complex genetic trait that is governed by many different genes affecting different levels of function including (but not limited to) tyrosine metabolism and transport, intracellular trafficking, and signal transduction. Mutations in any of these genes can result in problems in the production and distribution of melanin. Consequently, albinism appears in many different forms of varying severity and may be inherited in several different patterns.

The most severe form of albinism, known as *tyrosinase-negative oculocutaneous albinism*, is caused by a loss-of-function mutation in the gene for tyrosinase (*TYR*) on chromosome 11. This type of mutation strikes at the very root of melanin production and affects the pigmentation of both the skin and eyes. There are two subtypes of disease, depending on the location of the mutations in the gene—some mutations completely inactivate the enzyme, leading to zero tyrosinase activity (type 1A), whereas other mutations only reduce the activity of the tyrosinase enzyme (type 1B). Both are autosomal recessive disorders because the presence of even a single copy of the functional gene can compensate for the activity of the enzyme. Symptoms are fairly typical (i.e., extreme pallor accompanied by defects in vision and other functions of the eye). Other forms of oculocutaneous albinism that have similar clinical symptoms result from defects in other genes involved in the chain of melanin production, such as the *TYRP1* (tyrosinase-related protein 1) gene on chromosome 9.

Certain other forms of autosomal recessive oculocutaneous albinism map to mutations in genes for melanin transport proteins, including the *MATP* gene on chromosome 5 and the *OCA2* gene on chromosome 15. (These genes are discussed in Chapter 4.) In addition, mutations in the melanocortin hormone receptor *MC1R* on chromosome 16 cause albinism due to defects in regulating the production of melanin.

In addition to the forms of albinism just described, there are also forms of the disease where only the eyes are affected. These forms of the disease are called *ocular albinism*. Patients with these forms have normal skin and even eye (iris) color but lack pigment in the retina. An X-linked form of ocular albinism is caused by mutations in a gene called *GPR143* that encodes a G-protein-coupled membrane receptor that is expressed almost exclusively in the melanocytes that supply pigments to the eye tissues. The protein is presumably involved in the signaling pathways that mediate the transport of pigment to the iris and retina.

In addition to the specific conditions described earlier, partial or localized albinism may be part of a spectrum of symptoms in certain other complex diseases such as Waardenburg syndrome, certain forms of deafness, and bleeding disorders. Some of these examples are discussed in separate entries in this chapter.

Although there is no cure for the underlying genetic defects that cause albinism, certain measures can be taken to ensure the protection of the skin and eyes from sunlight. Sunburn risk can be reduced by avoiding the sun, by using sunscreens with SPF (sun protection factor) of 30 or even higher, and covering oneself completely with clothing when exposed to sun. Ultraviolet protective sunglasses may help relieve photophobia, and corrective lenses can help with myopia.

See also: Skin, Eye, and Hair Color (Ch. 4); Cancer—Melanoma, Waardenburg Syndrome.

ALCOHOLISM

The tendency of children to mimic their parents' drinking habits was documented thousands of years ago by the likes of Plato and Aristotle. But before entering a discussion of the modern-day thoughts on the connection between alcoholism and our genes, it may be wise to say something about the issue of the disease itself. Although it is true that these writings show an awareness of the existence of alcohol misuse, abuse, and even addiction, it must be noted that it was not until the twentieth century that alcoholism was formally recognized by the medical community as a disease rather than merely a social evil or behavioral problem.

Today, we define alcoholism as a chronic, progressive disease wherein individuals have lost control over their alcohol-drinking habits to such an extent that drinking interferes with their ability to conduct normal activities. Major symptoms include a craving for alcohol, an inability to stop drinking once having begun, a physical dependence characterized by withdrawal symptoms when deprived of alcohol over any period of time, and a gradually growing tolerance for the substance. By treating alcoholism as a disease, it becomes possible for physicians to diagnose the condition and devise strategies for treatment.

What about the inheritability of this disease? Although epidemiological studies show that alcoholism definitely runs in families, they have not been able to ascertain whether the recurrence of drinking problems in successive generations was a learned behavior due to the environmental influence or a characteristic built into the genes. The current stance is that alcoholism is a multifactorial and polygenic disorder (i.e., one whose outcome and inheritability depend on multiple genes, environmental factors, and the interactions between them).

A possible biochemical mechanism for the development of alcoholism appears to be linked to the nature of the metabolites of alcohol degradation that accumulate in the body. People more prone to alcoholism have higher levels of metabolites giving pleasurable effects, whereas those not prone to alcoholism have higher levels of metabolites giving unpleasant effects. Biochemical studies have shown that a moderate amount of alcohol results in higher blood acetaldehyde among alcoholics (as well as children and siblings of alcoholics) when compared with levels in a control population of nonalcoholics. Because acetaldehyde is the intermediary metabolite in alcohol's degradation into acetate, its accumulation in the blood indicates some stalling in the activity of the enzyme responsible for the reaction (e.g., a slow-acting *ALDH* gene would render the person more prone to alcoholism). In contrast, it has been shown that individuals—most often East Asians—with a mutant version of *ALDH2* are more prone to alcohol intoxication and consequently less prone to achieving tolerance and becoming addicted than people lacking this mutation.

The connection between alcohol and the body's pleasure centers suggests a host of additional factors that could have a complicating effect on the predilection to, and development of, alcoholism, but we will restrict ourselves to one example here. The activity of another enzyme, catechol-O-methyltransferase (COMT), which has a crucial role in the metabolism of dopamine, has shown links to alcohol addiction. A genetic epidemiology study conducted on a Finnish population showed that alcoholics had a markedly higher frequency of a slow-acting version of the *COMT* gene when compared with individuals with no drinking problem.

See also: Alcohol Metabolism—Acetaldehyde Dehydrogenase (ALDH) (Ch. 4).

ALPORT SYNDROME

First described in 1927 by the British physician for whom it is named, Alport syndrome is a genetic condition in which patients exhibit a combination of progressive hearing loss and kidney failure. It is one of the most common types of hereditary nephritis and accounts for 3 percent of all childhood cases of chronic kidney failure.

The mutation that causes Alport syndrome is found in a gene called *COL4A5* on the X chromosome. This gene codes for the alpha chain of a type IV collagen that is a component of the basement membranes of epithelium cells of different tissues. In the glomeruli of the kidneys the basement membrane containing collagen IV acts as the filter that moves the waste materials from the blood into the urine while retaining the useful materials. The basement membrane in the cochlea of the ear allows it to receive sound signals. Mutations in the *COL4A5* gene result in the formation of abnormal protein, which leads to the formation of abnormal basement membranes that are unable to maintain their integrity. As the membranes disintegrate, the kidneys lose their ability to retain proteins and cells in the blood, and these materials begin to leak into the urine. The abnormal collagen in the cochlea of the ear results in a progressive deafness that is characterized by the loss of high-tone sounds first.

For the most part, Alport syndrome is more severe (i.e., it proves fatal at a younger age) in men than in women. This is because of the location of the *COL4A5* gene on the X chromosome. Women, unless they are homozygous for the mutation, have one normal copy of the gene, which partially offsets (but does not override) the harmful effects of the abnormal protein. Because men have only one X chromosome, this partial correction mechanism is not available to them. The *COL4A5* gene has a very high rate of spontaneous mutation, and nearly 20 percent of diagnosed cases have no familial history of Alport syndrome. Genetic testing may be advisable in suspected cases because treatment of kidney failures by dialysis or transplants and treatment of deafness with cochlear implants may be feasible. Gene therapy for this disorder may also be possible in the future.

See also: Cartilage—Collagen (Ch. 4); Deafness.

ALZHEIMER'S DISEASE (AD)

Alzheimer's disease (AD) is one of several disorders that are characterized by the gradual loss of brain cells. The areas of the brain that control memory and thinking skills deteriorate first, followed by cell death in other regions of the brain. Consequently, some of the most frequently observed symptoms of the disease include a slowly progressing inability to remember facts and events and then later to recognize friends and family. The disease is named for the German physician Alois Alzheimer (1864–1915), who first described it in 1906. Although it was once considered rare, research has shown that Alzheimer's disease is in fact the leading cause of dementia, the umbrella term for a group of symptoms related to loss of thinking skills. About 4.5 million Americans suffer from AD, and the disease is the fourth-leading cause of death in adults. Famous among those affected by Alzheimer's is former U.S. president Ronald Reagan (1911–2004).

The incidence of Alzheimer's disease rises steeply with age. It is also diagnosed twice as often in women as in men. It tends to be inherited, suggesting that there is a strong genetic component in the risk of getting the disease. Scientists currently believe that mutations in at least four genes, found on chromosomes 1, 14, 19, and 21, play a role in the disease. AD1 is caused by mutations in the amyloid precursor gene on chromosome 21; AD2 is associated with the *APOE4* gene on chromosome 19; AD3 is caused by mutation in a chromosome 14 gene encoding a protein called presenilin-1 (PS1); and AD4 is linked to mutations in a gene on chromosome 1 that encodes a similar protein, presenilin-2 (PS2). Mutations in these genes appear to lead to the formation of lesions in the brain consisting of fragmented brain cells surrounded by protein fragments called "amyloids." Amyloid is actually a general term for protein fragments that the body produces normally. In a healthy brain, these protein fragments get broken down and eliminated. In AD, the fragments are not eliminated and instead accumulate to form hard, insoluble "amyloid plaques" that are characteristic of the disease.

The *PS1* and *PS2* genes are relatively well-characterized. Mutations in the *PS1* gene are involved in about 70 percent of inherited AD cases where the onset is early (i.e., at less than 60 years of age). Scientists believe that these mutations may also play a role in certain late-onset Alzheimer's cases. Because the presenilin proteins encoded by these genes were discovered only relatively recently, not much is known about their normal functions. But studies suggest that these membrane proteins may be involved in apoptosis, or programmed cell death. They seem to influence the secretion of an abnormal type of amyloid, called beta-amyloid, that gives rise to plaques.

The abnormal beta-amyloid is a fragment of the amyloid precursor protein (APP). This protein is found on the cell surfaces in many tissues and organs. The protein may help nerve cells grow and may also turn certain genes on and off, but researchers still do not know its exact function. Researchers have identified several mutations within exons 16 and 17 in the *APP* gene on chromosome 21, which lead to the formation of beta-amyloid. One of these mutations is believed to be responsible for 5–20 percent of all early-onset familial AD cases. The most common APP mutation is a substitution of the amino acid isoleucine for valine at position 717 of the peptide sequence.

Alzheimer's disease is evidently a complex, polygenic disease, wherein environment also plays a role in the outcome of the disease. Currently, scientists are exploring the interaction between the various gene loci and environmental factors that might affect a person's susceptibility to AD.

AMYOTROPHIC LATERAL SCLEROSIS (ALS)

Sports fans and physicists might find common ground in their knowledge of this disease, which counts among its famous patients the baseball player Lou Gehrig and cosmologist and *A Brief History of Time* author Stephen Hawking. Also known as Lou Gehrig's disease, ALS is a progressive degeneration of the nerve cells in both the brain and spinal cord that control voluntary muscles. As these nerve cells die, the afflicted individuals gradually lose control over their normal motor capabilities even while retaining their normal intellectual faculties.

Amyotrophic lateral sclerosis can strike at any age, although onset is usually in early to mid adulthood. The earliest symptoms caused by the weakening of voluntary muscles include clumsiness, a tendency to trip and fall, twitching muscles, fatigue, slurred speech, and difficulties in breathing. With the death of the nerve cells, the muscles they are supposed to control no longer receive the messages they need in order to function and begin to waste away with prolonged disuse. Eventually, the patients become paralyzed, and most do not survive for very long. (Hawking, diagnosed with ALS at 21 and who has survived past 60, is one of the rare exceptions.) According to the National Institute of Neurological Disorders and Stroke (NINDS), about 20,000 people are afflicted with ALS, and 5,000 new cases occur each year in the United States. It seems to occur about 1.5 times more frequently in men than in women.

Exactly what causes ALS is not completely understood. About 5–10 percent of all cases are familial (i.e., transmitted through the genes from one generation to the next), whereas the remainder seem to occur sporadically. The first major breakthrough toward understanding the genetics of ALS came in the early 1990s with the discovery that some 20 percent of the familial ALS cases were linked to mutations in the gene on chromosome 21 that codes for an enzyme called superoxide dismutase (*SOD1*). Widely distributed in cells throughout the body, superoxide dismutase is an important housekeeping enzyme whose function is to remove supercharged oxygen molecules from the cells by converting them to less harmful substances. Superoxide molecules are by-products of normal cellular activities. They are very reactive and can thus cause considerable damage if allowed to accumulate inside a cell. Exactly why a mutation in SOD should selectively harm the motor neurons when the enzyme is produced in all the cells is still a puzzle. Meanwhile, the reason that antioxidant drugs appear to benefit some ALS patients may be due to the fact that they help compensate for the defective superoxide dismutase and hence prevent nerve cells from further damage.

Amyotrophic lateral sclerosis has also been linked to other genetic loci besides *SOD1*. A particularly interesting example is a gene on chromosome 11 called *CNTF* (for ciliary neurotrophic factor). The normal gene product of *CNTF* is expressed only in nerve cells, where it promotes the synthesis of neurotransmitters and aids in nerve cell survival. This gene appears to be a modifier for ALS. This means that mutations in *CNTF* alone do not have any ill effects but modify the course of disease (for instance, by greatly accelerating its onset) when present in combination with *SOD1* mutations. In addition, ALS susceptibility is associated with the gene for a polypeptide of nerve cell filaments encoded on chromosome 22 and regions of chromosome 18. One theory for this is that a defect in the glutamate transport system allows glutamate to build up to levels that are toxic to nerve cells. The drug riluzole helps to reduce damage to the nerve cells by suppressing the release of glutamate.

ASTHMA

Asthma is a chronic inflammatory disease of the lungs that affects more than 5 percent (an estimated 17 million) of the total U.S. population. According to the National Institutes of Allergy and Infectious Diseases (NIAID) of the NIH, nearly 500,000 people are hospitalized and more than 5,000 succumb to asthma each year. Although asthma affects people of all ages and ethnicities, children, especially those living in inner cities and other environments with low air quality, are among the most susceptible to this disease. In the United States, African Americans are more frequently diagnosed and succumb more often than other ethnic groups.

The fact that it is an inflammatory disease indicates that the underlying problem in asthma is with the immune system. It is the result of an overzealous immune response to various environmental antigens that the body perceives as threats. There are many different potential triggers of asthma attacks—airborne irritants such as viruses, pollen, tobacco smoke, and animal dander, vigorous exercise, and even certain foods. Typical symptoms include coughing, wheezing, shortness of breath, and chest tightness, often in response to specific environmental triggers. Other signs of inflammation, such as hives or rashes, often accompany asthma. The physiological basis for the respiratory symptoms is a narrowing of the airways that is caused by edema (fluid in the intracellular tissue space) in the lungs and the influx of inflammatory cells into the walls of the airways.

Asthma is a complex, or polygenic, disease, which means that several genes can contribute to a person's susceptibility to the condition. Researchers have identified regions on at least seven chromosomes—5, 6, 7, 11, 12, 13, and 14—that are associated with asthma susceptibility. The roles of these chromosomal regions in asthma predisposition are still not clear, but one of the most promising sites for investigation is on chromosome 5. Mutations in a gene of unknown function called *BHR1* (for bronchial hyperresponsiveness) in this region are specifically linked to bronchial asthma. In addition, it is known that this region is rich in genes coding for key molecules in the inflammatory response seen in asthma, such as cytokines and their receptors. Scientists have discovered that the *PHF11* gene on chromosome 13 is linked to asthma susceptibility. This gene appears to regulate the blood B cells that produce Immunoglobulin E, the allergic antibody. The gene is a potential target for drugs that could turn off the production of IgE and control asthma. Another gene, *GPRA* (G-protein-coupled receptor for asthma susceptibility) on chromosome 7, also appears to be involved in the susceptibility to allergy and asthma. Mutations on yet another gene, *ADAM33* (for A disintegrin and metalloprotease) on chromosome 20, have also been found to play a role in the development of asthma.

Studies on a mouse model have revealed nearly 300 mouse genes, called "asthma signature genes," that appear to be involved in asthma pathogenesis. Homologs for these genes may be implicated in humans with asthma, and this knowledge could lead to asthma treatments tailored to an individual patient's disease. Of course, inheriting mutations in these genes does not necessarily lead to asthma. Susceptibility alleles, just one part of the puzzle, act together with factors in the environment and lifestyle to increase one's risk for developing asthma.

See also: Immunity (Ch. 4).

ATHEROSCLEROSIS

How many of us have a relative or know someone with a relative that has suffered from a heart attack? With more than 2,500 Americans dying from heart disease each day, the odds are that virtually everyone who reads this book fits that bill.

Perhaps the most direct cause of a heart attack is atherosclerosis, which is the thickening and hardening of the large arteries brought about by the buildup of fatty substances, cholesterol, cellular waste products—especially of the immune cells—calcium, and clotting factors such as fibrin in the inner walls of the arteries. Of these, cholesterol is perhaps the most common offender, although the deposition of other substances is often closely linked with atherosclerosis. According to the American Heart Association, over half of all deaths in the United States are a direct or indirect result of atherosclerosis. The buildup of materials in arteries is known as a plaque. As time progresses, arterial plaques can partially or totally block the flow of blood. This poses a double danger for the heart—not only does it respond to slowing of the blood flow by attempting to pump harder and thereby getting fatigued, but the impeded blood also reduces the oxygen available to its muscles, which also leads to exhaustion. When a blood clot forms on the plaque's surface, it can completely block the artery, resulting in a heart attack (due to a reduction in oxygen to the heart) or a stroke (by cutting off the oxygen supply to the brain).

Although some amount of arterial hardening is normal as people grow older, the progression to atherosclerosis is a pathological event that is influenced by many environmental and genetic factors. The disease often starts in childhood, and the arteries become clogged over many years, initially without symptoms. Key risk factors include elevated levels of cholesterol and triglycerides (fats) in the blood, high blood pressure, smoking, diabetes, and lack of exercise. So our daily habits and lifestyle, including diet and exercise, play an important role in determining whether or not we get this disease.

Recent studies have shown that several different genes and their protein products play a role in the development of atherosclerosis. Many of these genes appear to be related to the processing of cholesterol. For example, high levels of lipoprotein A, Lp(a), have been associated with atherosclerosis. Lipoprotein A is a plasma lipoprotein that consists of apolipoprotein A (ApoA) linked by a covalent bond with a low-density lipoprotein (LDL) particle. The Lp(a) fragments accumulate in atherosclerotic plaques, where they may promote the bursting of these plaques, a process known as thrombogenesis. The gene that encodes ApoA is on chromosome 6, but little is known about variations in the gene and their association with atherosclerosis.

Another apolipoprotein, ApoE, has also been found to be associated with atherosclerosis. Apolipoprotein E is a plasma lipoprotein that plays an important role in the degradation of particles rich in cholesterol and triglycerides and in carrying excess cholesterol back to the liver cells. It can exist in several different forms and is coded for by a gene found on chromosome 19. Defects in apolipoprotein E sometimes result in its inability to bind to receptors, which leads to an increase in a person's blood cholesterol and consequently his or her risk of atherosclerosis. However, scientists are still undecided about exactly how the various forms of apolipoprotein E affect atherosclerosis.

Mutations in selectin E (endothelial leukocyte adhesion molecule-1), which is encoded by a gene on chromosome 1, have been found to increase susceptibility to atherosclerosis. The normal protein is thought to be involved in the process of blood cells sticking to the inner lining (endothelium) of blood vessels in inflammatory events such as atherosclerosis. An A-to-C change at nucleotide 561 of the cDNA converts Serine to Arginine in the protein, and this polymorphism is associated with severe atherosclerosis and a higher risk of coronary artery disease (CAD).

Researchers have found a link between atherosclerosis and a gene called *ALOX5*, found on chromosome 10, which codes for the enzyme *5-lipoxygenase*. Normally, this enzyme generates leukotrienes, which are important mediators of inflammation. People with a certain mutant variant of this gene have been found to have a tendency to develop thickened artery walls—a symptom of atherosclerosis. The mutations tested involved deletions or additional nucleotides in a special transcription factor–binding part of the promoter of the gene. This mutation appears to produce relatively high levels of leukotrienes in the blood vessels, which is believed to contribute to atherosclerosis.

See also: Lipid Metabolism—Cholesterol (Ch. 4).

ATTENTION DEFICIT HYPERACTIVITY DISORDER (ADHD)

Attention deficit hyperactivity disorder is a common childhood behavioral or psychiatric disorder believed to affect about 5 percent of all children and adolescents. According to the National Institutes of Mental Health of the NIH, the average rate of incidence in the United States is about one child in a classroom of 25 to 30. Children typically begin to exhibit symptoms in preschool or early school and are unable to focus, get distracted easily, and often display a tendency to act impulsively.

In 1845, a physician named Heinrich Hoffman found that he had difficulties finding suitable reading materials for his 3-year-old son. He subsequently wrote a book of poems about children and their behaviors titled *The Story of Fidgety Philip* that closely describes the symptoms that we recognize as ADHD today. A more formal description of the condition appeared in medical journals for the first time in 1902, when a British physician published a study on a group of children with behavioral problems relating to impulsive behaviors and inattentiveness. Today, we tentatively classify both ADHD and a related disorder known as ADD (which differs mainly in lacking the hyperactivity associated with ADHD) as disorders related to certain inborn defects and do not regard them as the sole product of poor child-rearing practices. Although not all scientists agree on the authenticity of ADHD as a disorder, twin studies have suggested that it does have a strong genetic component. For instance, if one identical twin has the disorder, the likelihood that the other will as well ranges from 65 percent to 90 percent, a statistic that is comparable to those for other diseases such as schizophrenia and bipolar disorder, where genetics plays a partial role.

Genetic work has shown evidence of links between ADHD and certain genes of the dopamine system (see the entry on behavior in Chapter 4). In many cases, the mutations underlying the condition map to genes for dopamine receptors *DRD4* and *DRD5* on chromosomes 11 and 4, respectively. Although the mutations in these genes have essentially the same disease phenotype, their normal versions have quite the opposite activities. Whereas the *DRD4* product triggers a set of signaling pathways by inhibiting adenylyl cyclase upon binding with dopamine, D5 stimulates adenylyl cyclase and triggers a completely different type of signal transduction pathway in the brain. Furthermore, *DRD5* is expressed specifically in the limbic region of the brain, whereas *DRD4* is more widespread. The fact that both receptors are implicated in ADHD appears to indicate that it is the presence of dopamine in the cellular milieu rather than an interference with its signaling activities that is linked to behavior. This link seems to be borne out by the observation that a third gene whose mutations have been linked to ADHD—*DAT1* on chromosome 5—encodes a transport protein for dopamine. Mutations in *DAT1* seem to affect the attachment of dopamine to its receptors, thereby increasing its extracellular concentration.

Still another suspect in ADHD is a protein called synaptosomal-associated protein 25 (SNAP-25) that is expressed highly and specifically in the nerve cells and helps trigger the release of neurotransmitters from nerve cells. The gene for this protein is found on chromosome 20. The hunt for ADHD-related genes continues, with other possible candidates on chromosomes 5, 6, 16, and 17.

See also: Behavior (Ch. 4).

AUTISM AND PERVASIVE DEVELOPMENTAL DISORDERS (PDD)

Remember "Rain Man"? It was the nickname of the character played by Dustin Hoffman in the 1988 movie of the same name. In the film, Hoffman depicted a withdrawn autist, a man who had some brilliant abilities—for instance, he could memorize entire telephone books with one reading and accurately count matchsticks or toothpicks in large bundles—but who was almost incapable of performing some of the most mundane tasks of daily living without an unchanging environment—he only wore underwear from K-Mart and had to eat exactly eight fish sticks (no more, no less). In the movie, he was also obsessively afraid of flying and knew the crash records of virtually every airline company—he would only fly Qantas.

The specific combination of abilities and disabilities displayed by Rain Man may have been fiction, but the condition he portrayed—autism—is quite real. It is a developmental disorder that affects aspects of a person's social behavior. Derived from the Greek word "autos" (meaning self), the word autism is believed to have been first introduced by a Swiss psychiatrist, Eugen Bleuler, in 1911 to describe the self-exclusion of an individual from all social contact. Leo Kanner, a physician in Baltimore, used the term in 1941 to make the first formal diagnosis of children who displayed a marked inability to form social relationships with others. One year later, Dr. Hans Asperger in Vienna also described (independently, in a German-language publication) similar cases of developmental or behavioral problems. Today, we classify autism—or autistic disorder, to give the condition its correct medical name—as one in a spectrum of disorders of development and behavior collectively called the *pervasive developmental disorders* (PDD), although the terms are frequently used interchangeably. Other related disorders falling under the umbrella category of PDD are Asperger's syndrome, Rett syndrome, and childhood disintegrative disorder.

Characteristic features of PDD include an impaired ability to communicate, form relationships with others, and respond appropriately to the environment. Different people with PDD seldom, if ever, show the same symptoms or deficits. Distinctions among different categories may be quite subtle. However, sufferers do tend to share certain problems that affect their behavior in more or less recognizable and predictable ways. For instance, they often appear to live in a different, isolated world. They appear indifferent, remote, and unable to connect with others at an emotional level. Although autism is not a type of mental retardation per se, many people with this disorder display lower than normal intelligence, a fact that may be related to the lack of proper communication between the patients and those around them rather than any innate deficits of intelligence. They also often show delays in language development. Asperger's syndrome, according to the National Institute of Mental Health, is on the milder end of the spectrum, where patients are often high-functioning and even display certain peculiar talents (like Rain Man). Patients do not typically demonstrate language disabilities but are prone to repetitive behaviors and odd obsessions. Rett syndrome is predominantly a disorder of girls, although there are rare cases among boys also. Mental retardation is a common feature, and in fact Rett syndrome is one of the most frequent causes of mental retardation among girls in the United States. Signs of abnormal development—such as reduced muscle tone, wringing hand movements, autistic-like behavior, and seizures—begin to manifest some 8–10 months after birth. Childhood disintegrative disorder is a very rare form of PDD. Children are born and seem to develop normally for relatively longer periods of time (2–4 years) before their behavior suddenly "disintegrates" into a condition typical of autism. Language skills and social behaviors regress, and the children become gradually more indifferent to their surroundings.

So what role, if any, does genetics play in such diseases? Not surprisingly, there is no simple relationship between our genes and PDD, and environment plays a huge role in their development. Still, to date, researchers have identified regions on at least seven different chromosomes that may increase susceptibility to PDD, including chromosomes 2, 3, 7, 11, 15, 19, and X.

An interesting and curious example of the genetics of autism is the association between the disorder and the secretin system. Now, the best-known function for secretin is as an endocrine hormone of the digestive system. Produced by the intestinal mucosa, it stimulates the pancreas and liver to send their secretions—bicarbonates and bile salts—into the gut (see the entry on food and digestion in Chapter 4). But mutations in both the gene for secretin (*SCT* on chromosome 11) as well as the gene for its receptor (the *SCTR* gene on chromosome 2) have been implicated in manifestations of autism. Exactly how this happens is not understood, but the presence of the secretin receptors in brain cells indicates that secretin also has a neuroendocrine function and that it is a disturbance in this function that leads to the development of autistic symptoms.

Scientists have also discovered a definitive link between Rett syndrome and mutations of a gene called *MeCP2* on chromosome X. The normal protein encoded by this gene belongs to a family of nuclear proteins that specifically bind to methylated DNA in the promoter regions of genes and thus inhibit gene expression. The MECP2 protein is essential during embryonic development. No doubt it is puzzling to see an X-linked disease that is more common in females because in our earlier discussions we stressed the fact that X-linked diseases are seen much more commonly in males! Why then do we see fewer Rett syndrome boys than girls? One reason is that the absence of a normal copy in the genome is lethal for males, either during pregnancy or shortly after birth. But this reason alone does not explain Rett syndrome girls because in most cases the normal gene on the second X chromosome should compensate for the mutation (which is what happens in classical X-linked diseases). The answer to this conundrum is related to the phenomenon of X-inactivation (see Chapter 3 for the explanation of this phenomenon) in females. Because this process occurs randomly in each cell, a female that inherits the Rett trait will have both normal cells in which the chromosome with the mutant gene is inactivated and abnormal cells with the normal version of the gene that is inactivated. This gives rise to a situation where there is a small amount of the normal protein, which prevents the mutation from being completely lethal. The presence of the mutated version of the protein, however, leads to the development of disease symptoms.

Finally, we note the subjective nature of PDD. The fact that they affect behavior rather than some aspect of physiology or structure raises a somewhat troubling issue, namely the issue of whether differences should be treated as symptoms of a disease. Are we guilty of diagnosing people with autism as having a disorder merely because they are different from us and in a minority? If someone from a foreign country who speaks an entirely different language and knows a completely different culture were suddenly transplanted into, urban America, wouldn't he or she display "abnormal" and antisocial behaviors? And wouldn't we do the same or feel the same if one of us were suddenly transported into a foreign country amidst people who spoke no language known to us? Neither of these situations warrants the treatment of the "different" person as ill or abnormal. Is it then fair or right to impose a single set of behavioral norms on all individuals of the human race and deem autism a disease?

There may be no simple answer to such questions, but perhaps a look at history might help. Although autism was not named formally until the twentieth century, historians have uncovered documentary evidence showing that, even in the eighteenth century, the social behaviors that we associate with autism today were regarded as "abnormal" or aberrant. In their book entitled *Autism in History: The Case of Hugh Blair of Borgue*, published in 2000, historian Robert Houston and cognitive psychologist Uta Firth describe the case of the titular character, a man of no discernible occupation in eighteenth-century Scotland, who was charged with insanity by his brother in a property rights case. Using the descriptions of his behavior by different people who testified, these modern-day scholars deduced that Hugh Blair would have been diagnosed as autistic today. Although not constituting irrefutable proof, this case shows that, even in a different time and place, the yardsticks used by people to distinguish between normal and abnormal were similar to those we use today. The pivotal issue in deciding this matter appears to be the level at which people are able to interact with others (i.e., sociability) and recognize and respond to changes in their cultural and linguistic environments. Whereas "normal" people in the scenarios posed earlier would, after an initial period of adjustment, at least try to communicate with others around them (for instance, by attempting to learn the local language), patients with PDD would remain indifferent to their surroundings.

AUTOIMMUNE DISEASES

As we saw in the previous chapter, one of the key properties of the human immune system is its ability to distinguish between the body's own cells and molecules (namely "self" antigens) and those belonging to external, potentially harmful agents such as bacteria and viruses. When this system of recognition fails, different things can happen. On the one hand, the immune system may fail to recognize certain foreign antigens and mount a proper response to remove the threat. This is what happens, for example, in the case of certain types of cancer. The other possible outcome of this lack of recognition is an attack by the immune system on the cells, tissues, and organs of the very body it was designed to protect. The resulting disease condition is called an autoimmune disease.

Over the years, scientists and physicians have identified many different autoimmune conditions in human beings. Broadly speaking, these fall into two main groups: *systemic* autoimmune disorders, in which there is immunologically induced damage to many different tissues and organ systems of the body, and localized or *organ-specific* disorders, which, as their name indicates, are localized to a single organ, tissue, or cell type.

What the immune system decides to attack depends on what goes wrong with the system. In some instances, the immune system produces specific antibodies (called autoantibodies) against a normal body protein. In Hashimoto's thyroiditis and diabetes mellitus type 1 (see the entry on diabetes), for example, the immune system produces autoantibodies against the thyroid gland and islet cells, respectively. However, there are many diseases where the autoimmune attack is actually a consequence of a somewhat overenthusiastic immune response. Again, these inflammatory responses against oneself may result in either generalized or organ-specific disease. An example of a systemic disease is rheumatic fever (also called rheumatoid arthritis), an autoimmune disease that follows cases of streptococcal infection (e.g., strep throat), often many years after the initial infection. Instead of clearing the bacteria from the bloodstream, the antibody forms complexes with the bacterial antigens, which are then deposited at sites such as the joints or heart muscles. Various nonspecific "soldiers" of the immune system (e.g., complements and phagocytes) then proceed to attack these areas of the body, giving rise to the painful swelling of the joints typical of rheumatic fever. Guillain–Barre syndrome is an example of a disease where the inflammatory response is primarily directed against a single substance, specifically the outer myelin sheath that covers the nerves.

Many autoimmune diseases show some association with an underlying genetic defect or group of defects. Scientists have identified a gene called *CTLA4* (cytotoxic T-lymphocyte associated serine esterase 4) on chromosome 2 that seems to be implicated in many autoimmune diseases. This gene encodes a T-cell-specific serine esterase enzyme that normally serves as the "brakes" in an immune response, preventing the cells from turning on the body after destroying the pathogen. Mutations in the gene reduce the ability of CTLA4 to keep the immune system in check and increase general susceptibility to autoimmune disorders. Patients diagnosed with autoimmune thyroiditis, Graves disease, or diabetes, for instance, will often have mutations in *CTLA4* in addition to other mutations.

Autoimmune Polyglandular Syndrome (APS)

Autoimmune Polyglandular Syndrome is characterized by problems in multiple endocrine glands, most notably the parathyroid and adrenal glands, but also other glands such as the thyroid, gonads, and pancreas (insulin-producing islet cells). Chronic infections of *Candida,* a type of fungus, particularly of the skin and mucus membranes, are another common feature of this disease. In addition, people with APS tend toward complete baldness early in life, have poor teeth (especially badly developed enamel), and suffer from inflammation of the cornea and whites of the eyes, chronic hepatitis, diarrhea, and anemia.

Although very rare, APS is interesting to immunologists and geneticists because it is the only systemic autoimmune disease definitively linked to a defect in a specific gene. The affected gene, called *AIRE* (for Auto*immune Re*gulator), lies on chromosome 21 and encodes a transcription factor that is found in both the cytoplasm and nucleus of cells. Exactly how mutations in this gene target the endocrine glands for immunologic attack is not completely understood, but scientists hope that the identification of the gene and mutation will help them to understand the mechanisms underlying this disease, as well as autoimmune disorders in general, and eventually devise treatments.

Diabetes Mellitus Type I

See the entry on diabetes in this chapter for details about this disease.

Hashimoto's Thyroiditis

In Hashimoto's thyroiditis (named after the physician who first described this syndrome), the immune system attacks and destroys the thyroid gland. Antibodies are produced against proteins of the thyroid cells, including a globular protein called thyroglobulin and certain microsomal proteins. The net result of antibody action is an eventual deficiency of the hormones of the thyroid gland (hypothyroidism), whose normal function is to control the rate of metabolism in the body.

Hashimoto's syndrome is the most common cause of hypothyroidism. It appears most commonly in middle-aged women, although it can occur in both sexes and at any age. The deficiency or lack of the thyroid hormones makes the body unable to adjust its metabolism to changes in its environment. Consequently, people with hypothyroidism are subject to fatigue, high sensitivity to cold, weight gain, dry skin, muscular weakness, constipation, muscular cramps, and even depression. Women often experience increased menstrual flow and cramps. A goiter (enlarged thyroid) is another common symptom.

The prognosis for patients with Hashimoto's syndrome is good because of the development of hormone replacement therapies. A synthetic version of the thyroid hormone is available that the patient can take orally to reverse the effects of the underlying hypothyroidism. Because this disease is chronic, the treatment is a lifelong one, requiring periodic checkups to adjust the level of the hormone according to the body's changing needs. The flip side of the good prognosis is the danger in a failure to diagnose this disease—the absence of hormone replacement therapy can have serious, even fatal, consequences. Because thyroid function often appears normal in patients with Hashimoto's syndrome until a significant portion of the thyroid is destroyed, diagnostic tests must include tests for antithyroid antibodies in addition to thyroid hormone functions.

Hashimoto's syndrome has been linked to at least two distinct genetic loci. One is the *CTLA4* gene on chromosome 2 mentioned earlier. The second locus, on chromosome 8, has not yet been associated with any functional gene.

Systemic Lupus Erythematosus (SLE)

More commonly known as lupus erythematosus or just lupus, SLE is a chronic, inflammatory autoimmune disease that affects many organs and tissues, including the skin, joints, heart, lungs, kidneys, and nervous system. It typically begins in a single location and then spreads to others as the condition progresses. Some ways in which lupus manifests itself include joint pains, a butterfly-like rash across the nose and cheeks, accumulation of protein in kidney cells, inflammation in parts of the heart and lungs, the tendency to form blood clots easily, and, as the disease progresses, strokes and pulmonary embolisms, seizures, headaches, and psychoses.

Lupus occurs in 1 out of 2,000 Americans and in as many as 1 in 250 young, African American women. It affects nine times as many women as it does men and mostly affects people between the ages of 10 and 50 years.

The pathology of lupus is due to the production of autoantibodies. The most prevalent autoantibodies in SLE are directed against the nuclei of various cells and are called antinuclear antibodies. They form immune complexes that are widely deposited in the joints, nerves, skin, muscles, and connective tissues, making these tissues the target of various inflammatory responses. When only the skin is affected by the immune response, the resulting condition is a variant of lupus called *discoid lupus*. Although there is no cure for lupus, anti-inflammatory medicines can control the worst of its symptoms and organ damage.

Exactly what causes SLE is not known. Genes, viruses, ultraviolet light, and drugs may all play some role in setting off the autoimmune response. Genetic factors seem to increase the general tendency to develop autoimmune disease, and a number of different genes and other loci on the genome are linked to lupus susceptibility. People who have relatives with lupus are more likely to get the disease than those in the general population. Some scientists think that the immune system in lupus patients is more easily activated by external factors such as viruses or ultraviolet light. For instance, a brief exposure to the sun can worsen and in some cases even trigger symptoms of lupus, as can certain medications for other diseases. Doctors have also observed that the symptoms of woman patients with SLE worsen before their menstrual periods. This phenomenon, together with the higher incidence of lupus in women, seems to point to a role for female hormones in the disease. Currently scientists are investigating this relationship between lupus and female hormones.

Scientists have found a link between lupus and the failure of enzyme Dnase1 to perform its normal functions properly. Normally, this enzyme, which is encoded on chromosome 16 in humans, plays a role in keeping the cell free of extra DNA (i.e., DNA that is not part of the chromosome or the mitochondrial plasmids) by chopping it up into small pieces. When researchers turned off the Dnase1 gene in mice, they found that although the mice appeared healthy at birth, the majority of them showed signs of lupus after 6–8 months. Presumably, the accumulation of undigested DNA within the cells causes them to burst (lyse) and release their contents into the bloodstream. When this happens, the immune system mounts an attack against the cellular contents, including the DNA and other nuclear material. Thus, a genetic mutation that disrupts the body's cellular waste disposal appears to be involved in the beginning of lupus.

See also: Immunity (Ch. 4).

BLEEDING DISORDERS

When someone has a bleeding disorder, their blood has difficulties in coagulating properly, which causes the patient to have a tendency to bleed longer. Patients bruise easily because of internal bleeding, and wounds and cuts take longer to heal. Bleeding disorders can arise either from defects in the blood vessels or from abnormalities in the blood itself. Here we discuss bleeding disorders caused by genetic defects.

Factor I Deficiencies—Disorders of Fibrinogen

Factor I deficiencies constitute an extremely rare class of genetic bleeding disorder associated with defects in the genes for fibrinogen on chromosome 4. According to the National Hemophilia Association, there have been only 200 diagnosed cases of this disease since the first report in 1920. Mutations in the fibrinogen gene can result in afibrinogenemia (a complete lack of fibrinogen), hypofibrinogenemia (reduced levels of fibrinogen), or dysfibrinogenemia (the presence of dysfunctional fibrinogen). Afibrinogenemia has very low rates of embryo survival, and when the babies do survive proves fatal very early in life. It is inherited in an autosomal recessive manner because the mutations must be present on both genes for a complete absence of the protein to result. Dysfibrinogenemias are extremely varied and can affect any of the functional properties of fibrinogen. Clinical consequences of dysfibrinogenemias include hemorrhage, spontaneous abortion, and thromboembolism.

Factor V Deficiency

Also known as Owren's disease or parahemophilia, this bleeding disorder is the result of defects in the gene for the clotting factor V, which normally helps to accelerate the activity of thrombin. When levels of factor V are low, blood clotting is delayed or progresses slowly. People with this deficiency may have occasional nosebleeds, excessive menstrual bleeding, and bruising, although many have no symptoms. The first sign of this condition may be bleeding following surgery. Bleeding can occur almost anywhere in the body, and death from hemorrhage has occurred with this disorder. However, the outcome is good with proper diagnosis and treatment.

Hemophilia

Hemophilia means the inability to clot blood. It is perhaps the best-known of all bleeding disorders. *Hemophilia A*, also referred to as *classic hemophilia*, is a disorder resulting from a deficiency in the clotting factor VIII, a key component of the blood coagulation cascade. The normal function of factor VIII is as a cofactor, a molecule that lends a helping hand, in the activation of factor X to factor Xa, which activates prothrombin to thrombin (see the entry on blood clotting in Chapter 4). Individuals with deficiencies in factor VIII suffer joint and muscle hemorrhage, bruise easily, and have prolonged bleeding from wounds. If they are not careful, they could potentially bleed to death. Treatment for this disease involves infusion of factor VIII concentrates prepared either from human plasma or by recombinant DNA technology.

Hemophilia A can arise from a variety of mutations in the gene for factor VIII. In fact, some 150 different single-site mutations have been identified as the cause of hemophilia A. Depending on how these mutations affect the level of activity of factor VIII, the condition may be severe, moderate, or mild. Because it is an X-linked disease, hemophilia A primarily affects males, in whom it is inherited with a frequency of 1 in 5,000–10,000 in all populations. Most females are able to compensate for any deficiency in factor VIII due to the presence of a normal copy of the factor VIII gene on their second X chromosome. As a result, women are mostly carriers of the disease and pass the defective gene to the next generation. Cases of female hemophiliacs are very rare indeed.

Hemophilia has a long history, having been described first by Babylonian Jews more than 1,700 years ago. It was brought to widespread public attention as a royal or princely disease in the nineteenth century when Queen Victoria, a carrier, transmitted it to several European royal families. The most famous case was the heir to the Russian throne, Alexis Romanov, whose mother Alexandra, a cousin of Queen Victoria, was also a carrier.

Hemophilia B results from deficiencies in clotting factor IX, which is also encoded on chromosome X and so exhibits a sex-linked inheritance. The prevalence of hemophilia B is approximately one-tenth that of hemophilia A. Patients with hemophilia B have prolonged coagulation time and decreased factor IX clotting activity. As in hemophilia A, there are severe, moderate, and mild forms of hemophilia B that reflect the factor IX activity. At least 300 unique factor IX mutations have been identified, 85 percent of which are point mutations, 3 percent short nucleotide deletions or insertions, and 12 percent gross gene alterations.

Hemophilia B Leyden is a specific form of factor IX deficiency characterized by a gradual increase in factor IX levels after puberty. Whereas clinical manifestations of the prepubertal condition are essentially the same as in hemophilia B, the increase in factor IX after maturity elicits a corresponding lessening in bleeding episodes and other symptoms of hemophilia as the person grows older. In contrast with hemophilia B, which is caused by mutations in the functional regions (exons) of the gene, the Leyden version is specifically associated with point mutations in the promoter region. The promoter controls the levels of gene expression, and mutations in the region block the proper expression of the gene. Even with point mutations, however, the promoter is sensitive to hormonal changes and is "turned on" during puberty due to the activity of testosterone. This, in turn, leads to gene expression and the production factor IX. Named for the fact that it was first described in families of Dutch ancestry, hemophilia B Leyden accounts for a small percentage of all B hemophiliacs.

Von Willebrand Disease

Perhaps the most common inherited bleeding disorder, especially among women, von Willebrand disease (vWD) is due to a deficiency in von Willebrand factor (vWF). To give an idea of the relative incidence of this bleeding disorder in comparison with others, abnormalities in vWF can be detected in approximately 8,000 people per million on the basis of sensitive laboratory tests. Clinically significant vWD occurs in approximately 125 people per million. This is a frequency at least twice that of hemophilia A.

The von Willebrand factor is a complex protein that functions both as an anticoagulant factor and a mediator of adhesion between platelets to the collagen exposed on the walls of the blood vessels. Consequently, protein deficiencies result in defective platelet adhesion and a deficiency of factor VIII. The result is that vWF deficiency can cause bleeding that appears similar to that caused by platelet dysfunction or hemophilia. Frequently there is bleeding into the skin and from mucous membranes, which line the nose, gums, and the gastrointestinal and genitourinary tracts. Symptoms include frequent and prolonged nosebleeds, easy bruising, bleeding gums, and prolonged bleeding following dental work or surgery. Women with this condition also experience very heavy menstrual periods.

Von Willebrand disease is an extremely heterogeneous disorder that has been classified into several major subtypes. The underlying reason for this heterogeneity is probably the fact that the disease is caused by different mutations in the vWF gene, which lies on chromosome 12. Type I vWD is the most common and is inherited as an autosomal dominant trait. This variant is caused by a simple deficiency in the amount of vWF in the blood. Type 2 vWD is caused by the formation of a dysfunctional protein rather than merely a deficiency of the normal protein and is subdivided further according to whether the dysfunctional protein has decreased or, paradoxically, increased function in certain laboratory tests of binding to platelets. Type 3 vWD is the most severe form clinically and is characterized by a virtual absence of vWF and a recessive inheritance.

See also: Blood Clotting (Coagulation) (Ch. 4).

CANCER

We often talk of cancer as though it is one disease, but cancer is really a group of different diseases that can appear in any part of the body and take many forms. The one common characteristic shared by all cancers is an abnormal uncontrolled growth and multiplication of certain cells in the body. Normally, a cell knows when to stop growing, usually in response to the other cells around it. In many organs and tissues, cells also know when to die (see the entry on apoptosis, or programmed cell death, in Chapter 4). But sometimes cells lose all control over their normal development. They fail to mature properly or to die as originally intended, use up the body's resources, and disrupt the normal functions of the organs and tissues. The result is a cancer originating at the site where the cells first lost control. Often cancer cells break off from their original site and spread to other parts of the body, where they implant themselves and cause further damage. This spreading of cancer is called metastasis and is a sign of an advanced and usually nontreatable cancer.

Different tissues in the body give rise to distinct types of cancers. For example, cancers in tissues such as the skin, breast, and liver typically arise in the epithelial tissues of these organs and are called carcinomas. Cancers originating in soft tissues such as the connective tissues, muscles, and fat are sarcomas, blastomas are cancers of nervous tissue, and leukemias, lymphomas, and myelomas are the cancers of different blood-forming cells. The term "tumor" is used to describe cancers of solid tissues, including various carcinomas, sarcomas, and certain blastomas.

What causes a cell to lose control over its normal development and become cancerous? Just as there are many types of cancer, so also the triggers that set off cancer are numerous and as yet incompletely understood. As we have seen in previous chapters (see the entries on apoptosis, cell cycle, development, and signal transduction in Chapter 2 and apoptosis in Chapter 4) the cell cycle of is a complicated and highly regulated system of different proteins acting together. So a defect or disruption of any component of the system can cause the cell to lose control and become cancerous. As we will see, many cancers are linked to defects in the genes that regulate the cell cycle. As these genes in turn are susceptible to modifications and disruption from a wide variety of environmental factors, the spectrum of cancer causation is very wide indeed!

Breast Cancer

Cancers originating in the mammary or breast tissues are called breast cancers. Most of these are carcinomas (i.e., cancers of the epithelial tissue) and originate in the ducts that carry milk. Breast cancer is one of the most commonly diagnosed human carcinomas and one of the most heavily monitored cancers as well. The National Cancer Institute estimates that about one in 50 women in the United States will develop breast cancer by age 50. Around 180,000 people, the vast majority of them women, are diagnosed with breast cancer each year, and about 40,000 women die of the disease each year in the United States. (Although male breast cancer is very rare, men can also get this type of cancer.)

Breast cancer is usually detected as a lump or thickening of the skin in or around the breast tissue. Such signs appear only after the tumor is fairly advanced, however, and a mammography or breast X-ray can detect tumors at a much earlier stage. Following the observation of a suspicious lump, doctors need to do a biopsy to confirm the presence of cancer and determine its type.

Breast Cancer Susceptibility Genes

A number of genes have been found to be associated with breast cancer, although these genes currently account for only 5–10 percent of all breast cancer cases. Two genes, designated as *BRCA1* and *BRCA2*, have been identified thus far. There is evidence that other such genes exist, and scientists are actively searching for them.

BRCA1. *BRCA1* is a tumor suppressor gene located on the long arm of chromosome 17. Mutations in this gene have been associated with both breast and ovarian cancers. When a person inherits one mutant copy of *BRCA1*, their risk of developing breast cancer is over 80 percent, but the actual development of cancer requires the accumulation of more mutations in other genes, and at least one of these mutations must occur in the other copy of *BRCA1*.

BRCA2. *BRCA2* also functions as a tumor suppressor gene; however, its exact function has not been well-characterized. The gene is located on chromosome 13. Unlike *BRCA1*, *BRCA2* has not been linked to ovarian cancer. Also, mutations in *BRCA1* are usually insertions and point mutations, but *BRCA2* mutations seem to be mostly deletions of short DNA sequences.

Risk Factors for Breast Cancer

Most scientists and doctors believe that breast cancer is caused by a combination of genetic and nongenetic factors. Nongenetic factors that raise the risk of breast cancer include early puberty, menopause after age 55, first childbirth after age 30, obesity, alcohol use, and excessive radiation exposure. But most of these risk factors cause less than a twofold increase in the risk of breast cancer, so that each makes a relatively small contribution to breast cancer risk. Other potential risk factors include an imbalanced diet—high in fat and low in fiber, fruits, and vegetables—and the lack of exercise, both of which contribute to obesity, a previously mentioned risk factor.

Less than 1 percent of breast cancers are associated with genetic syndromes such as Cowden syndrome, Li–Fraumeni syndrome, Bloom syndrome, Peutz–Jeghers syndrome, and Werner syndrome. Interestingly, however, patients with these disorders, some of which have been described elsewhere in this book, often have a high risk of developing breast cancer.

There is also a possible risk of breast cancer associated with a number of common genetic variants. For instance, ataxia telangiectasia (AT) is a rare, neurodegenerative childhood disease that affects the nervous system. Children with AT are susceptible to childhood cancers. Epidemiological studies show that women who are heterozygous for a mutation in the *ATM* gene (i.e., they have one mutant copy and one normal copy of the gene) may have up to a fivefold higher risk for breast cancer. Another potential risk factor involves the metabolic enzyme N-acetyl transferase 2 (NAT2). People with one type of NAT2 are "fast acetylators," and those with another variant are "slow acetylators." Epidemiologists have found that among slow acetylators who also smoke, breast cancer risk was significantly higher than normal.

See also: Neurocutaneous Syndromes—Ataxia-Telangiectasia.

Proteins That Control the Spread of Breast Cancer

Scientists have found a protein called Endo180 in breast cancer cells. This protein, which is most likely encoded by the *PLAUR* gene on chromosome 19, plays a key role in the formation of tiny antennae, known as filopodia, on the cell surface, which help the cell move around. Scientists also found that cells with Endo180 are attracted to another protein, called u-PA, which is a serine protease involved in the degradation of the extracellular matrix. The gene for u-PA (*PLAU*) resides on chromosome 10. If either of these two proteins is missing, cancer cells can no longer move to other parts of the body. This is a very exciting discovery because containing cancer cells and preventing them from spreading is very important in treating the disease.

Burkitt Lymphoma

Burkitt lymphoma is a solid tumor of B lymphocytes and is most commonly seen among young people between the ages of 12 and 30 in Central Africa, although children in other parts of the world are also affected. The syndrome was named after Denis Parsons Burkitt, a physician who first mapped the distribution of the disease across Africa. Although relatively uncommon in the United States (although the incidence has risen since AIDS became widespread), it is endemic in Africa and other tropical countries around the equatorial belt, with a constant rate of 100 cases per million children.

The main feature of Burkitt lymphoma is an excess of small noncleaved, and hence undifferentiated, B cells in different tissues. Although it may spread diffusely throughout the body, Burkitt lymphoma has a special predilection for the lymph nodes, spleen, and bone marrow (i.e., tissues where B cells are synthesized). This results in the enlargement of the jaws (especially notable among African populations) and of the abdomen (due to splenic enlargement), although other organs may be affected. Based on its epidemiology, there are two main types of this disease, the African type, which is endemic, and a sporadic or spontaneously appearing version, which seems to be more widespread around the globe.

The genetic basis for Burkitt lymphoma appears to be a chromosome translocation of the *c-myc* gene on chromosome 8. In about 90 percent of the cases, the *c-myc* gene is moved close to the antibody heavy-chain enhancers on chromosome 14, and in other cases it is moved close to the enhancers of the antibody light-chain genes on chromosomes 2 and 22. The translocation puts the gene, whose product is a transcription factor essential for mitosis, under the control of the regulatory elements that are responsible for the vigorous synthesis of antibodies. Consequently, there is an overproduction of the transcription factor, which results in the loss of control of the cell's proper growth cycle and hence the development of malignancies. About 25 percent of Burkitt lymphoma cases are also associated with infections of the Epstein–Barr virus. The mechanism by which this virus causes disease is not fully understood, but it is believed that the lymphocytes have EBV receptors and absorb the virus, thus shielding it from the body's immune system. Consequently, these cells are more prone to excessive proliferation when chromosomal translocation occurs.

See also: Chromosome, Translocation (Ch. 2).

Chronic Myeloid Leukemia (CML)

Chronic myeloid leukemia is a cancer of the white blood cells that is characterized by the production of large numbers of abnormal blood cells—specifically the myeloid leukocytes—in the bone marrow. It usually arises in the pluripotent stem cells that give rise both to granulocytes (neutrophils and basophils) and platelets. The cancer has a long initial phase, known as the "chronic" phase, which can range from months to years and during which time there are few or no symptoms. The leukemic cells grow slowly during this stage of the disease, but they can be found in the circulating blood, and elevated levels can serve as an early warning sign. Eventually, the chronic phase gives way to a rapid acceleration phase during which the abnormal cells proliferate very actively and disease becomes apparent. Symptoms include fevers, fatigue, anorexia, pain in the bones,

Normal Chromosomes

Translocated Chromosomes

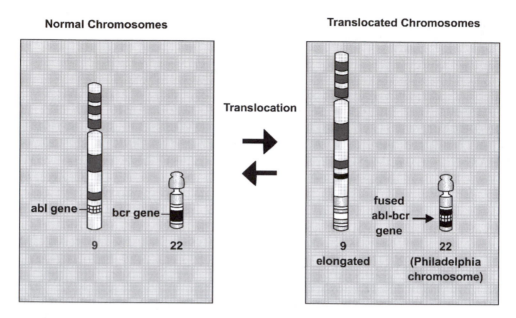

Figure 5.1. The Philadelphia chromosome (translocation). [Ricochet Productions]

an enlargement of internal organs, notably the spleen and liver, and increased susceptibility to bruising and bleeding. Although these are somewhat nonspecific signs, their development will call for blood tests, which will give a definitive diagnosis.

Chronic myeloid leukemia is the result of a specific genetic event that occurs in the stem cells. This event is the reciprocal translocation between chromosomes 22 and 9, which results in a chromosome 9 that is longer than normal and a shortened chromosome 22, called the "Philadelphia chromosome" (see Figure 5.1) for the city where scientists first made the discovery in 1960. The translocation occurs at very specific locations of both chromosomes: in the middle of the *BCR* (breakpoint cluster region) gene on chromosome 22 and the *ABL1* gene on chromosome 9. Although scientists have studied the Philadelphia chromosome extensively because of its harmful effects—in addition to CML, it has also been implicated in certain acute leukemias—the exact function of the normal *BCR* gene product is still unknown. It is known to possess both a serine/threonine kinase activity and GTPase activating abilities. These activities suggest that it is a signaling molecule, but the signal transduction pathways in which it normally participates remain unknown. *ABL1* is a proto-oncogene that encodes a tyrosine kinase (see the entries on Signal Transduction in Chapters 2 and 4 for details) that has been implicated in processes of cell differentiation, cell division, cell adhesion, and stress response. (*ABL* stands for Abelson, which is the name of a leukemia virus that also produces a similar protein.)

As discussed in Chapter 4, a balanced translocation can lead to the formation of new genes at the sites of chromosome breakage and reunion. In the Philadelphia chromosome, the fusion of the *BCR* and *ABL1* genes results in the formation of a new gene that has the 5' section of *BCR* fused with most of *ABL1*. This gives rise to a protein that has the tyrosine kinase activity of *ABL1* but lacks the regulatory switch that keeps the normal ABL protein under tight control. As a result, the fusion protein is continuously active in the cell, which causes the unchecked stimulation of the different signal transduction pathways controlled by this gene product, which in turn leads to the development of cancer.

The treatment of CML is aimed at reducing the growth of abnormal cells in the bone marrow. The chronic phase may be kept in check with chemotherapy, but more drastic measures, such as bone marrow transplants, are required once the cancer progresses to the accelerated phase. A drug called imatineb mesylate (or Gleevec) is used to treat patients who have not responded well to transplantation therapy. This drug blocks the active site of the BCR–ABL protein and prevents ATP from binding there so BCR–ABL cannot then phosphorylate other proteins in the signaling cascade. It acts directly on the leukemic cells to inhibit their growth. Another therapeutic approach that seems to be giving encouraging preliminary results is the use of short segments of antisense DNA, which aims to suppress the formation of the leukemic cells while not affecting normal bone marrow development.

Colon Cancer

Cancers of the large intestines (i.e., the colon and rectum) are among the most widespread types of inherited cancers known and the third most common cancers to affect people overall in the United States. According to estimates from the American Cancer Society, there were nearly 100,000 new cases of colon cancer and 50,000 new cases of rectal cancer diagnosed in the United States in 2003.

Several factors seem to contribute to the development of colorectal cancers. Scientists have established links with mutations in at least three distinct genes: *MSH2* and *MSH6* on chromosome 2 and *MLH1* on chromosome 3. Interestingly, all three genes encode proteins that participate in DNA repair mechanisms. The proteins made by *MSH2* and *MLH1*, for instance, are mismatch repair enzymes, whereas *MSH6* codes for a protein that recognizes and binds specifically to GT mismatches. In the Chapter 4, we mentioned how the breakdown of these functions could lead to uncontrolled cell growth and division. However, these proteins are expressed virtually everywhere in the body and are not confined to any particular tissue, so there is no obvious reason for them to cause cancers specifically in the large intestine when they are mutated. Their lack of association with cancers of other tissues is still something of a mystery.

Researchers have discovered that mutations in the gene *AXIN2* are associated with the development of colorectal cancer. This gene, which sits on chromosome 17, codes for a protein belonging to an important biochemical pathway that includes several other proteins, such as the one encoded by the adenomatous polyposis coli gene (*APC*) on chromosome 5, whose mutations have also been linked to colorectal cancer and familial adenomatous polyposis. The AXIN2 and APC proteins, like other tumor suppressors such as p53, control the expression of genes critical in the cell division process.

In addition to these genes, there are other risk factors for colorectal cancers. The most significant environmental factors are age and diet. Although both men and women are at equal risk, most but not all cases occur in people over the age of 50. A diet that is high in fat and low in fiber poses a high risk for developing these cancers. Other bowel diseases, such as ulcerative colitis (see the entry on inflammatory bowel disease) and intestinal polyps, also increase the chances of colorectal cancers.

Although the death toll for this type of cancer still remains high—nearly 57,000 people in the United States died in 2003, for example—the overall death rate has gone down steadily over the past 15 years. Newer, better techniques for detecting these cancers at the earlier stages and improved treatments have contributed to this encouraging trend.

See also: DNA Synthesis/Replication and Repair (Ch. 4); Inflammatory Bowel Disease, Intestinal Polyps (Familial Adenomatous Polyposis).

Lung Cancer

Lung cancers are the leading cause of cancer deaths among both men and women both in the United States and in other countries, such as Canada and China. They are among the most malignant of cancers, with a five-year survival rate of less than 15 percent.

Most lung cancers are carcinomas, namely tumors originating in the epithelial tissues, although on occasion certain sarcomas may also arise in the lungs. They begin in the tissues lining the airways and have usually spread to adjacent and surrounding tissues before they are detected. Lung carcinomas are classified into two main categories based on the appearance of the cancer cells under the microscope—the small-cell and the non-small-cell carcinomas. Although these groups are not necessarily homogeneous with respect to origins or appearance, the approaches to treatment are consistent within each group.

Mutations in several different genes, in response to many different triggering factors, are known to cause lung cancers. They often activate oncogenes, which have also been implicated in other cancers. Examples of genes affected in lung cancer include:

1. A tumor suppressor gene called *SLC22A1L* located on chromosome 11. Alterations in this gene have also been associated with breast and ovarian cancers.
2. The gene for p53 tumor antigen on chromosome 17, which has been found at elevated levels in a wide variety of transformed or malignant cells.
3. Oncogene homologs of certain mouse sarcoma viruses found on chromosome 12 (*RASK2*) and chromosome 7 (*BRAF*).

Small-Cell Lung Carcinomas

Accounting for about 15–25 percent of all lung cancers, the small-cell carcinomas start in the hormonal cells of the lung, usually in the large airways—the bronchi—of the lungs, in the central part of the chest. This type of cancer has also been called oat-cell cancer because the cancer cells look like grains of oats when viewed under an ordinary light microscope. The cancer cells are very small and spread (i.e., metastasize) easily because of the constant flow of blood and lymph through the lungs. This has two consequences: These carcinomas are usually detected only after the cancer has spread and thus cannot be surgically removed. However, they have proven somewhat susceptible to chemotherapy and radiotherapy (the latter only if the cancer is detected at a relatively early and localized stage). Regardless of treatment, prognosis is poor, and the five-year survival rate is about 6 percent, which means that 6 people out of every 100 diagnosed with lung cancer survive beyond 5 years.

Genetically, small-cell lung carcinoma has been shown to arise due to a specific chromosomal abnormality—the deletion of a locus called *SCLC* on chromosome 3. The deletion has been shown to occur in response to certain environmental stresses, most notably prolonged exposure to cigarette smoke.

Non-Small-Cell Lung Carcinomas

The non-small-cell carcinomas of the lung are a heterogeneous group of cancers that are grouped together because they can all be treated surgically if caught in the early stages, but unlike small-cell carcinomas, they respond poorly to chemotherapy. There are three main subcategories of non-small-cell carcinomas, depending on the origin of the tumor. They include:

1. *Adenocarcinoma*—Arising in the mucus-producing cells of the lungs, most often near the edge of the lungs, this is the most common type of lung cancer, accounting for 30–50 percent of all lung cancers. Lung adenocarcinomas have been linked to smoking but are also the most common type of lung cancer to affect nonsmokers.

2. *Squamous cell carcinoma*—This type of cancer originates in the reserve cells, whose normal function is to replace damaged epithelial cells in the bronchi of the lungs. Squamous cell tumors are usually found near the center of the lungs, grow to large sizes, and metastasize easily to other locations, including the bones, brain, liver, and adrenal glands. This type of cancer is almost always caused by smoking and accounts for 25–40 percent of all lung cancers.

3. *Large-cell lung carcinoma*—This subcategory makes up about 10 percent of lung cancer cases and serves as a catch-all category for non-small-cell carcinomas that are not identifiable as either squamous cell carcinomas or adenocarcinomas.

Lung Cancer and Smoking

In no other case is the link between cancer and a specific environmental factor or mutagen as well-established as that between lung cancer and smoking. Cigarette (tobacco) smokers are about 13 times more likely to develop lung cancer than nonsmokers, although health experts believe that nonsmokers who inhale cigarette fumes from smokers (called second-hand smoke) are also at increased risk for developing cancer. All told, cigarette smoke may cause up to 87 percent of all lung cancer cases. The reason for the risk is that smoke, and particularly tobacco smoke, contains chemicals that are proven carcinogens (cancer-causing substances), the constant exposure to which causes cells to develop mutations in their genes. Because the lungs are the organs that are most exposed during smoking, the cells of these organs are most likely to develop mutations and become cancerous. The link of tobacco to cancer is also borne out by the observation that people who chew rather than smoke it are at a high risk for developing cancers of the throat and esophagus. Other risk factors for lung cancer include radon gas (which is formed from the natural decay of uranium deposits in the ground and can seep into the surrounding water and air), asbestos, and industrial dusts containing substances such as uranium, arsenic, and vinyl chloride.

Melanoma

Melanomas originate in melanocytes, cells that produce the pigment melanin, which is responsible for the color of our skin, hair, and eyes (see Chapter 4). Consequently, melanomas are typically cancers of the skin and the pigmented portion of the eyes (namely, the iris). Although less common than other types of skin cancer, which involve the basal or middle layers of the epithelium, melanoma is the most aggressive type. Unfortunately, its incidence has been steadily increasing, and today it stands as the leading cause of death from skin disease as well as the biggest cause of cancer death in women between the ages of 25 and 30. According to estimates from the American Cancer Society, more than 50,000 new cases of melanoma are diagnosed per year in the United States, and the annual death toll is at nearly 8,000. Ocular melanoma is the most common form of eye tumor in adults.

Virtually anyone can develop melanoma, although the cancer is typically more common in lighter-skinned people, especially those with a tendency to freckle or get sunburn easily. People with a large number of moles or birthmarks also seem to be at higher risk. Family history is also a strong indicator of risk. Melanomas may appear anywhere on normal skin but often begin as a mole. Physicians currently classify the cancer into four main types:

- *Superficial spreading melanoma*—This is the most common type of melanoma both in general and especially in Caucasian populations. The cancer begins with a flat, irregular brown-to-black lesion and spreads over larger areas of the skin.

- *Nodular melanoma*—This form usually starts as a raised area (like a mole) that is dark blackish-blue or bluish-red, although some may lack color.

- *Lentigo maligna melanoma*—This version usually affects elderly people, who characteristically develop it in areas of the skin most exposed to the sun such as the face, neck, and arms. The cancer appears as large, flat areas of tan-colored discoloration, often intermixed with darker brown patches.

- *Acral lentiginous melanoma*—The least common form of melanoma, this form of the cancer is seen more frequently among African American populations than others. It usually originates on the palms, soles, or under the nails.

Germ-line mutations in genes for proteins such as the tumor suppressor p16 (encoded by *CDKN2* on chromosome 9) and cyclin-dependent kinase 4 (on chromosome 12) that are involved in regulating cell division are associated with familial forms of cutaneous (skin) melanomas. Genes whose signaling proteins are specifically involved in melanin synthesis and transport, such as those described in Chapter 4 (see the entry on skin, eye, and hair color) may also be implicated in certain forms of this cancer. There is also a family of genes known as the melanoma antigens that are clustered on the X chromosome. These genes seem to be expressed only in the cells of melanoma patients or in vitro by melanoma-derived cell lines, and their normal function, if any, is unknown.

A study of people with inherited and noninherited forms of the disease has found a potential link between the gene *NRAS*, mapped to chromosome 1, and exposure to UV light, which is strongly associated with melanoma incidence. Scientists are speculating that UV exposure may cause mutations in *NRAS*, which have been implicated in various cancers, including melanoma.

Treatment for melanomas is only effective if the cancer is detected at the very early stages and the lesions are shallow. The cancerous skin cells and a portion of the normal surrounding skin usually need to be removed by surgery, and a skin graft may be necessary afterward especially if the area affected was large. Follow-up in the form of lymph node biopsies is advisable to monitor the spread of cancer. Radiotherapy, chemotherapy, or immunotherapy may be used to supplement surgery and to prevent the spread of cancer but cannot replace the primary surgical treatment.

Retinoblastoma

Retinoblastoma is a malignant tumor of the eyes in young children. Although its incidence in the general population is low compared with other human cancers, it is the most common eye tumor diagnosed in young children and the third most common cancer in children overall. About 90 percent of the cases occur before the age of 5, and children may even be born with the disease.

Common signs of retinoblastoma include leukokoria—which is the appearance of a white (rather than black) reflecting pupil—and squinting. Somewhat less commonly, the child's vision may show signs of early deterioration, and the eyes may become inflamed and red. There are two forms of retinoblastoma: a familial or inherited form of the disease, occurring in 30–40 percent of all cases, and a sporadic form that is not passed along to the next generation. Inherited tumors tend to occur at a younger age, more often than not originate in both eyes (i.e., they are bilateral), and arise in many different spots of the retina at once (i.e., they are multicentric). Sporadic or spontaneous retinoblastomas develop somewhat later (sometimes even into adulthood) and tend to occur as single tumors in one eye (unilateral) only.

The underlying cause for both familial and sporadic retinoblastomas is the same, namely mutations in the retinoblastoma gene (*Rb*), a tumor suppressor gene on chromosome 17 (see the entry on tumor suppressor genes in Chapter 4) that makes a protein involved in regulating the cell cycle. This protein is largely expressed in the retina and in certain bone tissues, which leads to the development of retinoblastomas and osteosarcomas (tumors in the bone tissue) when mutations occur. The cancer-causing mutations of *Rb* are recessive, which means that both copies of the gene must be defective or absent for tumors to develop. As long as even one copy of *Rb* is normal, we will be fine because the normal protein will compensate for the loss of function caused by defects in the allele. In familial retinoblastoma, a child inherits a copy of a defective *Rb* gene from one of its parents and damages or loses the second, normal copy of the gene early in development, usually before birth. This means that nearly all of the child's cells will contain two damaged *Rb* genes, which leads to the development of tumors in the retina. Sporadic retinoblastoma cases develop due to the occurrence of mutations in both copies of the *Rb* genes of a single retinal cell during development. Consequently, the tumor occurs in just one spot.

Untreated, retinoblastomas are almost always fatal, but they respond to treatment if detected early enough. In fact, this tumor has one of the best cure rates of all childhood cancers. If the cancer is fairly advanced, treatment may entail the removal of the eye (or both eyes) altogether, but with improved technologies both for early detection and treatment, the prognosis for preserving both vision and life has greatly improved since the 1980s or so. Consequently, early detection through proper eye exams and genetic screening is important.

See also: Tumor Suppressor Genes (Ch. 4).

CHARCOT–MARIE–TOOTH DISORDER

Named for three physicians who first identified this degenerative neurological disease, Charcot–Marie–Tooth disorder (henceforth CMT) is one of the most common neuropathies diagnosed in the world, affecting one in 2,500 people according to U.S. statistics. Two French physicians—Jean-Marie Charcot and Pierre Marie—and an Englishman, Howard Henry Tooth, independently described the condition in 1886. The disease is characterized by a slow progressive degeneration of the muscles of the feet, lower legs, hands, and forearms and a mild loss of sensation in the limbs, fingers, and toes. One of the first signs of CMT is the development of a high-arched foot or an observable disturbance in gait. Other symptoms include abnormalities of the bones of the feet such as high arches and hammer toes, problems with hand function and balance, occasional lower leg and forearm muscle cramping, the loss of some normal reflexes, occasional partial sight and/or hearing loss, and the development of curvature of the spine. Most of these symptoms occur by the age of 30, and although the symptoms contribute significantly to a lowered quality of life, CMT is not fatal.

Perhaps one of the reasons for the high incidence of CMT is its genetic heterogeneity. This means that the disease symptoms are caused by different mutations on different genes. This variety also lends CMT diverse and often confusing patterns of inheritance. One of the most common forms of the disease, labeled CMT type 1A, maps to a gene on chromosome 17 that encodes a membrane protein (called PMP22 for peripheral myelin protein) that is a major component of myelin, the substance that forms the fatty sheath that protects nerves and helps with conducting impulses in the peripheral nervous system. The specific type of mutation in this gene is a gene duplication that results in the overproduction of PMP22, which leads to abnormalities in the normal structure and function of myelin. The inability of myelin to conduct impulses properly, in turn, causes the wasting of muscles serviced by these nerves. Point mutations in the genes of another protein component of myelin (myelin protein zero), encoded on chromosome 1, also cause similar symptoms inherited in a similar autosomal dominant fashion (CMT type 1B). An autosomal recessive form of the disease has been mapped to mutations in the gene for lamin, also on chromosome 1, which is also known to be associated with various types of muscular dystrophy. Yet another recessive form of CMT has been mapped to the gene for a protein called ganglioside-induced differentiation-associated protein 1 (GDAP1) encoded on chromosome 8 that plays a role in nerve development. There is also an X-linked form of CMT syndrome that maps to mutations on a gene for connexin, a gap junction protein that mediates the communication between peripheral nerves.

Finally, it should be mentioned that the same proteins involved in the type 1A and type 1B CMT are also involved in a disease called *Dejerine–Sottas syndrome* (DSS) (also sometimes known as CMT type 3), which is a more severe form of CMT that may manifest as early as infancy. Research into understanding the pathogenesis of CMT, through the use of animal models for the disease, should also give insight into DSS and may lead to therapies for both diseases.

See also: Muscular Dystrophy.

COLOR BLINDNESS AND ACHROMATOPSIA

Color blindness is the name for a group of vision defects characterized by a deficiency in the ability to tell the difference between certain colors, such as red and green or blue and yellow. The word "blindness" for this disorder is actually a misnomer because there is no loss of visual acuity or any degeneration of eye tissue. Rather, the deficiency in color vision is caused by genetic defects in the genes for the photoreceptors that are present in different populations of cone cells in the retina.

The most common form of color blindness is the difficulty in distinguishing between red and green, particularly the blue-green shades. Also known as red–green color blindness, this is an X-linked disorder that is known to affect some 10 million American men, which is 7 percent of the total male population in the United States. In contrast, only about 0.4 percent of the female population is affected. Mutations causing red–green color blindness affect one of two adjacent genes for either the green (*OPN1MW*) or red (*OPN1LW*) light receptors or opsins (see the entry on the senses in Chapter 4). About 75 percent of all red–green deficiencies are caused by mutations in the former (i.e., green cones), whereas red cone mutations seem somewhat less frequent and account for only 25 percent of the total cases. As indicated by the X-linked nature of the condition, the mutations are recessive and only result in symptoms when there is no compensation by a normal copy. This is also the reason why blue–yellow color blindness is so rare. The gene for the opsin that absorbs blue light lies on chromosome 7, and mutations are thus inherited in a classical recessive pattern. Only individuals with two copies of the defective gene will exhibit difficulties in distinguishing between blue and yellow.

In very rare instances, a person may lack the ability to perceive any color at all and view the world in shades of grey, black, and white. This condition is called *achromatopsia*. The incidence in the United States is about 1 in 33,000. People with the disorder have no real understanding of the concept of color—red may be perceived as dark grey and yellow as a lighter shade of grey. Unlike red–green and blue–yellow color blindness, achromatopsia is accompanied by other defects in vision, including decreased visual acuity and sensitivity to light.

Although very rare in most parts of the world, achromatopsia is known to affect nearly 6 percent of the total population of a small Western Pacific island called Pingelap. Genealogical studies on this population led to the discovery of the genetic cause of the condition. Scientists found that the affected individuals had mutations in a gene called *CNGB3* on chromosome 8. This gene encodes a nucleotide-gated ion channel that participates in the signal transduction pathways stimulated by the photoreceptors (opsins) in cone cells. When the protein is defective because of mutations, it fails to transduce signals properly and does not process the information about different colors, leading to achromatopsia. A related gene that has been similarly implicated in European populations is *CNGA3* on chromosome 2.

See also: The Senses—Vision (Ch. 4).

CYSTIC FIBROSIS

Cystic fibrosis (CF) is a genetic disease—attributable to mutations in a single gene—in which normal body fluids such as mucus, tears, saliva, and pancreatic juices become viscous and sticky. Instead of their normal lubricating activities, these secretions start to clog up the passageways in the organs where they are produced and interfere with their proper functioning. The effect is especially marked in the lungs and pancreas—patients have difficulties in breathing properly and digesting food because of the accumulation of mucus that blocks the respiratory passages and the pancreatic ducts. Unless the mucus is cleared away consistently, the patient will die. Indeed, cystic fibrosis is still regarded as a terminal disease, with respiratory failure as the most common cause of death.

Cystic fibrosis is perhaps the most widely diagnosed fatal inherited disorder, affecting over 30,000 Americans. It occurs most commonly among Caucasians (whites)—at a frequency of 1 in every 3,000 births—and less so among people of African and Asian descent (1 in 15,000 people and 1 in 31,000 people, respectively). An inherited condition, it manifests itself very early in life. As recently as the 1990s, the life expectancy of children born with this condition was very low, and most did not survive beyond their teens. Although there is no established cure for the disease, the situation is now considerably improved, and many CF patients now live into their thirties and beyond and also enjoy a considerably better quality of life.

One of the major reasons for these improvements was the identification and cloning of the *CFTR* gene. The normal protein product of this gene, on chromosome 7, is an ion transporter (see the entry on channel proteins and cross-membrane transport in Chapter 4) that specifically carries sodium and chloride ions (i.e., salt) across membranes in epithelial cells. When the salt concentrations are unbalanced because of improper transport, these cells are unable to produce their secretions (mucus, saliva, etc.) at their normal viscosity. Since the identification of this gene, several hundred different mutations have been identified in different CF patients. The severity of the disease symptoms of CF is directly related to the characteristic effects of the particular mutation(s) that have been inherited by the sufferer. Because the disease is caused by the deficient functioning of a gene, CF is an autosomal recessive disease that can be inherited from non-symptomatic carriers of the mutated gene.

Because CF is a single-gene disease, it would be an ideal candidate for gene therapy (see the entry on gene therapy in Chapter 2). In fact, gene therapy at an early age is the only way to cure the disease. However, such a cure is still many years away from realization. Currently, treatment for CF involves little more than easing the symptoms and the patient's discomfort. Physical drainage, antibiotics, and other treatments, such as bronchodilators and decongestants, help to clear the lungs of mucus. Patients with advanced disease may need a lung transplant. Digestive problems are usually milder than respiratory ones and are usually managed through diet control and vitamin supplements. Intestinal obstructions can be controlled with the use of enemas.

DEAFNESS (CONGENITAL/HEREDITARY)

Deafness, or the lack of hearing, is an extremely widespread problem that can happen to anyone and at any age. A human being's ability to hear depends on the cochlea, situated in the inner ear. More specifically, it is the hair cells located in this region (about 16,000 cells per ear) that detect noises and carry the information to the brain for processing. Most hearing loss is related to problems with this region (although there may be a few instances where the problems may lie with the processing of auditory information rather than its reception). There are thus many possible causes of deafness (e.g., trauma to the inner ears, acute ear infections, very loud noises, certain antibiotics, tumors in the region of the ear and brain, and also genetic defects). According to the NIH, about 1 in every 1,000 infants born in the United States has some sort of profound hearing defect, of which about half are of genetic origin. In this section, we will consider some of the genes and mutations that are implicated in deafness.

Genetic deafness can be the singular result of a gene defect or be syndromic; that is, one of the symptoms in a syndrome with a wider spectrum of defects or abnormalities, such as Alport syndrome (characterized by kidney problems), Pendred syndrome (thyroid enlargement), and Waardenburg syndrome (abnormal skin pigmentation). In some instances, the hearing loss may result from a combination of genetic and environmental factors. This section looks at some of the genes implicated in nonsyndromic hearing loss.

Many cases of hereditary deafness appear to be linked to mutations in genes for proteins that are either known or suspected ion channels. For instance, one of the most common causes among Caucasian (European) populations is a mutation in the connexin26 (*Cx26*) gene. The connexins, as we saw in the previous chapter (see the entry on channel proteins and cross-membrane transport), are membrane proteins that form channels that enable ions to flow from one cell to its adjacent cell at the gap junction between the two cells. *Connexin-26*, which is on chromosome 13, codes for a protein that seems to be expressed in the cells of the inner ear, where it regulates the flow of potassium ions. Potassium ions are important for the proper functioning of the inner ear, and hence mutations in *connexin-26* lead to defects in hearing. A second gene linked to cases of both congenital deafness and progressive hearing loss later in life is *TMC1* on chromosome 9. This gene is expressed specifically in hair cells of the cochlea of the ear and is key to the function of the cells. Although the specific function of the protein is yet to be determined, scientists know from its structure that it is a transmembrane protein and suspect it also may be an ion channel.

Still another form of deafness (estimated to account for 20 percent of all nonsyndromic cases) has been traced to a combination of a mutation and a specific environmental factor. The mutation, which is a base substitution of an A with a G at position 1555 of a mitochondrial gene for a ribosomal RNA subunit (rRNA 12), does not cause deafness by itself. However, when people with this mutation are treated with aminoglycoside antibiotics, such as streptomycin, gentamycin, or kanamycin, they suffer severe and even irreversible hearing loss. Because the gene is encoded in the mitochondria rather than in any of the chromosomes, its inheritance is strictly maternal. This means that offspring of both sexes will inherit the mutation if their mother carries it but are in no danger of inheriting this susceptibility from their father.

Early screening for genetic deafness is advisable for many reasons. Congenital hearing defects can appear in children with no familial history of deafness due to the acquisition of new mutations during pregnancy or the absence of environmental cues. Left unchecked, deafness can have devastating effects on a child's development of speech and social and even cognitive skills. Although it may not be possible to completely restore a person's hearing, especially that of someone who has been deaf from birth, the identification of deafness-related genes early in life can markedly improve the ways in which the deafness might be managed and also prevent or stall progressive deafness. For instance, in regions of the world where aminoglycoside antibiotics are used widely, patients should be screened for the A1555G mutation in order to prevent the development of deafness.

DIABETES

Do you know someone who has diabetes? Or someone whose relative or friend has diabetes? According to the National Institute of Diabetes and Digestive and Kidney Diseases (NIDDK) of the NIH, your answer to these questions is very likely "yes." And if it happens to be "no," they may even go so far as to say that you do only you don't know it yet! By their estimates, some 18 million people—more than 6 percent of the total U.S. population—suffer from some form of diabetes. Of these, 13 million are diagnosed, whereas about 5 million (nearly one-third) are not yet diagnosed. Each year, by NIDDK's accounting, some 1 million Americans aged 20 or older are diagnosed with this disease.

Diabetes, like cancer, is an umbrella term for more than one disease, albeit a smaller group of disorders with a more cohesive set of clinical manifestations. The name derives from the Greek word "diabainein," which means "to pass through," and reflects the symptom of excessive urination that is distinctive of the condition. Versions of diabetes have been known since ancient times: It is alluded to in Hippocratic treatises, and there is archaeological evidence to suggest that the ancient Egyptian pharaohs may have also suffered from it. The vast majority (well over 90 percent) of modern-day diabetics suffer from one or the other version of diabetes mellitus, or "sugar diabetes." A much smaller group suffers from "water diabetes," which is also known as diabetes insipidus. As it is the less common type of diabetes, we have left the discussion of this condition for the end of this entry.

Diabetes Mellitus (Sugar Diabetes)

The main feature of diabetes mellitus is the inability of the body to metabolize sugar, specifically glucose. This results in the accumulation of sugar in the blood (i.e., hyperglycemia) and tissues, as well as its excretion via the kidneys. This last characteristic is the reason for the full name of the disease: The term "mellitus" comes from the Greek word for sweetness and honey, and the whole term literally means "the passing of sweet water."

Hyperglycemia inflicts damage to the body in two ways. First, as we know, our cells depend on glucose as the main source of chemical energy for performing the various tasks of sustaining life. The accumulation of sugar in the blood means that the cells are being starved of this resource. In the face of sustained starvation, they will simply shut down or die. Over time this will lead to organ failure, a frequent occurrence in untreated diabetes. Indeed, diabetes mellitus is the leading cause of kidney damage, loss of eyesight, and leg amputations. The second effect of hyperglycemia is caused by the changes in homeostasis and an upset in the electrolyte balance of the body. The kidneys work overtime attempting to flush out the excess sugar, which results in excessive urination. Unchecked, this will eventually lead to kidney failure, loss of consciousness, and death. Chronically high glucose also causes significant damage to the blood vessels, which may be why diabetes often goes hand in hand with hypertension and cardiovascular disease.

There are three main forms of diabetes mellitus, which are classified according to their underlying cause and their response to different treatments.

Type 1 Diabetes Mellitus

Also known as insulin-dependent diabetes mellitus (IDDM), type 1 diabetes mellitus is an autoimmune disease caused by the immunological destruction of the insulin-producing cells of the pancreas known as the islets of Langerhans. Insulin, as we saw earlier (see the entries on homeostasis and hormones and the endocrine system in Chapter 4), controls blood sugar levels by telling cells to utilize it for energy. When there is no insulin, the cells do not receive the signal they need to use the sugar, which results in the accumulation of sugar outside the cells, (i.e., in the blood and tissues). The dangerous thing about this type of diabetes is that most of the insulin-producing cells are destroyed long before the patient develops any of the characteristic symptoms of diabetes such as fatigue, frequent urination, and increased thirst. Although people usually develop type 1 diabetes mellitus before the age of 30, it can strike as early as the first month of life. Close control of sugar levels decreases the rate at which these events unfold.

Type 1 diabetes mellitus is a complex, or polygenic, disease, which means that it is linked to mutations in a number of different genes. Scientists have found at least 10 different loci that appear to contribute toward a susceptibility to this condition. Examples include genes for insulin on chromosome 11, the enzyme glucokinase on chromosome 7, and a transcription factor called hepatic nuclear factor (HNF1) on chromosome 12. Specific MHC types show a higher degree of association with type 1 diabetes mellitus than with other types of sugar diabetes. There is also a region on the short arm of chromosome 6 called the insulin-dependent diabetes mellitus (IDDM) locus that seems to harbor at least one gene (not yet identified) that confers disease susceptibility.

One of the most widespread autoimmune diseases, type 1 diabetes occurs in 1 out of 800 people in the United States. Individuals at high risk for developing the disease can be identified within a group of those who have a close relative with the disease. Efforts are now under way to evaluate prevention strategies for these family members at risk.

Although there is no cure for the underlying immunological condition, the adverse effects of IDDM can be controlled with hormone replacement therapy. A synthetic form of insulin is available as an injection. The regular administration of this hormone has markedly improved the life expectancy and quality of life of diabetics worldwide.

Type 2 Diabetes Mellitus

As widespread and common as it is, type 1 diabetes mellitus nevertheless accounts for only 5–10 percent of the total cases diagnosed. The overwhelming proportion (90–95 percent) of diagnosed cases belong to this second category, type 2 diabetes mellitus. This disorder is also known as non-insulin-dependent diabetes mellitus (NIDDM) because its underlying cause is not related to insulin production. Here, there is no autoimmune attack of the pancreas, and the islet cells produce normal levels of insulin. The problem lies in the inability of the other cells of the body to respond to insulin signals and thus use sugar efficiently. In other words, these cells ignore insulin, a condition known as insulin resistance. The result of this resistance is the same as a lack of insulin production, namely hyperglycemia and eventual organ failure.

Many of the telltale symptoms of type 2 diabetes are the same as those for type 1 diabetes: frequent urination, excessive thirst and hunger, dramatic weight loss, irritability, weakness and fatigue, and nausea and vomiting. Some other symptoms may include recurring or hard-to-heal skin, gum, or bladder infections, blurred vision, tingling or numbness in the hands or feet, and itchy skin. Symptoms appear gradually, and often in older people, as compared with the onset and appearance of insulin-dependent diabetes. Some people may have such mild symptoms that the disease might go unnoticed for several years.

Type 2 diabetes mellitus is a complex disease that is linked to many genes on different chromosomes as well as numerous environmental factors. Genetic factors include genes for specific proteins, as well as some that appear to confer susceptibility to the disease (e.g., on chromosome 20). Often the mutated proteins associated with type 2 diabetes are signaling molecules involved in the signaling pathways of sugar metabolism. Examples include an insulin receptor substrate (IRS2) on chromosome 13 and the glucagon receptor on chromosome 17. Among the nongenetic factors influencing the occurrence of this disease are older age, obesity, physical inactivity, familial and personal medical history, and ethnicity. About 80 percent of diabetics are overweight, and this is likely because excess fat in the body prevents insulin from working properly. The disease occurs at a higher frequency among African American, American Indian, and Hispanic populations.

Although insulin therapy is not an appropriate therapy for NIDDM, there are treatments in the form of diet and exercise regimens that help control the disease once it has been properly diagnosed. People with a familial history of diabetes, obesity, and hypertension should be especially vigilant and controlled in their habits. With the proper precautions, it is possible for a diabetic to lead a normal, productive life.

Gestational Diabetes

This is the name given to a condition that develops during pregnancy and exhibits the characteristic features of type 2 diabetes (i.e., improper sugar metabolism, insulin resistance, hyperglycemia, and excessive urination). It affects about 4 percent of all pregnant women, and about 135,000 cases are reported in the United States every year. It is not an autoimmune disease. Possibly, the insulin resistance that underlies the development of symptoms is triggered by the altered fat and sugar metabolism that a woman experiences during pregnancy. Consequently, doctors treat it the same way as they do type 2 diabetes, by controlling diet and exercise rather than by administering insulin. Although there is no well-defined genetic link, gestational diabetes is most common among women with a family history of type 2 diabetes. Conversely, it is also a risk factor for the disease—women suffering from this condition have a 20–50 percent higher than normal risk of developing NIDDM later in life. The ethnic distribution is also similar to that of type 2 diabetes mellitus.

Diabetes Insipidus (Water Diabetes)

Although it may be confused with diabetes mellitus due to the common symptoms of excessive urination and constant thirst, diabetes insipidus (DI) is a completely different disease (or group of diseases) that is caused by defects in water homeostasis and not sugar metabolism. Both the similarity of symptoms and the distinguishing features were evidently observed a very long time ago, as reflected in the name the Greeks gave this syndrome. "Insipidus" means tasteless and was likely chosen by the ancient Greeks to describe the quality of the urine in contrast with the "sweet" urine of diabetes mellitus. The kidneys do not function properly in removing the waste material and concentrating the urine, leading to the passing of large volumes of very pale, almost colorless fluid with very low ion concentrations. Patients with water diabetes often drink much larger volumes of fluids, especially water, than normal. They may also exhibit other symptoms of incontinence such as bedwetting and nocturia (waking up frequently at night to pass urine).

One of the worst consequences of water diabetes is dehydration, severe incidents of which can end in death. Oftentimes, the water drunk by DI patients is insufficient to keep up with the amounts lost in the urine. Water loss is hyperosmolar, which means that the urine passed is very dilute and the relative concentration of ions and plasma proteins in the blood rises to very high levels. If this happens too quickly or is prolonged for some reason, there is considerable damage to various organs, particularly the brain. Additional complications may arise from the expansion of the ureters and bladder caused by the prolonged passage of high volumes of urine.

There are four types of diabetes insipidus that are fundamentally different both in cause and treatment approach. Only the categories attributable to genetic defects will be covered in this section.

Neurohypophyseal Diabetes Insipidus

Also known as pituitary DI or central DI, neurohypophyseal diabetes insipidus is the most common type of water diabetes and is caused by a lack of the antidiuretic hormone vasopressin. (See the entries on homeostasis and hormones and the endocrine system in Chapter 4.) Mutations in the gene for vasopressin, which is located on chromosome 20, can lead to this disease, although it may also be caused by other factors, such as injuries or tumors in or around the pituitary gland, which produces the hormone. The development of symptoms of excessive thirst and urination early in childhood, as well as a family history of disease, are clues that a physician may look for in diagnosis. The disease is incurable, but its symptoms are treatable with synthetic forms of vasopressin.

Nephrogenic Diabetes Insipidus

As indicated by its name, the defect in nephrogenic diabetes insipidus lies in the kidneys, making them unable to conserve water. Mutations leading to nephrogenic diabetes insipidus map to two distinct genes. The more common form of the disease is X-linked and is linked to mutations in the gene for a vasopressin receptor of kidney cells. When this gene is defective, the kidney cells are unable to respond to the signals from vasopressin, which leads to excess urine flow. The second gene associated with the nephrogenic condition is on chromosome 12 and makes a membrane protein called aquaporin, a family of proteins whose function is to regulate the passage of water in and out of cells. The specific aquaporin involved here is selectively expressed in the collecting tubules of the kidneys, and mutations in the gene thus interfere with the ability of the kidneys to control the flow of water.

See also: Homeostasis—Water Homeostasis (Ch. 4).

DOWN SYNDROME

Most fetuses with trisomy (an extra chromosome caused by defects during meiosis or fertilization) do not survive, but one type does and is well-characterized: those with Down syndrome. In this condition, there is an extra copy of chromosome 21, the smallest of the autosomes. The presence of this extra chromosome gives rise to a wide range of symptoms. Many children with trisomy 21 suffer from slow development and mental retardation, heart defects, epilepsy, hypothyroidism, respiratory problems, susceptibility to infection, and obstructed digestive tracts. Many adults with trisomy 21 develop Alzheimer's disease. The outward physical effects of Down syndrome often include a flattened nose bridge, single creases on the palms of the hands, decreased muscle tone, and the characteristic epicanthal folds over the eyes. The syndrome is named for John Langdon Down, the English physician who first described the condition in detail in 1866 and who

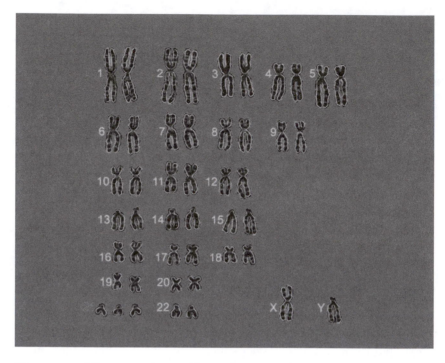

Karyotype of Down syndrome. Note mismatch (trisomy) of chromosome 21 pair. [© J. Cavallini/Custom Medical Stock Photo]

also coined the term "Mongoloid" to describe it. (This term is no longer used because of its racially derogatory implications.)

Why does the presence of the extra chromosome lead to disease? Scientists believe that it disrupts normal gene expression patterns and leads to the overexpression of several proteins whose genes are present in triplicate. This excess appears to have deleterious effects on the organism. Research has suggested that only a small portion of the chromosome is actually needed in triplicate in order to see the effects of Down syndrome. As it turns out, this region—labeled as the Down Syndrome Critical Region—is not confined to a single isolated part of the chromosome but includes genes scattered across the chromosome. Only about 20 to 50 genes of the 200 to 250 genes on chromosome 21 are believed to be involved. The identity of these genes is not confirmed, but there are some likely candidates:

1. *Superoxide dismutase*—SOD1 is a copper- and zinc-containing metalloenzyme that normally destroys toxic free radicals that are produced as by-products of metabolism within various cells. Specifically, this enzyme catalyzes the conversion of superoxide anions to hydrogen peroxide, which is then broken down by glutathione peroxidase (Gpx) and catalase (encoded on other chromosomes) to less harmful products. If SOD1 is overexpressed without a corresponding increase in Gpx and catalase, these enzymes are unable to keep up with the hydrogen peroxide, which

is damaging to the cells and induces cell death. This is the underlying basis for the signs of premature aging and impaired immune system function associated with Down syndrome.

2. *COL6A1*—This gene encodes part of collagen VI and is normally expressed in certain parts of the heart. Overexpression of this type of collagen may be associated with congenital heart defects. Several polymorphisms and functional variations in the 3' end of the coding sequence of *COL6A1* have been found to be associated with heart defects in Down syndrome patients.

3. *ETS2*—This is a proto-oncogene that encodes a transcription factor found in many different cell types. Studies in mice have shown that the gene is highly expressed in newly forming cartilage in the fetus, and it is likely that in humans *ETS2* has a role in skeletal development. An excess of the gene product may lead to the skeletal abnormalities associated with Down syndrome. *ETS2* also appears to regulate the transcription of beta-amyloid precursor protein (beta-APP), a protein involved in plaque formation in the brains of patients with Alzheimer's disease (AD). Research suggests that overexpression of *ETS2* may also play a role in the development of the brain abnormalities seen in Down syndrome as well as being the factor that links AD and Down syndrome together.

4. *CAF1A*—This gene codes for human chromatin assembly factor-1, a protein that is thought to play the role of assembling chromatin in DNA synthesis, possibly by helping to assemble histones onto replicating DNA. The overexpression of the protein may be detrimental to DNA synthesis in Down syndrome patients.

5. *DYRK*—This gene expresses a "dual-specificity tyrosine-regulated kinase," a signaling protein that seems to play a significant role in the pathways that regulate cell proliferation. When overexpressed, it may cause mental retardation.

6. *CRYA1*—The formation of cataracts in patients with Down syndrome may be related to trisomy of this gene, which normally expresses the lens protein crystallin.

7. *GART*—This gene actually codes for three enzymes that synthesize three different purines (nucleic acid building blocks) in vertebrate animals. It is a "housekeeping" gene that is probably expressed in all dividing cells, and its overexpression disrupts DNA synthesis and repair in Down syndrome patients.

8. *IFNAR*—This is the gene for interferon, overexpression of which likely interferes with the proper functioning of the immune system as well as other organ systems.

9. Transient myeloproliferative syndrome is a leukemia-like reaction that occurs in some newborn infants with Down syndrome. A search for a

gene on chromosome 21 that causes this syndrome revealed a gene that is now labeled *TAM*. The exact function of this gene is not completely understood.

Other genes that are also implicated in Down syndrome include those that express amyloid precursor protein (*APP*), glutamate receptor (*GLUR5*), a phospho-fructokinase (*PFKL*) expressed in the liver, and cystathione beta synthase (*CBS*). However, no gene has yet been linked for certain to any feature associated with Down syndrome.

The symptoms in Down syndrome vary widely from patient to patient, and it is not always possible to predict the outcome. Although the exact cause for the abnormal meiosis is unknown, the incidence is higher for mothers over the age of 35, especially those getting pregnant for the first time. Prenatal diagnosis and genetic counseling are highly recommended for prospective parents to make informed decisions and plan adequately. Diagnostic tests for Down syndrome include amniocentesis and chorionic villus sampling (CVS).

See also: Chromosome (Ch. 2); Triple X Syndrome, Turner Syndrome.

DWARFISM

Dwarfism is a general term for a large number (over 80) of genetically transmitted medical conditions in which one of the most prominent phenotypic features is exceptionally short stature (i.e., a dwarf-like appearance). The conditions differ from a simple lack of height (shortness) in several respects, most notably disproportional body parts and the development of progressive deformities, which eventually lead to disabilities. Dwarfism is often part of a constellation of symptoms that develop in a number of different genetic defects of genes specific for the muscles, bones, and connective tissues. In the most severe cases, affected children may lose the ability to walk, and other functions, such as breathing, are also profoundly affected. Gradual paralysis and even death may result, especially if there is no medical intervention.

We now know the underlying bases for many forms of dwarfism. Although individual conditions are often rare, the numbers add up because of the sheer variety of genes and mutations that can be affected. Even though there are no cures as such, we can prevent or lessen the disabilities caused by the bone deformities that develop in these diseases. Consequently, early diagnosis and treatment (even prenatal diagnosis and genetic testing) can save a child from becoming disabled. We discuss some examples of genetic syndromes associated with dwarfism.

Achondroplasia

Achondroplasia is one of the most common forms of dwarfism and perhaps one of the earliest recorded birth defects in history, as indicated by depictions in ancient Egyptian art. Individuals with achondroplasia typically show abnormal body proportions (e.g., markedly shortened limbs in contrast with a relatively normal-sized torso), a large head (which may or may not be caused by the accumulation of excess fluid in the brain) with a prominent forehead, and a nose that is somewhat flattened at the bridge. The middle and ring fingers diverge (are separated), giving the hand the appearance of a trident. Adults are typically less than 4 feet in height.

What causes achondroplasia? Although its name is derived from the Greek word meaning "without cartilage formation," individuals with the disease do have cartilage. In fact, they may even have an excess of this tissue because the processes that normally convert the fetal cartilage tissue into bone during development are defective, especially in the long bones of the upper arms and thighs. As a result of this defect—specifically caused by mutations in the gene for fibroblast growth factor receptor 3 (*FGFR3*), situated on chromosome 4—the rate of cartilage conversion is slowed down considerably. Interestingly enough, about 80 percent of all diagnosed cases of achondroplasia seem to occur via new sporadic mutations rather than a mutation inherited from a parent. Inheritance of the defective gene follows an autosomal dominant pattern. The most common type of mutation giving rise to this condition is a point mutation that causes the substitution of arginine for glycine in the receptor. The result of this mutation seems to be a gain of function, an increase in activity of FGFR3. Whereas the normal function of this receptor protein is to slow down the formation of bone by inhibiting the proliferation of chondrocytes (via signaling pathways), the mutant FGFR3 appears to be much more active and severely limits bone growth.

See also: Cartilage, Growth Factors (Ch. 4).

Achondrogenesis

Achondrogenesis is a rare type of dwarfism characterized by defects in the formation (genesis) of cartilage and bones. The mutations causing this disease occur in a gene on chromosome 5 that makes a sulfate transport protein that carries sulfate molecules in and out of cells. As we saw in Chapter 4, sulfates are important parts of the cartilaginous proteins chondroitin and collagen. Consequently, if they are deficient during development, the tissues are not formed properly.

Achondrogenesis is a congenital defect in that it is almost always caused by de novo mutations during development and is seldom inherited. The reason for this is evident from the fact that most children do not survive for long beyond birth, if indeed they make it that far. Those that are born have very short limbs, a narrow chest, a large head in contrast with the rest of the body, and a narrow jaw. The ribs especially are very thin and tend to fracture easily.

Rather confusingly, another disease, attributable to a mutation in an entirely different gene, has also been labeled as achondrogenesis. This condition, also known as the Langer–Sardino type, is characterized by the virtual absence of ossification (i.e., the conversion of cartilage to bone) in the vertebral column, hips, and pubic bones. The affected gene in this case is on chromosome 12 and encodes a collagen protein (designated as COL2A1). Again, mutations always occur de novo and are expressed in an autosomal dominant fashion.

See also: Cartilage—Collagen (Ch. 4).

Cockayne Syndrome

Cockayne syndrome is a rare inherited form of dwarfism in which people have a high sensitivity to sunlight and show signs of premature aging. It is named for a London physician of the early twentieth century who concentrated on investigating hereditary diseases of children.

There are two main forms of Cockayne syndrome. In the classical form (also called type I Cockayne syndrome), the symptoms are progressive and typically become apparent after the age of 1 year. An early-onset or congenital form of Cockayne syndrome (type II) is apparent at birth. Both diseases are linked to mutations in genes that code for proteins involved in DNA repair, a process that normally takes place in response to DNA damage caused by exposure to UV radiation, of which sunlight is a major source. People with Cockayne syndrome are unable to perform a certain type of DNA-repairing activity known as "transcription-coupled repair," which normally occurs when coding sequences of DNA are being replicated. Unlike most other diseases involving DNA repair, Cockayne syndrome is not linked to cancer.

Specifically, type I Cockayne syndrome is caused by mutations in an excision-repair gene on chromosome 5, called excision repair cross-complementing gene 8 (*ERCC8*), and the type II form is caused by *ERCC6*, which is on chromosome 10. Both proteins are known to interact with components of the transcriptional machinery and with DNA repair proteins. More is known about ERCC8, a DNA-binding protein with some ATPase activity that is believed to promote DNA–protein complex formation at the site of repair.

See also: DNA Synthesis/Replication and Repair (Ch. 4).

Diastrophic Dysplasia (DTD)

Diastrophic dysplasia is another rare form of dwarfism, in which patients are clubfooted and develop malformed joints in the hands. It is caused by defects in the same gene as that responsible for achondrogenesis, namely the sulfate transport protein on chromosome 5. In fact, this gene is linked to other forms of dwarfism as well, although the mechanism by which different diseases are caused is not properly understood. It is very likely that different mutations affect either the structure of the protein itself or when the gene is expressed. For instance, the achondrogenesis just described is a developmental disorder, but DTD is specific to certain locations, namely the hand and foot joints. The negatively charged sulfate-containing glycopeptides contribute to the shock-absorbing properties of chondroitin and are therefore important for the maintenance of the skeletal joints. Mutations associated with DTD evidently affect the structure of the protein, leading to deformed joints, which are most evident in the hands and feet.

See also: Cartilage—Chondroitin (Ch. 4).

Ellis–van Creveld Syndrome

Also known as "chondroectodermal dysplasia," this form of dwarfism is characterized by the development of short limbs, polydactyly (the formation of additional fingers or toes), the malformation of the bones of the wrist, dystrophy of the fingernails, partial hare lip, cardiac malformation, and often prenatal eruption of the teeth. Congenital heart defects occur in about 50 percent of patients. The most common heart problem is an atrial septal defect (i.e., a hole in the heart between the upper two heart chambers). Generally quite rare, Ellis–van Creveld syndrome has, however, been observed in a higher than normal frequency among people of the Old Order Amish community, a small and rather genetically isolated population in Lancaster County, Pennsylvania.

Ellis–van Creveld syndrome has been mapped to autosomal recessive mutations in two nonhomologous genes, designated as *EVC* and *EVC2*, that are found lying head-to-head on the short arm of chromosome 4. Changes in the *EVC* gene are also responsible for a related syndrome, *Weyers acrofacial dysostosis*. Individuals with this disorder have mild dwarfism, polydactyly, and dental abnormalities. It is inherited in an autosomal dominant manner. An individual with one abnormal *EVC* gene has Weyers acrofacial dysostosis, and a person with two changed *EVC* copies has Ellis–van Creveld syndrome. The gene *EVC2*, also known as limbin, is normally expressed in the growth plates of long bones. The normal functions of the proteins encoded by either gene remain unknown, and their determination is important for trying to understand the mechanisms of disease.

EPILEPSY

Known since ancient times to afflict humankind and imbued with all manner of cultural meanings—the Hippocratic corpus has a treatise on it called "The Sacred Disease," and other societies, even today, regard epileptic seizures as evidence of divine visitations—epilepsy is one of the common neurological diseases that affect human beings. The characteristic symptom of this disorder is the occurrence of repeated seizures or "fits," which are caused by the abnormal signaling or firing of neurons in the brain. Seizures vary greatly in type and severity, as well as frequency and the age of onset. Although there is no known cure for the disorder, symptoms are also known to go away over time in some cases.

Any event that causes a disturbance in the normal patterns of neuronal activity in the brain has the potential to cause epilepsy. An estimated 30 percent of all cases of epilepsy are caused by head injuries, tumors, stroke, or infection. Although the cause of many forms still remains unknown, there are at least 12 different forms of epilepsy that show evidence of linkage to mutations in specific genes. Although each of these forms is quite rare, together they account for a significant proportion of the total incidence of epilepsy. Some examples follow.

- A particularly aggressive form of epilepsy also known as *LaFora disease* has been linked to mutations in the *EPM2A* gene on chromosome 6. This gene encodes a regulatory phosphatase protein called laforin that is expressed primarily in neurons. When mutated, it induces the accumulation of glycogen-like masses—known as LaFora bodies—in the brain cells that may play a role in disrupting the normal electrical flow of these cells and causing seizures. Mutations in a neighboring gene, called the malin gene, also induce similar symptoms.

- Mutations in a number of different genes for the subunits of ion channels cause a generalized type of epilepsy that is characterized by febrile seizures (i.e., seizures with a simultaneous rise in body temperature, especially in children under the age of 6). The affected genes include three genes for subunits of voltage-gated sodium channels: *SCN1A* and *SCN2A1* on chromosome 2, *SCN1B* on chromosome 19, and a gene on chromosome 5 for a chloride channel that is also a receptor for the ligand GABA, an important inhibitory neurotransmitter in the brain (known as *GABRG2* or *CAE2*).

- In addition to the generalized form just described, a very severe form of infantile epilepsy is also caused by mutations in either *SCN1A* or *GABRG2*. Interestingly, these seizures seem to be caused by de novo mutations rather than those that are inherited from either parent.

- Another chloride channel that is encoded by the *CNCL2* gene on chromosome 3 has been linked to both a generalized and a juvenile form of epilepsy.

The prognosis for epilepsy is mixed. Although seizures do not necessarily cause further damage, and most epileptics can lead outwardly normal lives, it is not uncommon for people with epilepsy, especially children, to develop behavioral and emotional problems. In addition, people with epilepsy may be more prone than others to sudden unexplained death. Some treatment for epileptic seizures is available in the form of anticonvulsive drugs, although not all patients respond to them in the same way. By focusing on the genetic basis for epilepsy, scientists hope to develop more effective anticonvulsive treatments and possibly gene replacement therapies for this and other seizure disorders.

See also: Channel Proteins and Cross-Membrane Transport (Ch. 4).

FRIEDREICH'S ATAXIA

Named for the German physician Nikolaus Friedreich, who first described this condition in 1863, Friedreich's ataxia is a rare inherited disease marked by a progressive ataxia (i.e., the loss of control over the voluntary muscles) together with an enlargement of the heart. Ataxia in this case results from the degeneration of spinal tissue and the motor nerves of the arms and legs. Usually appearing between the ages of 5 and 15, Friedreich's ataxia begins with symptoms of difficulties in otherwise simple motor functions such as walking. Over time, muscles begin to waste away and weaken, leading to the development of deformities, especially of the hands and feet. A gradual loss of sensation, beginning in the extremities and later spreading to other parts of the body, is also typical. Most patients develop a condition known as scoliosis, which is a curving of the spine to one side and can impair proper breathing. Disease progression is variable but often ends with the patient confined to a wheelchair some 15–20 years after the symptoms first begin to appear. Many do not survive, as they succumb to cardiac problems caused by the enlargement of the heart. Incidence in the United States is about 1 in 50,000, which makes this disease one of the most prevalent inherited forms of ataxia. Men and women are affected equally.

Friedreich's ataxia is caused by a specific type of mutation known as a trinucleotide expansion in a specific gene, *FRDA*, on chromosome 9. This gene encodes a mitochondrial protein, labeled as frataxin, which functions in the regulation of iron transport and respiration in the mitochondria. The disease-causing mutation is an expansion of the trinucleotide sequence GAA within the gene—where normally people have 8–30 copies of this triplet, Friedreich's ataxia patients may have as many as 1,000. Evidently, the expansion is a loss-of-function mutation because the disease is inherited in a recessive manner. It has a tendency to expand over successive generations, which means that the phenomenon of anticipation is observed in this disease. The larger the number of GAA repeats, the earlier is the onset of disease, the more severe its symptoms, and the quicker the decline of the patient. The frataxin protein is very similar to a protein called YFH1 in yeasts, and much of our information about the function and activity of the former has actually come about due to studies on the latter.

Although there is no effective cure for Friedrich's ataxia or indeed any other neurodegenerative disease, measures can be taken to alleviate symptoms and prevent complications (e.g., heart failure) to help patients lead close to normal lives for as long as possible. Physical therapy, for example, may help prolong the use of the limbs. Genetic testing and neonatal diagnoses may help some people to prepare themselves. Scientists hope that advances in understanding etiology and genetics will lead to breakthroughs in finding treatments for this disease.

GLAUCOMA

The National Eye Institute of the NIH defines glaucoma as a group of diseases that can damage the optic nerves and cause progressive vision loss and blindness. It is one of the leading causes of blindness in the United States, where about 3 million people have the disease and roughly 50 million more are at risk.

Damage to the optic nerve in glaucoma is caused by the buildup of pressure due to the improper flow of the fluids that nourish the ocular (eye) tissues and help maintain normal pressure (called intraocular pressure) within the eyes. This fluid is produced just under the eyes and performs its function by flowing continuously through the anterior chamber of the eye—the space between the cornea and iris—at a constant rate. The fluid leaves the anterior chamber through a mesh-like area at the open angle where the iris and cornea meet. By far the most common subset of glaucoma, *open-angle glaucoma* is the result of improper drainage of the fluid through these meshes. Symptoms develop rather slowly, beginning with the loss of side or peripheral vision and gradually progressing to total blindness.

Typically, glaucoma develops later in life, although certain forms may manifest as early as adolescence. Although there is no way to restore lost eyesight, preventive treatment (e.g., by alleviating intraocular pressure) is possible, and consequently an early diagnosis is very important in preventing blindness. The NIH estimates that about 3 million Americans suffer from open-angle glaucoma, although about half this population is unaware of the condition. Although the disease can strike any person of any ethnicity, some groups, such as African Americans over the age of 40 and Mexican Americans over 60, show a much higher predisposition (about 3 to 4 times more than whites) for glaucoma. People with a family history of glaucoma are also at higher risk of developing the condition.

The genetics of glaucoma is quite complicated—causal links have been established with mutations in about 20 different genetic loci, not all of which are associated with known genes. One of the most common causes—affecting nearly 100,000 individuals in the United States—is a mutation in a gene called *MYOC* (also known as *GLC1A* or *TIGR*) on chromosome 1. This gene codes for a protein called myocilin that is expressed specifically in various ocular tissues and has cytoskeletal functions. It would appear that mutations in the myocilin gene cause increased pressure in the eye by obstructing the flow of ocular fluid due to the formation of a faulty protein that leads to structural defects in the meshwork through which the fluids drain out in the anterior chamber.

Examples of other genes and genetic loci implicated in glaucoma include:

1. The optineurin (*optic neuro*pathy *in*ducing) gene on chromosome 10 that encodes a protein that interacts with transcription factors to prevent cell lysis.

2. A gene on chromosome 2 that makes one of the members of the cytochrome 450 family that functions in the metabolism of steroids, especially those participating in signaling pathways in the eye tissues.

3. Genetic loci named *GLC1C, GLC1F*, and *GLC1D* on chromosomes 3, 7, and 8, respectively. Although these loci are not part of any known functional genes, specific variants have been linked to adult-onset open-angle glaucoma.

HEMOCHROMATOSIS

Although iron plays certain essential roles in the human body, such as in oxygen transport and the synthesis of key intracellular enzymes, it is actually toxic to most tissues and organs. An excess buildup of iron will eventually lead to the failure of these organs to perform their functions properly. Consequently, there is a need to regulate the amount of iron in the body and maintain a balance between the body's need for it and its toxicity. Under normal conditions, the body maintains this balance at the level of uptake, absorbing only as much iron as it needs from dietary sources and excreting the rest because it does not have any natural means for getting rid of excess iron from the blood and tissues.

Hemochromatosis is an inherited disease in which the body absorbs and stores too much iron. Symptoms associated with this disease reflect the damage done to various organs—notably the liver, heart, and pancreas—where the iron accumulates. Often, but not always, the symptoms begin to manifest in middle age, beginning with nonspecific complaints such as fatigue, heart palpitations, joint pain, stomachaches, impotence, and loss of menstrual periods, and later developing into more serious problems, including gray or bronze skin pigmentation, liver cirrhosis and cancer, heart failure, joint disease, diabetes mellitus, chronic pituitary and sexual dysfunction, chronic abdominal pain, severe fatigue, and a heightened susceptibility to infections. Various environmental factors also influence the clinical course of hemochromatosis; excess alcohol use, for instance, exacerbates the disease, whereas menstruation or blood donations—which rid the body of iron, at least for a time—may help lessen severity. The disease is quite easily treated, especially if detected early, before the accumulated iron begins to exert its harmful effects.

One of the most common hereditary diseases known in the United States, hemochromatosis occurs in about 5 in every 1,000 people and is especially high in the Caucasian population. One possible reason for the high incidence is that the disease phenotype can result from mutations in different genes. This pattern of expression reflects the fact that iron absorption in the body is a multistep process that is controlled by different proteins and that the disruption of any of these steps can lead to iron accumulation. At least five different types of hemochromatosis have been identified, mapped to genes on different chromosomes (see Table 5.1).

- The classic syndrome is associated with mutations in the *HFE* gene on chromosome 6. The protein product of this gene is a membrane-bound protein that seems to regulate the interactions between transferrin (the iron-carrying molecule in blood—see the entry on blood in Chapter 4) and its receptor molecules on the cell surface. This syndrome is more prevalent among males.

- A mutation in the gene for a transferrin receptor molecule on chromosome 7 is another cause of hemochromatosis, called HFE-3.

- Juvenile hemochromatosis, also called type 2 HFE, is characterized by a younger age of onset and has been linked to two distinct mutations. The first is a mutation of a gene (*HAMP*) for an antimicrobial peptide produced in the liver and encoded on chromosome 19. The second mutation has been found to have a specific locus on chromosome 1 that is not part of any known gene.

- A fifth form of the disease (HFE-4) has been linked to a mutation in the gene on chromosome 2 that codes for an iron transport protein, ferroportin, produced in the duodenum. Whereas the first four forms of hemochromatosis mentioned follow an autosomal recessive pattern of expression and inheritance, the mutation on the ferroportin gene expresses as a dominant phenotype.

See also: Blood—Globulin (Ch. 4).

Table 5.1. Hemochromatosis genotypes

Disease type	Mutated gene	Location	Protein product/function	Inheritance
Classic or type 1	*HFE*	Chm. 6	Membrane protein regulating transferrin-receptor interaction	Recessive
Juvenile hemochromatosis (type 2A)	Unknown	Chm. 1	No known gene at locus associated with disease	Recessive
Juvenile hemochromatosis (type 2B)	*HAMP*	Chm. 19	Antimicrobial peptide in liver	Recessive
Type 3	*TFR*-2	Chm. 7	Transferrin receptor molecule on cell surface	Recessive
Type 4	*SLC11AE*	Chm. 2	Ferroportin	Dominant

HEREDITARY HEMORRHAGIC TELANGIECTASIA (HHT)

Hereditary hemorrhagic telangiectasia is an inherited bleeding disorder in which excessive bleeding is caused by the malformation of the blood vessels rather than by abnormalities in the clotting cascade (see the entry on bleeding disorders). In people born with this syndrome, the capillaries—the small thin-walled blood vessels that normally form connections between the veins and arteries and deliver blood to the interior parts of different organs—do not develop properly. This gives rise to direct connections between the arteries and veins. When this happens, the arterial blood, which is under high pressure, flows directly into the veins without being dispersed through a capillary network that normally serves to relieve this pressure. The high pressure leaves the connections between the arteries and veins very fragile and prone to easy rupture and bleeding. Patients with HHT thus suffer from bouts of chronic and excessive bleeding at the site of the abnormalities. Common external symptoms include the incidence of chronic or frequent nosebleeds (in about 95 percent of all HHT cases) and the appearance of red to purplish spots or lacy red veins in the skin of the hands, face, and mouth due to malformed blood vessels in these areas (in 90 percent of cases). Blood loss through the gastrointestinal tract is another common symptom. Among the internal organs, the lungs, brain, and liver seem most susceptible to HHT-related malformations and consequent bleeding. Lesions in these organs can have other serious consequences, such as embolisms (traveling blood clots), stroke, and brain abscesses caused by infections.

The incidence of HHT is about 1 in every 5,000 people. It has no particular predilection for either sex or any ethnic group. Although there is no cure for HHT, individual symptoms can be treated, and hence screening is useful. Typically inherited as an autosomal dominant disease, HHT occurs in multiple forms that are caused by mutations in different genes. Type I HHT is caused by mutations in the *ENG* gene on chromosome 9 and shows a high frequency of lesions in the lungs. The *ENG* gene normally produces a membrane glycoprotein that is expressed primarily in endothelial cells (i.e., by the blood vessels), where it forms part of the receptor for the transforming growth factor beta. Mutations in the gene affect the proper development of the blood vessels. Mutations in a second gene for a TGF-beta receptor known as activin-like kinase (*ALK1*) on chromosome 12 are associated with type II HHT. This receptor is expressed in high levels in the placenta and the lungs, although the frequency of capillary abnormalities in this form of HHT is relatively low. Scientists have also identified a third form of HHT associated with a high frequency of liver abnormalities that is not linked to mutations in either the *ENG* or *ALK1* genes. The gene for this HHT (type III) has not been identified as yet. Finally, defects in the tumor suppressor gene *MADH4* on chromosome 18 cause HHT along with juvenile polyposis (i.e., the formation of intestinal polyps).

See also: Growth Factors (Ch. 4).

HIRSCHSPRUNG'S DISEASE

Hirschsprung's disease, also known as congenital megacolon, is a complex inherited disorder in which nerve fibers are absent in segments of the bowel, resulting in severe bowel obstruction. The disease affects about 1 in 5,000 people and is about four times more common in males than females. Mutations in various genes result in the failure of the development of nerve cells in the wall of the colon of a fetus. The affected segment of the intestine lacks the ability to relax and move bowel contents along. This constriction causes the bowel in the neighboring region to dilate, producing megacolon (dilation of the colon). The disease usually affects varying lengths of bowel segment, most often involving the region around the rectum.

Dominant mutations of the *RET* gene on chromosome 2 seem to be responsible for anywhere from 7 to 41 percent of all cases and up to 50 percent of familial cases. The gene expresses a tyrosine kinase receptor (proto-oncogene tyrosine–protein kinase receptor *re*arranged during *t*ransfection). It is expressed in neural precursor cells shortly after they leave the neural plate and colonize the gut of the developing fetus. In addition, mutations in the endothelin-3 gene (*EDN3*) on chromosome 20 as well as the gene for its receptor—the endothelin receptor type B gene *EDNRB* on chromosome 13—are also implicated in this disease. These proteins are components of a signaling pathway that probably operates late in the neural crest colonization process. Deletions of *RET*, *EDN3*, and *EDNRB* also lead to the disease. In addition, regions on chromosomes 3, 10, and 19 have been found to be associated with Hirschsprung's disease and together explain its complicated inheritance pattern. People with Down syndrome (trisomy 21) are 100 times more likely to have Hirschsprung's disease, although the nature of this association is not fully understood.

Type 4 Waardenburg syndrome, also known as Waardenburg–Hirschsprung disease (WS4), shares features with Hirschsprung's disease. People who have WS4 often suffer from anomalies in the colon. There is no evidence that *RET* mutations cause WS4, but mutations in *EDN3* and *EDNRB* have been implicated in WS4. In general, individuals with a heterozygous mutation in *EDNRB* or *EDN3* have HSCR or occasionally features of Waardenburg–Hirschsprung disease (WS4), whereas those with homozygous mutations in either gene are more likely to have severe manifestations of WS4.

Hirschsprung's disease can be largely cured through surgery in which the diseased, nonfunctioning segment of the bowel is removed to restore normal bowel function.

See also: Waardenburg Syndrome.

HUNTINGTON'S DISEASE (HD)

Huntington's disease (HD) is a debilitating neurological disease in which symptoms typically do not appear until mid-life. The disease was first identified and described as a "hereditary chorea" by George Huntington in 1872. People who suffer from the disease have problems controlling physical movements (hence the name "chorea"—or "dance"), intellectual functioning, and emotion. Early signs of HD include mood swings, irritability, depression, memory loss, and uncontrolled movements. As the disease progresses, it becomes progressively more difficult to walk and speak, and memory and intellectual functions continue to decline. Patients generally die of heart failure or pneumonia.

Huntington's disease is caused by mutations in a gene (on chromosome 4) that encodes a protein that was named "huntingtin." This gene was first discovered in 1993 by Elena Cattaneo at the University of Milan. The normal huntingtin protein is believed to up-regulate the transcription of brain-derived neurotrophic factor (BDNF), a chemical found in the brain that is necessary for the correct functioning of nerve cells in the basal ganglia, the part of the brain that controls movements. When the huntingtin protein is mutated, production of BDNF drops. This leads to the death of the neurons of striated muscles and the symptoms of HD.

Like certain other neurological diseases (see the entries on Friedreich's ataxia and muscular dystrophy—myotonic dystrophy), Huntington's disease is caused by a specific type of mutation, known as a trinucleotide expansion, involving the triplet sequence CAG. These repetitive mutations are believed to arise during DNA replication. When the DNA unzips to begin replication, a large number of CAG codons must be replicated. The longer the strand of CAG repeats, the more likely it is for there to be an error in its replication because the copying machinery might miscount the exact number of CAG units.

The appearance of disease symptoms is strongly correlated with the number of times this codon is repeated—the greater the number of repetitions, the more likely it is for an individual to experience the disease. Most of us have 10–15 units of this sequence in our huntingtin gene and show no symptoms. It seems that the disease is inevitable in people with more than 40 CAG repeats, and the outcome is uncertain in the range of 30–40 repeats. The number of repeats tends to increase from one generation to the next, which gives rise to the phenomenon of anticipation (see Chapter 4), an increase in severity and earlier onset of the disease over successive generations. For example, if somebody has over 60 CAG repeats, he or she is likely to be affected as early as the twenties by juvenile Huntington's disease.

Rather than eliminating transcription of the gene, the CAG expansion appears to confer new properties on the mRNA or the protein product. This type of gain-of-function mutation gives the disease its autosomal dominant character. In addition, scientists believe that the extra CAG repeats contribute to HD by involving the glutamine amino acid residues encoded by this codon. These glutamines are responsible for forming cross-links within and between proteins through a reaction catalyzed by enzymes known as transglutaminases (TGases). TGases may be involved in neurodegeneration by increasing glutamine cross-linking that could induce the formation of aggregated protein structures, with consequent neuronal death.

There is still no cure for Huntington's disease, but the individual symptoms (such as depression, involuntary movements, and others) can be treated with drugs.

The huntingtin gene is also implicated in another disease, called the Wolf–Hirschorn syndrome. In contrast with HD, this disease is the result of a loss of gene function, rather than a gain, caused by the deletion of the huntingtin gene. The affected individual is born with disfigured facial features, suffers from mental retardation, and rarely survives beyond early childhood.

INBORN ERRORS OF METABOLISM

In this section, we will be looking at inherited disorders that are caused by defects in a single gene. The defect or mutation causes a particular enzyme involved in some metabolic pathway to be either defective or missing altogether. As metabolic reactions often proceed in a stepwise or cyclical manner, an interference in even one step can throw an entire system off balance and result in disease. Some examples of inborn metabolic diseases are phenylketonuria (PKU) and Tay–Sachs disease.

Inborn errors of metabolism affect about 1 in every 5,000 babies born and are usually diagnosed at birth. Obvious physical features such as malformations of the skeleton or abnormal hair usually suggest the presence of a metabolic condition. Diagnosis is confirmed by checking the blood or urine for one or more of the compounds involved in the specific metabolic pathway affected. There may be too much of a substance or not enough of another. For example, a blood test confirms the presence of abnormally high levels of phenylalanine in infants that suffer from PKU.

In most cases of these diseases, the mutated gene is autosomal recessive in character. In other words, the gene is on one of the 22 autosomes, and there must be two copies of the faulty gene, one inherited from each parent, in order for the enzyme defect to exist. The genetic mutations are present from the time of conception, and nothing can be done to prevent them. However, clinical symptoms of the diseases caused by the inborn error may be preventable in some cases. For instance, if the mutation results in a particular substance in certain foods not being broken down, the affected individual might be able to stave off clinical symptoms by avoiding those foods. An example is phenylketonuria, which may be controlled by removing the amino acid phenylalanine from the diet. Genetic counseling and family planning are also important in managing these diseases. A couple who know that they are both carriers of a gene responsible for an inborn error may choose not to have children at all or to take advantage of recent advances in assisted pregnancy or prenatal diagnosis. This strategy is often used for conditions such as Tay–Sachs disease.

Gaucher Disease

Gaucher disease is the result of the accumulation of harmful quantities of a fatty substance called glucocerebroside in the spleen, liver, lungs, bone marrow, and, in rare cases, the brain. This fatty acid is a by-product of the recycling of blood cells and certain nervous tissue components such as myelin. Normally it is degraded via a metabolic pathway beginning with the activity of an enzyme called glucocerebrosidase. Mutations in the glucocerebrosidase gene may cause the enzyme to partially or completely lose function, which results in the accumulation of the lipid material and the development of disease.

Disease symptoms are related to the loss of function of the cells in which the fatty acids accumulate. Although they can all be traced to mutations in the same gene, there are actually three types of Gaucher disease (types 1–3), with different symptoms predominating. The three types of disease are caused by mutations in different parts of the gene.

In type 1 Gaucher disease, which is the most common, the recycling of macrophages in the bone marrow is affected. This interferes with the normal bone marrow functions such as platelet and blood cell formation, which gives rise to symptoms of anemia, fatigue, easy bruising, and splenic and liver enlargement. The presence of lipids in the cells seems also to trigger the loss of minerals in the bones, causing them to weaken. Type 1 Gaucher disease does not affect the brain. Disease symptoms may appear at any age. Type 2 and type 3 Gaucher disease are both different from type 1 in their early onset and the involvement of the brain. Type 2 Gaucher disease becomes apparent by 2–3 months of age, with symptoms of liver and spleen enlargement and brain damage. Patients seldom survive beyond 2 years. The type 3 form is most frequently associated with seizures. Liver and spleen involvement seem to be more variable than in type 2.

There is no permanent cure for Gaucher disease, but treating patients with normal glucocerebrosidase has proved to be effective in reducing the symptoms of type 1 disease as well as type 3 disease if the latter is detected early enough. There is, however, no therapy for the neurological symptoms of types 2 and 3.

Gaucher disease is the most common example of a category of metabolic disorders called lipid-storage diseases, which also includes Tay–Sachs syndrome (discussed later) and Niemann–Pick disease. Although it may affect anyone, Gaucher disease is most prevalent among the Ashkenazi Jews of Eastern and Central European ancestry, affecting about 1 in every 450 people, in contrast with a rate of 1 in 100,000 people in the general public. It is also the most common inherited disease among the Ashkenazi population.

Gyrate Atrophy of the Choroid and Retina

Gyrate atrophy of the choroid and retina is an example of a deficiency of an enzyme of the urea cycle, which oddly enough results in damage to the tissues of the eyes. People with this disorder suffer from a progressive loss of vision, eventually leading to blindness between the ages of 40 and 60 caused by the gradual degeneration or atrophy of the choroid—the thin layer of cells coating the eye— and retina. Mutations that cause gyrate atrophy occur in the gene for an enzyme called ornithine ketoacid aminotransferase (OAT) on chromosome 10, which converts the metabolite ornithine into the amino acid glutamate. Ornithine levels are elevated in the plasma of patients with gyrate atrophy, and in an earlier era before genetic testing, the detection of this substance in the blood was likely the first indication of this disease. A reduction of dietary sources of the amino acid arginine—which is the precursor to ornithine in the urea cycle—has proven useful in slowing the progress of the degeneration of the ocular tissues.

Menkes Disease

Menkes disease, also known as kinky hair disease because of the peculiar appearance of kinky, brittle, and colorless or steel-colored hair, is an inborn defect of metabolism that markedly affects the body's ability to absorb copper. People with Menkes disease show signs of severe brain damage and abnormalities of the walls of their arteries early in their infancy and seldom survive beyond 10 years of age.

Menkes disease is caused by mutations in a gene for a copper-metabolizing protein called ATPase. The normal enzyme plays an important role in transporting copper ions to tissues such as the brain and bones during development. A deficiency or absence of enzyme activity leads to an uneven distribution of copper in the body—sites such as the brain and liver have abnormally low levels, whereas organs such as the kidneys and intestinal lining accumulate higher than normal amounts of copper. Both of these situations result in undesirable consequences. The lack of copper in the tissues where it is needed (e.g., the developing nerve and bone cells) causes degeneration, and its accumulation in other sites proves to be toxic to those organs.

Unlike the other examples discussed here, Menkes disease is X-linked rather than autosomal. The gene is present on the long arm of the X chromosome and has no corresponding allele on the Y chromosome, so it is seen much more frequently in male infants than among females.

The prognosis for Menkes disease is very poor. As mentioned earlier, children born with this disease usually die within a few years. Copper supplements such as copper histidinate, given in the form of subcutaneous or intravenous injections, have shown some effectiveness in delaying death if the disease is diagnosed very early, but the success rate has been limited.

Niemann–Pick Disease

Niemann–Pick disease is a group of diseases related to genetic errors in lipid metabolism. The first description of this class of diseases came from a German pediatrician, Albert Niemann, who in 1914 described the case of a child with brain and nervous system impairment. Then, in the 1920s, another physician, Luddwick Pick, studied the tissues of children who had died of this type of disease and found it to be a lipid-storage disease with a pathology that was distinct from previously described storage diseases such as Gaucher disease and Tay–Sachs syndrome.

Today we know of at least three separate disorders that are classified as Niemann–Pick disease. Types A and B are caused by mutations in the same gene, *SMPD1* on chromosome 11. This gene encodes a lysosomal enzyme called *sphingomyelin phosphodiesterase*, which is normally involved in the metabolism of nervous tissue glycolipids. The type A disease is an acute form that manifests early in infancy and typically ends in death before the child is 3 years of age. It is characterized by the accumulation of structural lipids—predominantly sphingomyelin—in the cells of the bone marrow, blood vessels, and nervous tissues. Type B Niemann–Pick disease, which is also called the visceral disease, is less common than type A. It is a chronic, less severe form of the disease in which patients show only limited nervous impairment and survive into adulthood. Genetic studies on samples from both types of patients indicate that small deletions or nonsense mutations that result in truncated gene product and missense mutations that render the enzyme noncatalytic cause type A Niemann–Pick disease, whereas missense mutations that produce a defective enzyme with some residual activity cause the milder type B disease.

The third category of Niemann–Pick disease—type C—is caused by defects in cholesterol metabolism. About 95 percent of the cases mapped to mutations in a gene called *NPC1* on chromosome 18. This gene encodes a protein whose normal structure is very similar to a group of membrane proteins that play a role in cholesterol homeostasis and metabolism. The protein is believed to play a role in regulating the intracellular transport of cholesterol from the lysosomes—the intracellular compartments where they are metabolized—to other destinations in the cell. Mutations in the protein seem to affect this transport function because people with Niemann–Pick disease type C tend to accumulate cholesterol within their lysosomes. Clinical features of this disease are very variable but include a spectrum of neurological symptoms beginning 1–2 years after birth and eventually causing the death of the patient. The remaining 5 percent of type C disease cases are associated with another gene involved in cholesterol transport, called *NPC2*, on chromosome 14. In the course of studying the genetics of type C Niemann–Pick disease, scientists also found a less severe form known as type D (also called the Nova Scotia type), which appears to occur exclusively among people of Acadian ancestry from Nova Scotia in Canada. This variant is associated with specific mutant alleles of the *NPC1* gene.

There is currently no effective treatment for this group of diseases. Scientists have shown that inserting copies of the *NPC1* gene into cultured skin cells from NPC patients compensates for the abnormal cholesterol buildup, suggesting that gene therapy might eventually be able to cure the disease.

Phenylketonuria (PKU)

Phenylketonuria (PKU) is a rare metabolic disease caused by the absence or deficiency of an enzyme called phenylalanine hydroxylase (PAH) that breaks down the amino acid phenylalanine in the body and occurs in about 1 in 10,000 Caucasians in the United States. Although a normal part of many proteins, the amino acid phenylalanine by itself is toxic to many tissues, particularly nerve cells, in the body and hence must not be allowed to accumulate in the tissues. Normal PAH is a vital enzyme in the metabolic pathway to convert phenylalanine to another amino acid, tyrosine (which the body is able to utilize), and keeps the levels down to about 1 mg/dl of blood. When the enzyme is missing or inactive, phenylalanine and closely related substances (products of its metabolism or breakdown) rise to very high levels in the body, usually > 30 mg/dL. These compounds strip the protective covering of nerve cells, which is made of the protein myelin. Once exposed, nerve cells die and the affected individuals suffer severe mental retardation and usually do not live beyond their thirties.

Some of the accumulated phenylalanine in the bloodstream is broken down by alternative biochemical pathways that convert it to phenylketone and related products, which are then excreted in the urine. In fact, it was the detection of these compounds in the urine in certain mentally retarded patients that led the Norwegian biochemist Asbjørn Følling to first make the connection between the disease and its cause and to name the disease phenylketonuria in 1934.

Appearing shortly after birth, the most obvious symptoms of this disease are severe mental retardation, organ damage, and abnormal posture. Other symptoms might include vomiting, irritability, skin rashes, and a "mousy" odor to the urine. Some people may also have subtle signs of problems with nervous system function, such as increased muscle tone and more active muscle tendon reflexes. Because phenylalanine metabolism is related to melanin biosynthesis, many infants with PKU also tend to have lighter skin coloring than their unaffected siblings. Retardation sets in after birth if the infant is not diagnosed and treated in time.

Although it is a genetic condition that cannot be cured, PKU is completely treatable by cutting out phenylalanine from the diet very early in life. Today, all newborn babies are tested for PKU with a simple blood test and if the test is positive put on a lifelong phenylalanine-restricted diet. The Guthrie test for PKU was the first screening test developed for a genetic disease. It was adopted across the United States in the 1960s.

Phenylketonuria may be caused by different types of mutations—deletions, insertions, truncations, and single-site mutations—of the *PAH* gene, which is located on chromosome 12 and encodes a protein of 454 amino acids. Mutations in the gene may lead either to the incorrect formation of the protein or to malfunctions at the enzyme's active site. Certain PAH mutations (single-nucleotide polymorphisms) occur in certain populations at higher frequencies than in others. For example, those of Irish descent have higher frequencies of mutations and are susceptible to PKU (with an incidence as high as 1 in 2,000 people), whereas the Chinese have a very low incidence of 1 in 1,000,000. The differences in the frequencies of these mutations have been valuable to population geneticists for tracking migrations and movements of people throughout history.

Tay–Sachs Disease

Tay–Sachs disease is a fatal genetic metabolic disorder in which harmful quantities of a fatty substance called ganglioside GM2 accumulate in the nerve cells of the brain. The accumulation of this substance results in a loss of function of the nerve cells that is associated with a wide range of severity in symptoms. In infants, symptoms include paralysis, dementia, blindness, and early death. There is also a chronic adult form of Tay–Sachs disease in which the affected individual suffers from nerve dysfunction and psychosis.

Tay–Sachs disease is most commonly, but not exclusively, associated with Ashkenazi Jews. French Canadians of southeastern Quebec and Cajuns of southwestern Louisiana also seem to be at increased risk for this disease—both groups have about 100 times the rate of occurrence of other ethnic groups.

The disease is caused by mutations in a gene called *HEX A* on chromosome 15. The gene, which is active in nerve cells, codes for part of an enzyme called β-hexosaminidase A. This enzyme is found in lysosomes—special compartments in cells where large molecules are broken down for recycling—where it helps to break down a lipid called ganglioside GM2. In people who have Tay–Sachs disease, mutations in *HEX A* result in low levels or the loss of β-hexosaminidase A. This results in the accumulation of GM2 in the nerve cells, which then interferes with normal cellular function. In the developing fetus, this leads to the abnormal development of nervous tissue, which is responsible for the symptoms of retardation and paralysis described earlier. When the disease manifests later in life, the gradual loss of nerve cells leads to nervous dysfunction.

Couples who are planning to have children can take a blood test to see whether they are Tay–Sachs mutation carriers. This test measures the amount of β-hexosaminidase A in blood. Because they have one functional *HEX A* gene, Tay–Sachs carriers have about half as much of the enzyme as noncarriers, which is sufficient for their own needs. Scientists have recently developed genetic tests that look for known mutations in *HEX A* that cause Tay–Sachs disease.

INFLAMMATORY BOWEL DISEASE

Inflammatory bowel disease (IBD) is the name given to a group of diseases characterized by abdominal pain and discomfort, diarrhea, and bleeding (in the gut) caused by an overreaction of the immune system in the alimentary canal. There are two main types of disease that affect different parts of the digestive tract and to different extents—ulcerative colitis and Crohn's disease. *Ulcerative colitis*, as the name indicates, affects just the large intestine (colon) and involves only the innermost lining (i.e., the mucosa). It typically progresses to the formation of ulcers and causes rectal bleeding. In contrast, *Crohn's disease* can occur anywhere along the digestive tract from the mouth to the anus (although it is most commonly seen in the small intestine) and involves the deeper layers of the gut wall. The combined prevalence of these diseases is about 200–300 cases per 100,000 people in the United States. Although the outward symptoms of these two forms of IBD are quite similar, a differential diagnosis can be useful in deciding the best course of treatment and also to prevent the progression to ulcer formation in cases of ulcerative colitis.

A number of factors, including genetics, diet, and race, all appear to play a role in causing IBD. It is evidently a complex trait because many different genes on at least eight different loci in the human genome appear to contribute to its development. Scientists have mapped a susceptibility locus for Crohn's disease to chromosome 16 that contains several genes for proteins known to be involved in the inflammatory response. By far the most definitive link to the disease is with *CARD15*, a gene that makes an intracellular protein involved in apoptosis. Other candidate genes found in this region of chromosome 16 include CD19, a cell-surface protein involved in B-lymphocyte function; sialophorin, which is involved in leukocyte adhesion; the CD11 integrin cluster, involved in cell adhesion; and the interleukin-4 receptor. The last is particularly interesting because IL-4-mediated functions are altered in inflammatory bowel disease. Mutations in a gene for an intestinal mucin—glycoproteins of mucus produced at various sites in the body—called *MUC3A*, on chromosome 7, have been linked to ulcerative colitis. Using new DNA microarray technology, scientists have identified nearly 170 genes that are directly or indirectly linked to this disease, further complicating our understanding of it.

INTESTINAL POLYPS (FAMILIAL ADENOMATOUS POLYPOSIS)

An intestinal polyp may be best described as a cellular growth that forms in the innermost, mucus-producing layer of the intestinal (typically the colon) wall. Although such a polyp—which is called an adenomatous polyp because the tissue of origin is epithelial tissue—is initially benign, it has the potential to develop into a malignant cancer. In *familial adenomatous polyposis* (FAP), the affected individual has literally hundreds to thousands of these polyps throughout the colon. The average age for the appearance of the polyps is 16, although they may appear much earlier in childhood or not develop until an individual is in his or her thirties. Unless treated for the syndrome, these patients will almost certainly develop colon cancer.

The culprit in FAP is a tumor suppressor gene called *APC* (for adenomatosis polyposis coli) on chromosome 5. The function of the normal protein is to promote apoptosis in the cells of the colonic epithelium by controling the localization and activity of a growth-stimulating transcription factor called beta-catenin. Mutations in *APC* result in the formation of truncated proteins that are unable to perform their function properly. Without normal APC to rein it in, beta-catenin activity is uninhibited, which causes cells to proliferate rather than die. As these cells grow in number, they form polyps. Although the colon (large intestine) is the primary site for these polyps, patients with FAP may also develop polyps in other parts of the digestive tract such as the small intestine and stomach.

Familial adenomatous polyposis is inherited in an autosomal dominant fashion, which means that the affected individual need have only a single copy of the mutated gene in order to develop colonic polyps. Apparently, the normal *APC* copy provides enough of the normal protein to control just enough beta-catenin activity to keep the cells from becoming malignant. However, this gene is highly susceptible to inactivation by sporadic mutations, which increases the chances that the adenomas will (polyps) develop into malignant tumors.

Although there is no complete treatment for FAP, there are certain measures that may be taken in order to minimize discomfort and forestall the development of malignant colon cancer. A colectomy—the surgical removal of part or all of the colon—is often necessary once the polyps emerge. The postoperative use of certain nonsteroid, anti-inflammatory drugs has proven useful in causing polyps to regress and preventing the formation of new ones. Because of the high rate of colon cancer among FAP patients, many doctors recommend a genetic evaluation of families at risk.

See also: Apoptosis, Tumor Suppressor Genes (Ch. 4); Cancer—Colon Cancer.

KLINEFELTER SYNDROME

Also known as the *47-XXY syndrome*, Klinefelter syndrome is a chromosomal abnormality in which men are born with an extra X chromosome. Strictly speaking, it is not a genetic disease that is passed on from one generation to the next but rather the result of a random aberration during the development of the gonads and embryo. It is a trisomy of the sex chromosomes and, as in all trisomies (see Chapter 3 for a complete explanation), it occurs as a result of the improper segregation of chromosomes during meiosis. In this case, it is the sex chromosomes that fail to segregate (i.e., separate) properly. Klinefelter syndrome is the outcome when an abnormal XY sperm fertilizes a normal egg or when an abnormal egg containing two X chromosomes is fertilized by a Y-containing sperm. The resulting embryo has a total of 47 chromosomes instead of 46 because of the extra X. (Other combinations give rise to different disorders, such as the triple X syndrome.)

Perhaps not surprisingly, given the underlying cause of an extra X chromosome, many of the symptoms of this disorder are related to the development of sexual characteristics. Infants with Klinefelter syndrome often appear normal at birth and begin to exhibit most symptoms around the age when boys normally show signs of puberty. Although symptoms vary greatly in severity, some telltale features include a delay or failure in the development of secondary sexual features, smaller than normal sexual organs, and diminished facial, axillary, and pubic hair. The mildest cases may go virtually undetected, with no abnormalities other than infertility. Other possible features include complete sexual dysfunction, gynecomastia (development of breast tissue), abnormal height, disproportionate body parts, and learning disabilities.

The syndrome is relatively common and occurs in 1 out of 500 male births. Although the exact causes leading to abnormal meiosis are unknown, there is a higher risk of occurrence when the mothers are older in age. This is true of other trisomies such as Down syndrome and triple X syndrome as well. In rare instances, some patients will have more than one extra X chromosome, for a total of 48–50 chromosomes. These males show more severe symptoms, such as more severe retardation, deformities of bony structures, and even more disordered development of male features.

Karyotyping is the best method of diagnosis of this syndrome and may be useful for determining treatment. There is no treatment for the infertility associated with Klinefelter syndrome, but administering testosterone beginning at the age of puberty will help in the development of the secondary sexual characteristics. Consequently, genetic testing, particularly of children of older parents, is useful. Counseling may be necessary to help patients cope with emotional problems arising from sexual dysfunction.

See also: Chromosome, Meiosis (Ch. 2); Triple X Syndrome.

LONG Q-T SYNDROME

Long Q-T syndrome (LQTS) is a relatively uncommon hereditary disorder of the heart's electrical rhythm that results from underlying structural defects in the ion channels of the heart. Although the cause is genetic and thus present from the time of birth, LQTS is usually not seen until childhood or adolescence. It manifests very suddenly and dramatically in response to certain types of stressors that cause sudden disturbances in the heart's rhythm (also known as arrhythmia) and lead very quickly to loss of consciousness and death.

Normally, when the heart muscles contract, they emit a series of electrical signals in a characteristic regular pattern known as a waveform. When a physician asks for an electrocardiogram (ECG or EKG), he or she is actually looking at a record of the waveform to make sure that it is normal. The Q-T interval represents the portion of the waveform between the activation and inactivation (contraction and relaxation) of the muscles of the heart's chambers. In people with LQTS, this interval is prolonged beyond the normal period, which leaves the individual afflicted with the condition extremely vulnerable to conditions such as hard exercise or deep emotional trauma, which cause the heart to beat abnormally rapidly. Faced with this sudden arrhythmia, the heart is unable to contract properly, and the blood supply to the brain and other parts of the body is stalled. Oxygen starvation in the brain will very quickly cause the person to faint. Meanwhile, if the heart does not regain its normal rhythm, it will go into spasms, which will cause ventricular fibrillation unless the patient is given emergency treatment. Thus, seemingly normal and healthy teenagers may succumb very unexpectedly and quickly to some stress if they have undiagnosed LQTS.

The genetic defects or mutations that have been detected in cases of LQTS map to different genes for ion channels. As we know from Chapter 4 (see the entry on channel proteins and cross-membrane transport), humans have hundreds of genes that encode different ion channel subunits in different tissues and organs. Not surprisingly, the genes linked to LQTS are those that are expressed most in the cardiac muscles and tissues. Most of the disease-causing mutations occur in genes for K$^+$ channels, but other channels have also been implicated. For example, mutations in the *KCNQ1* gene on chromosome 11—which encodes a voltage-gated K$^+$ channel that is required for the repolarization of the cardiac action potential—cause an autosomal dominant version of the syndrome classified as LQTS type 1, and those in the *KCNH2* gene on chromosome 7 (for a K$^+$ channel of a different family) cause LQTS type 2. A sodium channel subunit encoded by the *SCN5A* gene on chromosome 3, which is expressed primarily in cardiac muscles, causes LQTS type 3 when it is defective. Mutations in still another gene, *ANK2* on chromosome 4, which encodes an integral membrane protein that is not strictly speaking an ion channel at all, have been implicated in LQTS type 4.

Long Q-T syndrome is often fatal because of the rapid onset of symptoms, but it can be treated and the arrhythmia prevented if the existence of the condition is known. Drugs known as beta-blockers have been used successfully to treat disease symptoms in some cases. They work by affecting the response to certain nerve impulses in some parts of the body and help the heart to beat more regularly. In addition, commonsense measures, such as avoiding strenuous physical activity and other stresses, are also effective preventive measures. Research on how different genes interact with each other should also encourage the development of new treatments for LQTS.

LQTS and Sudden Infant Death Syndrome

There has been some evidence linking LQTS-causing mutations in certain ion channels to sudden infant death syndrome (SIDS), a mysterious condition in which seemingly healthy infants (usually between the ages of 2 and 4 months) die in their sleep from no apparent cause. In some of these cases, scientists found that the death may well have been due to an episode of arrhythmia caused by defective potassium channels (mutations in *KCN1Q*) or sodium channels (*SCN5A*), both of which are linked to LQTS. Meanwhile, scientists have discovered certain other genetic links for SIDS as well, including two mitochondrial genes, *MTTL1* and *MTND1*, which encode a transfer RNA for leucine and a respiratory chain subunit, respectively, and a gene on chromosome 17 (*SERT*) for a neurotransmitter transporting protein.

The unpredictable nature of SIDS is a haunting prospect that strikes fear into the hearts of parents everywhere, but there is hope. Since the 1990s, researchers have identified simple measures parents can take to greatly reduce their child's risk, and the incidence has certainly dropped since then. One of the most important precautions is to place infants on their backs for sleep instead of on their stomachs. With the identification of the genetic risks for SIDS, parents are in a better position to take the necessary precautions.

See also: Channel Proteins and Cross-Membrane Transport (Ch. 4).

MALE PATTERN BALDNESS

Male pattern baldness is a progressive loss of hair (alopecia) that follows a characteristic pattern on the scalp. Seen much more commonly in men than in women, male pattern baldness is characterized by the loss of hair beginning at the temples and the top of the head, causing a receding hairline and a bald spot. As it progresses, these bald areas continue to grow and eventually join, leaving a horseshoe-shaped line of hair around the sides and back of the head. Roughly 40 million men in the United States, and perhaps up to 50 percent of men worldwide, are affected. In contrast, the worldwide frequency among women is estimated at 20 percent. Balding may begin at any age from puberty onward, although it occurs more often with advancing age. Currently there is no way to prevent or cure this type of hair loss, although drugs such as finasteride (sold as Propecia) may be effective in slowing hair loss. In certain cases, doctors may prescribe the drug minoxidil (which readers may recognize from television commercials as Rogaine), which replaces lost hair rather than prevent hair loss. This treatment has usually proven more effective when applied at the early stages of hair loss.

Strictly speaking, male baldness is not a disease because it poses no physical risks to the individual and is not associated with pain or any other symptoms such as rashes, itching, redness, and so that accompany inflammation. However, the fact that there are drugs available for stalling or correcting the symptoms shows that many people perceive it as a problem. It has been included in this chapter because it represents a deviation from the perceived norm (of more complete hair growth) that appears to run in families and shows some degree of association with certain genes.

The medical term for male pattern baldness is androgenetic alopecia, which indicates that it is influenced by a combination of hormone—specifically androgen—activity and genetics. Although the androgens testosterone and dihydrotestosterone (DHT) are responsible for stimulating the growth of facial and underarm hair during puberty (by increasing the size of hair follicles in these regions), they play quite the opposite role in the scalp, especially later in life. Here they cause the hair follicles to become more sensitive to degradation and shorten the growing phase of the hairs, which leads to hair loss. Scientists have discovered a link between male pattern baldness and the activity of the enzyme 5-alpha reductase. This enzyme, whose gene is on chromosome 5, catalyzes the conversion of testosterone to DHT (its more potent form), which then stimulates the development of the sexual characteristics during puberty. Dihydrotestosterone is produced in the prostate gland, various adrenal glands, and the scalp. Actually, there are two nearly identical types (isoforms) of 5-alpha reductase, called SRD5A1 and SRD5A2, respectively. Although first isolated from the prostate gland, the first isoform is much more prominent in the scalp, whereas SRD5A2, whose gene is on chromosome 2, is more abundant in the prostate gland itself. Scientists have linked disturbances in the *SRD5A1* gene with variations in hair growth—not just baldness but also the other extreme, namely excess hair growth, or hirsutism. None of these traits are inherited in a Mendelian fashion, although the exact genetic events that lead to the different outcomes are not properly understood.

MARFAN SYNDROME

Marfan syndrome is an autosomal dominant inherited disorder of the connective tissue that is most typically characterized by the development of unusually long limbs and often compounded by problems in many parts of the body, including the skeleton (bones), lungs, heart, eyes, blood vessels, and nerves. Symptoms typically worsen with age. At an estimated occurrence of 1 out of every 5,000 Americans, it is not an uncommon disorder and shows no special preference for either sex or any ethnicity or geographic region. Medical experts today believe that Abraham Lincoln may have been afflicted with Marfan syndrome.

The underlying cause of Marfan syndrome is a defective gene for the protein fibrillin (*FBN1* on chromosome 15), an essential component of the elastic fibers found in connective tissue. Because connective tissue's function is to hold the body together and provide a framework for growth and development, a defect in one of its essential components has serious consequences in many different organs and tissues. The mutated *FBN1* gene shows variable expression, which means that two people, even from the same family and with identical gene defects, are often not affected in the same way. Also, even though only 5 percent of the cases are caused by spontaneous mutations and the remaining 95 percent by defects inherited from a parent, the syndrome is not always evident in newborn babies. Indeed, it can manifest itself at any age from infancy to late adulthood. The exact reason for the unpredictable nature of the way in which *FBN1* mutations express themselves is unknown.

Although there is no cure for Marfan syndrome, there are a range of supporting therapies, offered by different specialists, depending on the organ systems that are most affected. For example, beta-blockers may be used to control some of the cardiovascular symptoms, and braces and other structural supports may be necessary in people with skeletal defects. Unfortunately, even if the mutations are detected, it is not easy to predict the course that the syndrome will take. Nevertheless, prenatal or early genetic testing can prove useful because it helps in the early detection and alleviation of symptoms, markedly improving the quality of life of the people with Marfan syndrome.

MONILETHRIX

Monilethrix is a rare, dominant genetic disease in which hair fibers lose their smooth appearance and look like strings of beads. The disease has been linked to mutations in the genes for type two keratins, specifically the basic hair keratins 1 and 6 located on chromosome 12, which code for keratins of the hair cortex. As a result of these mutations, the cuticle of the hair does not develop properly or uniformly along the length of the hair shaft. This results in some parts with the cuticle (the so-called beads) and other interspersing portions without this outer covering. The lack of the cuticle leaves the hair shaft very brittle, and the hair of affected people frequently breaks off at these constrictions. People with monilethrix tend to suffer from moderate to excessive hair loss. This loss occurs most frequently in the back of the head and neck while leaving the front quite unaffected, although monilethrix may also manifest itself all over the scalp as well as other parts of the body. In some cases, the beaded hair is accompanied by defects of the hair follicles as well as the nails and teeth (enamel), and in some cases it has been associated with cataracts. The age of onset is relatively early, and the condition is actually known to improve over the years. Limited treatment with topical applications of retinoic acid or glycolic acids is possible.

See also: Keratin (Ch. 4).

MUSCULAR DYSTROPHY (MD)

Muscular dystrophy (MD) refers to a group of genetic disorders whose major symptom is a progressive degeneration, and hence weakening, of muscular tissue over time. The primary targets in these diseases are the voluntary skeletal muscles, although the degeneration of tissue may also extend to the cardiac and smooth muscles and even other organs in certain types of MD. There are over 40 different types of muscular dystrophies identified thus far, which are attributable to mutations in a variety of genes and the consequent deficiencies or defects in their products.

Duchenne/Becker Muscular Dystrophies

Originally identified separately because of the difference in severity and the age of onset, the Duchenne and Becker muscular dystrophies were actually found to be forms of the same disease. Both forms are caused by deletion mutations in the gene for dystrophin, a protein of the muscle fibers whose main task is to maintain the strength of the cells by anchoring the internal cytoskeleton to the cell membrane. When dystrophin is deficient, lacking, or dysfunctional, the cell membrane becomes permeable and leaky, which causes lysis. In the initial stages, the body is able to regenerate muscle cells, but in the face of continual degradation over prolonged periods of time, this ability is exhausted, and muscles begin to waste away. The damage is also exacerbated by immune reactions against the cellular contents that leak into the tissue.

Duchenne MD is the more severe form of this dystrophy. It is also the most common muscular dystrophy among children worldwide. Onset is early in life, between the ages of 2 and 6, and the disease first affects the pelvis, upper arms, and upper legs, eventually progressing to all of the voluntary muscles. Children begin walking on their own later than average, and even after they do they begin to have difficulties such as falling down frequently or having trouble rising from the ground or going up steps. Survival beyond the twenties is rare. The symptoms and progression of the Becker dystrophy are nearly identical, albeit less severe. Muscular degeneration can manifest as late as 16, and many patients survive well into middle age.

Differences between the two forms have been traced to the nature of the deletion mutations causing the disease. Surprisingly enough, the milder Becker dystrophy was found to be associated with much larger deletions than the Duchenne disease. As researchers discovered, what was important was not so much the amount of coding sequence that was deleted as *where* it occurred and whether it resulted in a frameshift (a shift in how the triplet code gets read for translation to protein) or not. (See the entry on mutation in Chapter 2 for a fuller explanation of the nature of these mutations.) The deletions associated with the Becker form are large but leave the reading frame of the gene intact, resulting in the synthesis of a partially functional version of dystrophin. The smaller deletions causing Duchenne dystrophy, however, cause a shift of the reading frame, resulting in the production of "nonsense" transcripts, which are highly unstable and quickly degraded in the cell. Consequently, there is a nearly complete loss of dystrophin function, causing a much more severe disease.

Duchenne and Becker muscular dystrophies exhibit a pattern of X-linked recessive inheritance. This means that the disease is much more prevalent in males and often transmitted from one generation to the next by female carriers. Incidence is estimated at 1 in 3,500 boys worldwide.

See also: Muscle—Dystrophin (Ch. 4).

Myotonic Dystrophy

Myotonic dystrophy is the most common form of adult muscular dystrophy. In addition to the muscular degradation and wasting characteristic of all muscular dystrophies, myotonic dystrophy is characterized by symptoms of myotonia, which is a loss of the ability of muscles to relax after contractions. The disease shows an early pattern of muscle wasting that is unique among the major muscular dystrophies. As opposed to the hip and shoulder muscles in the other dystrophies, the first muscles to be affected in this case are those of the face, neck, hands, forearms, and feet. In addition, myotonic dystrophy may be associated with mental deficiency, hair loss, and cataracts.

There are two types of myotonic dystrophy, based on which gene is mutated. The cause of the more common *type 1* (*dystrophia myotonia*) is the mutation of a gene on chromosome 19 for a protein kinase that is expressed in skeletal muscle. Specifically, there is an amplification of a triplet nucleotide (CTG) in the gene, which means that this triplet gets repeated many times in the gene. Disease severity varies with the number of repeats: normal (unaffected) individuals have from 5 to 30 repeat copies of AGC, mildly affected persons from 50 to 80, and severely affected individuals have 2,000 or more copies. Type 1 myotonic dystrophy exhibits one of the most dramatic cases of anticipation, which is the increase in severity of the disease and the onset at an earlier age over successive generations, due to the amplification of the trinucleotide repeat from one generation to the next. Interestingly, although the disease is autosomal, the most extreme amplifications are transmitted almost exclusively through the female line, and anticipation is noted only in the offspring of affected women.

Nearly identical symptoms of myotonic dystrophy are also caused by the amplification of a tetranucleotide unit (CCTG) in the first intron of a gene for a zinc-finger protein (ZNF-9) encoded on chromosome 3. This disease is also called *dystrophia myotonica 2*. The normal protein appears to function as an RNA-binding protein or transcription factor protein that influences the transcription of DNA into mRNA.

Facioscapulohumeral (FSH) Muscular Dystrophy

Named for the muscles that are selectively weakened—the face, shoulders (scapulae), and upper arms—FSH muscular dystrophy is the third most commonly occurring disorder of this category. Syptoms typically appear during the second decade of life and include a gradual weakening of the muscles in the areas mentioned that manifest in difficulties with speech and facial expressions, a "winging" or protrusion of the shoulder blades, the reduction of muscle mass in the shoulder area, and a weakened ability to bend and straighten the elbows. The observed weakness is often asymmetrical, in contrast with the even or symmetrical wasting symptoms of other dystrophies.

Facioscapulohumeral muscular dystrophy has been mapped to a deletion of a repeated nucleotide unit on a specific locus on chromosome 4. Researchers have identified a gene in the vicinity of this deletion temporarily known as the FSH region gene 1. The function of this gene is unknown, and the DNA sequence bears no homology to any known genes. Nevertheless, the fact that it has been conserved in evolution *and* has related sequences on several other chromosomes indicates that the FSH region gene is important.

Emery–Dreifuss Muscular Dystrophy

Emery–Dreifuss muscular dystrophy is another X-linked degenerative muscular dystrophy characterized by muscle weakness and wasting without the involvement of the nervous system. The most commonly affected areas include the shoulder, upper arm, and shin muscles. Joint deformities are also common. Often confused with Becker MD because of the similarities in age of onset, symptoms, and progression, Emery–Dreifuss MD may be distinguished from Becker MD on the basis of symptoms of cardiac involvement and mental retardation.

The underlying genetic basis for Emery–Dreifuss MD is the mutation of the gene for *emerin*, a nuclear membrane protein in skeletal and cardiac muscle cells that anchors the nuclear membrane to the cytoskeleton. Mutations in the gene result in the synthesis of a nonfunctional protein that destabilizes the membrane and makes muscles more susceptible to lysis and degradation.

In addition to the X-linked diseases caused by mutations in emerin, there is also an autosomal dominant version of this disease caused by certain mutations in the gene for *lamin,* which is on chromosome 1. This protein is a member of a highly conserved family of nuclear membrane proteins that are thought to play a role in nuclear stability, chromatin structure, and gene expression. The importance of lamin is further evidenced by the fact that different mutations of the gene cause a number of genetic diseases besides Emery–Dreifuss muscular dystrophy, including familial partial lipodystrophy, limb girdle muscular dystrophy, dilated cardiomyopathy, and Charcot–Marie–Tooth disease.

Congenital Muscular Dystrophies

Attributable to multiple genetic causes, the congenital muscular dystrophies are characterized by their onset at infancy, indeed even in utero, with babies lacking the muscle strength to move freely enough in the womb. Symptoms at birth often include floppiness, poor head control, muscle weakness, delayed motor milestones (e.g., when a baby first stands while holding onto something or learns to crawl or walk), and tightness in the ankles, hips, knees, and elbows. The baby's hips may be dislocated at birth.

A large proportion of congenital muscular dystrophies are caused by mutations on chromosome 6 for a gene for *merosin*, a membrane protein of the basement membranes in the placenta, skeletal muscles, and Schwann cells. This protein plays a role in the attachment, migration, and organization of cells into tissues during embryonic development by interacting with various extracellular matrix components. When this protein is mutated, these functions are hampered, leading to defective development of the muscles in the embryo and the consequent dystrophy.

Another congenital MD is *Fukuyama congenital muscular dystrophy*, which was first described in 1960 by a Japanese physician who reported more than 200 cases diagnosed in Japan alone. The genetic basis for this congenital disorder is the mutated gene for fukutin (chromosome 9), a regulatory protein that controls the migration and assembly of the neurons of the brain's cortex. This form of the disorder is prevalent in Japan.

As indicated by the examples cited, the primary symptoms of muscular dystrophy can arise from defects in a variety of genes and proteins, including those involved in embryonic development, signal transduction, and, obviously, muscle function. This heterogeneity underscores the importance of the muscles to the survival of humans and the intricate manner in which they are linked to other vital systems and processes in the body.

See also: Muscle (Ch. 4).

NEUROCUTANEOUS SYNDROMES

The neurocutaneous syndromes are a set of genetic disorders that share common characteristics of neurodegeneration (i.e., a decay of the tissues of the brain, spine, and peripheral nerves) accompanied by tumors in these tissues as well as the skin (hence the descriptor of cutaneous), internal organs, and skeletal bones. They are believed to arise from the abnormal development of embryonic cells caused by mutations in one or more genes that are involved in this process. Some of the common conditions follow.

Ataxia-Telangiectasia

Also known as the Louis–Bar syndrome, ataxia-telangiectasia was named for two early and characteristic, although by no means exclusive, symptoms—"ataxia," which means a lack of muscle control, evidenced by a wobbliness or lack of balance and slurred speech, and "telangiectasia," which is the appearance of tiny red spider veins in the corners of the eyes or on the surface of the ears and cheeks when exposed to sunlight. The spider veins are essentially harmless, but their appearance in conjunction with ataxia led to the naming of this disease.

Children born with ataxia-telangiectasia (A-T) appear normal at birth and usually begin to display symptoms of ataxia during their second year. This marks the beginning of the progressive degeneration of the cerebellum of the brain. Ataxia worsens over time, and the patients gradually lose control of various muscles, leading to confinement to a wheelchair and progressive difficulties in eye movement. Other symptoms include severe immunodeficiency and a proneness to respiratory infections, a high sensitivity to radiation, and the predisposition to cancers, especially lymphomas and leukemias. (Children with A-T tend to develop malignancies of the blood system almost 1,000 times more frequently than the general population.) At a microscopic level, the chromosomes are highly susceptible to breakage. Therapy or management is symptomatic, and there is as yet no known cure for A-T. Treatment is especially complicated, as radiation, which is one of the most widely used therapies for cancers, is not an option. Most patients die quite young either from cancer or respiratory illnesses.

Ataxia-telangiectasia is an autosomal recessive disease caused by mutations in the gene for a protein known as the ATM (for Ataxia-telangiectasia mutated) protein. Although the disease was mapped to chromosome 11 in 1988, it was not until 1995 that the actual gene was cloned and the mechanism of disease understood. The normal ATM protein is a member of a family of signal transduction proteins called phosphatidylinositol-3-kinases, whose function is to repair damaged DNA by adding phosphate groups to key substrate molecules. Mutations in this gene compromise the cell's ability to repair its DNA in response to damage, such as that induced by radiation. The variety of symptoms seen in A-T is a reflection of the diversity of cellular signaling pathways induced by ATM.

Epidemiologists have estimated the frequency of A-T at 1 in 40,000 to 100,000 live births in the United States, although the actual incidence may be higher because of improper diagnoses of young children who succumb early in life. The disease does not appear to have any geographic or ethnic predilections, and worldwide incidence probably parallels that in the United States.

See also: DNA Synthesis/Replication and Repair, Signal Transduction—Kinases (Ch. 4).

Neurofibromatosis

Neurofibromatosis affects the development and growth of neural (nerve) cell tissues, causing tumors to grow on nerves and producing skin changes and bone deformities. About 30 to 50 percent of new cases arise through spontaneous mutation, and once this change has taken place, the mutant gene is passed on to the next generation. There are two types of neurofibromatosis, called types 1 (NF1) and 2 (NF2).

Neurofibromatosis type 1, also known as von Recklinghausen's disease, is one of the most common genetic diseases, affecting about 1 in 3,000 individuals. It is an autosomal dominant disorder caused by mutations in the *NF1* gene on chromosome 17. Symptoms of NF1, which typically include skin discolorations called "café-au-lait spots" and fibromatous tumors, are often evident at birth or during infancy and early childhood. The *NF1* gene expresses a large and complex protein called neurofibromin. A part of this protein is similar to a family of proteins called guanosine triphosphatase-activating protein (GAP). GAP proteins appear to play a role in tumor suppression in certain cancers, suggesting that neurofibromin may have a similar role in the development of neurofibromas. Mutations in the *NF1* gene may lead to lower quantities of normal protein so that tumor development is not suppressed. A number of point mutations and gene deletions have been identified in affected individuals. Genetic testing for NF1 only tests for protein truncation and does not detect other types of mutations. In most cases, symptoms of NF1 are mild, and patients live normal and productive lives. In some cases, however, NF1 can be severely debilitating.

Neurofibromatosis type 2 is characterized by tumors on the auditory nerves. The tumors are bilateral (i.e., they occur on both sides of the body), and their formation often creates pressure on and damages neighboring nerves. Patients may suffer from tinnitus (a ringing noise in the ear), hearing loss, and poor balance. Headaches, facial pain, or facial numbness, caused by the pressure from the tumors, may also occur. The *NF2* gene is on chromosome 22 and encodes a putative tumor suppressor protein. The normal function of this gene is incompletely understood, but it is expressed at high levels during embryonic development. Gene expression in adults occurs almost exclusively in the nervous tissue. The protein encoded by the gene is similar to members of a family of proteins known to be involved in regulating the cytoskeleton and ion transport. Mutations in the gene interfere with the function of the protein and account for the clinical symptoms observed in NF2 patients. Like NF1, NF2 is an autosomal dominant genetic trait.

There is presently no cure for the neurofibromatoses, but treatment is available to control symptoms. Surgery can also reduce the effect of some bone malformations and painful or disfiguring tumors; however, the tumors may grow back. If the tumors become malignant (3–5 percent of all cases), the patient may undergo surgery, radiation, or chemotherapy.

Tuberous Sclerosis

Tuberous sclerosis is characterized by a combination of the following symptoms with varying degrees of severity: benign nodules in the brain and/or retina, wart-like lesions of the skin, seizures, and mental retardation. In the early 1990s, scientists correlated these symptoms with mutations in two genetic loci designated as tuberous sclerosis 1 (*TSC1*) and 2 (*TSC2*) on chromosomes 9 and 16, respectively. Like the neurofibromatosis genes, the *TSC* genes are involved in regulating the cell cycle through signal transduction pathways. *TSC1* turned out to be a tumor suppressor encoding a protein that the discoverers named hamartin because of its similarity to a yeast protein of the same name but of unknown function. *TSC2* codes a protein called tuberin that regulates GTPase-activated signaling pathways.

Von Hippel–Lindau Syndrome

Named for physicians who first described this disorder in the early twentieth century, von Hippel–Lindau (VHL) syndrome is an autosomal dominant neurocutaneous disorder caused by the abnormal growth of blood capillaries, which form knot-like angiomas instead of branching out through the organs as they normally do. Common sites for angiomas include the brain, retina, spinal cord, adrenal glands, and skin.

The genetic basis of VHL is a germ-line mutation in a tumor suppressor gene (*VHL*) on chromosome 3 that acts by regulating a transcription factor in a manner similar to the retinoblastoma gene (see the entry on tumor suppressor genes in Chapter 4). Specifically, *VHL* acts on HIF, a transcription factor that is made of two subunits and plays a role in oxygen-regulated gene expression. The VHL protein appears to be unrelated to any known family of human proteins, but scientists are learning more about it by using rats and mice as model organisms.

Although the angiomas (knotted capillaries) in VHL and some of the other neurocutaneous syndromes are benign insofar as they do not metastasize like cancers, they can and do interfere with the proper functioning of the sites where they occur.

With early detection, they can be removed to minimize structural and functional damage to various organs. Consequently, genetic screening for the presence of the VHL mutations is useful in families with a history of this disease.

OBESITY

Obesity can be quite simply defined as an excess of body fat. It is indeed true that the range of "normal" body weight in healthy humans is very wide and that some people are naturally larger than others. However, there is a limit beyond which excess fat causes serious health problems. Currently, the consensus among physicians is that men with more than 25 percent body fat and women with more than 30 percent body fat are obese. Obesity has been linked to several chronic conditions, including heart disease, diabetes, high blood pressure, stroke, and even certain forms of cancer.

The physiological reason for obesity is a prolonged imbalance of energy in the body. In other words, the amount of energy (or number of calories) a person consumes is not equal to the amount that the body uses for its various activities. The excess energy is converted into fat, which is stored in special cells called adipocytes throughout the body.

Now, based on this information, it may seem obvious that the way to avoid obesity is simply to maintain a balance between our food intake—food is by far the most direct and abundant source of energy for humans—and energy expenditure through physical activity or exercise. And indeed it is true that sensible eating and exercise habits go a long way toward maintaining body weight and preventing obesity. However, there is ample evidence to suggest that maintaining a proper energy balance depends on more than just food and exercise. First, the body does not simply use up the fat in adipose tissue when it needs energy. It always uses up the most readily available sources first—usually sugars and other carbohydrates. In general, fats and sugars are more difficult forms of energy for the body to burn, so the fat that is stored in the adipocytes tends to accumulate over time. Furthermore, the energy equation of consumption and utilization in the body involves many more factors than just food on one side and exercise on the other. Energy and body weight are key players in the process of homeostasis or balance described in Chapter 4. Consequently, obesity is related to a host of causes, including physiology, behavior, psychology, the environment, and, last but not least, our genes.

Our main interest in this section is the genetic component in obesity. Obesity is polygenic, which means that many genes appear to influence the condition. This is evident from the fact that there is an entire network of signals and molecules that contribute to weight homeostasis. Still, a relatively small proportion of obesity cases in the population have been linked to mutations in single genes (see Table 5.2). By understanding the action of these defects, scientists hope to gain a better understanding of the processes of fat storage and energy consumption. As

Table 5.2. Single genes linked to obesity

Gene/product	Chromosome	Normal function	Inheritance
Leptin	7	Homeostasis; controls appetite and stimulates energy expenditure	Recessive
Leptin receptor	1	Membrane receptor mainly in cells of the hypothalamus that transmits leptin signals into cells	Recessive
Proopiomelano-cortin	2	Pituitary hormone involved in stimulating the adrenal cortex and melanocytes	Recessive
Prohormone Convertase 1	5	Enzyme involved in regulating insulin synthesis in the body by processing pro-insulin	Recessive
Melanocortin-4 receptor	18	Membrane receptor that transmits signals from the hormone melanocortin in brain tissue	Dominant
SIM1 (Homolog of "single-minded" gene in *Drosophila*)	6	Transcription factor found primarily in human fetal kidney cells; "single-minded" regulates nerve cell formation in fruit flies	Likely dominant

you can well imagine, the potential payoff from this research in the form of effective weight-reducing therapies is enormous!

The molecule that has caused the most excitement with respect to its potential in weight reduction, as evidenced by experiments on mouse models, is *leptin*. As we discussed earlier, this hormone is a major player in various aspects of homeostasis. As shown in Table 5.2, obesity in humans has been linked to mutations in either the human leptin gene itself or in the gene for a leptin receptor. The result is the same in both cases. Either because of its own defects or because of those in its receptor, the leptin molecule fails to control the appetite properly, leading to increased food intake and fat storage.

PARKINSON'S DISEASE (PD)

Parkinson's disease (PD) is a progressive neurodegenerative disorder caused by damaged cells in the portion of the brain that controls movement. It is named after

On Capitol Hill, boxing legend Muhammad Ali (left) and actor Michael J. Fox joke around before the start of a Senate subcommittee on Labor, Health, Human Services and Education hearing on Parkinson's disease, May 22, 2002, Washington, D.C. The celebrities, who have Parkinson's, asked the panel for more funds for Parkinson's research. [AP/Wide World Photos]

James Parkinson, an English physician, who described it in an 1817 publication titled "Essay on the shaking palsy." However, this disease appears to have been with mankind since ancient times (like diabetes mellitus and epilepsy) because descriptions of conditions with its symptoms are to be found in ancient Indian and Chinese medical texts. Today, Parkinson's disease is seen all over the world and shows no special predilection for any geographic region or ethnic group. It is the second most common neurological disorder, second only to Alzheimer's disease. It is estimated that at least 500,000 Americans are affected, with some 50,000 new cases reported annually. Boxing champion Muhammad Ali and actor Michael J. Fox are two famous celebrities who have been diagnosed with this disease.

The physiological basis for Parkinson's disease is the death of neurons in a part of the brain known as the substantia nigra. These cells are responsible for producing the neurotransmitter dopamine, which is needed for carrying messages to limbs for conducting normal movements. Parkinson's disease is characterized by a severe shortage of dopamine. It is this deficiency that causes the symptoms of the disease. Among the most characteristic symptoms of Parkinson's disease are the uncontrollable tremors of the body. Eventually, brain damage spreads and the affected person experiences rigidity of limbs and muscles, gradual dementia (loss of mental faculties), depression, loss of the sense of smell, and inability to control the heart rate and blood pressure. This wide range of symptoms is hardly surprising, given that the neurons all over the brain that control many different functions of the body gradually cease to send or interpret nervous signals.

By and large, Parkinson's disease is a disease of age, and symptoms typically appear around the age of 60, with incidence and prevalence increasing as the population grows older. However, there is a less common form of the disease—the type of PD affecting Michael J. Fox—known as "early-onset" Parkinson's disease, which accounts for about 5–10 percent of all cases. There are subtle differences in symptoms between the two types.

Treatment options are available to help control the symptoms of the disease to a certain extent. The most effective treatment is to replace dopamine in the brain through levodopa (L-Dopa), which nerve cells are able to convert to dopamine. Other medications work by substituting dopamine with a different substance that "mimics" the action of dopamine. The advantage of this is that the detrimental effects of excess dopamine in the brain are less likely to occur. Other drugs act by helping nerve cells release more dopamine into the synapse. Still others help restore the dopamine/acetylcholine balance, which helps relieve tremors and muscle stiffness.

Parkinson's disease is a complex disease in which both genes and the environment seem to play a role. Scientists have found that close relatives of patients were more than twice as likely to have signs of the disease as the rest of the population. In familial PD, the pattern of inheritance is usually difficult to pin down. In cases where a pattern can be established, autosomal dominant patterns occur more often than autosomal recessive patterns. Several gene loci implicated in various forms of Parkinson's disease have been identified.

Researchers have linked susceptibility to *early-onset Parkinson's disease* to variations in genes on two main chromosomes: 4 and 6. The first specific gene to be identified was *SNCA* (also known as *PARK1*), which expresses a protein called alpha synuclein that appears to have a role in signaling and membrane trafficking in nerve cells. However, mutations in this gene only relate to early-onset Parkinson's disease in a few cases. Only a few families (from Italy and Greece) have been reported to have a mutation in this gene, which is inherited in an autosomal dominant manner.

A second gene, called *PARK2*, has been identified on chromosome 6. This gene encodes a protein called parkin whose exact function is still unknown but that may be involved in ubiquitin-related functions in the nerve cells. Variant forms of this gene have been found in families where Parkinson's disease is inherited in an autosomal recessive pattern. The protein ubiquitin expressed by the gene *CH-L1* on chromosome 4 has also been implicated. These genes only account for a small fraction of the early-onset cases. There are probably many more variations in genes still waiting to be discovered.

Genes for *late-onset Parkinson's disease*, the more common form of the disease, have proven more difficult to pin down. The strongest link found until now is with a region of chromosome 17 near the *TAU* gene (also called *MAPT*), which encodes a protein that helps to maintain the structure of brain cells and is known to be involved in Alzheimer's disease. Although this gene does not seem to be directly involved in Parkinson's disease, it seems to increase susceptibility. Defects in the enzyme N-acetyl-transferase 2 (encoded by *NAT2* gene on chromosome 8), which influences the body's ability to break down toxins, also appear to be linked to late-onset Parkinson's disease. There are two major variants of the protein: "fast processing" and "slow processing." The slow variant was found twice as often in persons with familial Parkinson's disease as compared with people without Parkinson's disease. In 2001, scientists successfully mapped a genetic locus on chromosome 1 that contributes to late-onset Parkinson's disease. Other suspected regions include the *PARK2* gene that was linked to early-onset Parkinson's disease and the long arm of chromosome 2, although no specific genes have been identified as yet.

In addition to the genes and regions mentioned, this disease has also been linked to variations in loci on chromosomes 5, 8, 9, 13, and X. However, to date, none of these discoveries have been useful for developing genetic tests that a physician can conduct for a patient to determine risk. Scientists are still actively looking for other genes for the various types of Parkinson's disease.

PAROXYSMAL NOCTURNAL HEMOGLOBINURIA (PNH)

Identified more than a century ago, paroxysmal nocturnal hemoglobinuria (henceforth abbreviated as PNH) is a rare disorder of the red blood corpuscles that is characterized by intermittent or paroxysmal episodes of massive destruction of these cells (hemolysis) and the consequent release of hemoglobin into the bloodstream and urine (hemoglobinuria). The latter is usually seen after an affected individual is asleep, because of which the label of "nocturnal" hemoglobinuria was used to describe the disease. The bone marrow is unable to compensate for the massive loss of RBCs and thus the patient suffers from anemia, which may in some instances progress to acute myelogenous leukemia. Paroxysmal nocturnal hemoglobinuria is also associated with a high risk of thrombosis, especially of the large intra-abdominal veins. Although the disease outcome is variable, depending on the timeliness of diagnosis and the response to treatments, the disease can be fatal. Most PNH-related deaths occur due to thrombosis. Therapeutic interventions for PNH are mainly aimed at treating symptoms. For example, anemia is addressed with folic acid and iron supplements, hemolysis is checked with regulated doses of steroids (prednisone), and thrombolytic agents are administered in the later stages. Blood transfusions with normal RBCs (and normal *PIG-A*) are also given to compensate for the defective cells.

At the cellular level, PNH is the result of a defect in the RBC membranes that leads to an increased sensitivity of these cells to complement-mediated lysis. This sensitivity has been traced to defects in a gene called *PIG-A* located on the X chromosome, which codes for an enzyme involved in the biosynthesis of cellular anchors whose function is to keep various membrane proteins anchored properly to the cell surface. Among these proteins are some that protect normal RBCs from destruction by the complement pathway. Mutations in *PIG-A* result in the destabilization of these proteins, leaving the RBCs highly susceptible to hemolysis by complement.

Paroxysmal nocturnal hemoglobinuria is an example of an *acquired* genetic disorder, namely a condition that arises from a mutation in a gene but that is not passed along from one generation to the next. It has been traced to a number of different mutations in the *PIG-A* gene, including single base substitutions, deletions, or insertions, as well as deletions of larger segments, indicating that any disturbance of the anchor proteins can increase RBC susceptibility to hemolysis. Exactly how the different mutations are induced remains unknown. Evidently, the mutation responsible occurs in a stem cell (blood cell precursors), which passes the defective gene on to *its* descendants—the blood cells of the adult organism. Because other cells of the body, including the sex cells, are not affected by this mutation, the disease is not typically inherited by the offspring of an affected person.

See also: Mutation (Ch. 2); Immunity—Complement Factors (Ch. 4).

PENDRED SYNDROME

Pendred syndrome is a form of syndromic hereditary deafness in which a congenital hearing deficiency is usually accompanied by goiter (enlarged thyroid gland) caused by a defect in the production of the thyroid hormones. An English physician, Vaughan Pendred, described the disease in 1896, and it was later found to account for as much as 10 percent of all hereditary causes of deafness. In addition to these symptoms, Pendred syndrome patients may also exhibit problems with balance, show retarded mental abilities, and be more prone to thyroid cancers.

In 1997, National Human Genome Research Institute (of the NIH) researchers identified a gene on chromosome 7 that they named the *PDS* gene, mutations in which were seen to cause Pendred syndrome. The inheritance is autosomal recessive, which indicates that the mutation causes a loss of function. The normal protein (pendrin) is a member of a family of anion-carrying (usually but not always sulfate) proteins called solute carrier proteins that are very similar to one another in structure but extremely tissue-specific in their expression. These genes appear to play roles in the development of the tissues where they are expressed. For instance, *PDS* belongs to the solute carrier 26 (SLC26) gene family and is produced specifically in the thyroid tissue and in the tissues of the inner ear. Mutations in PDS lead to the deficient development of these tissues and the consequent symptoms. (Mental retardation, for example, often results from severe thyroid deficiency during development.)

With the discovery of *PDS* and pendrin, scientists are finally beginning to understand the etiology of a disease that had confounded doctors for over a century. For instance, they found that pendrin was specific for chloride and iodide ions rather than sulfate because of slight differences in the ion-binding site of the protein. This sheds further light on why only certain tissues are affected—chloride and iodide ions are important for the proper functioning of the ears and thyroid, respectively, while sulfur plays a larger role in cartilage tissue. Scientists may also be able to confirm their suspicions that the lack of a goiter in some cases of Pendred syndrome results in a misdiagnosis of the cause of congenital deafness and that the incidence of the disease is actually higher than originally thought.

POLYCYSTIC KIDNEY DISEASE (PKD)

Polycystic kidney disease is a disease where large, fluid-filled cysts grow in the kidneys, gradually replacing normal kidney tissue, interfering with their normal excretory functions, and eventually leading to organ failure. With about 500,000 people in the United States estimated to have this disease, PKD is the fourth-leading cause of renal failure in the nation.

There are two main types of inherited PKD. About 90 percent of all cases diagnosed fall into the first category, which is an autosomal dominant form of the disease. The age of onset is typically between 30 and 40 years of age, for which reason this disease is also called adult PKD (APKD), although symptoms may begin much earlier. About half the people with APKD progress to acute renal failure, also known as end-stage renal failure. The second category, which is much rarer, is an autosomal recessive disease that begins very early in life, even while the baby is still in the womb. It is also known as polycystic kidney and hepatic disease-1 because the formation of renal cysts is typically accompanied by cyst formation in the liver as well. The difference in the two forms of PKD is due to the specific genes that are affected.

The dominant form of the disease is caused by mutations in the genes on chromosomes 16 and 4 for proteins called the *polycystins* (*PKD1* and *PKD2*). These proteins are integral membrane proteins that are expressed primarily in the kidney cells and are involved in interactions among cells as well as those between the cell surface and its matrix. Thus, the polycystins seem to participate in numerous signal transduction pathways. Scientists believe that they also play a role in the development of renal tubules. Interactions between these two polycystins enable the movement of cations such as calcium across the cell membrane. Defects in either protein evidently prevent their interaction, which in turn perturbs ion movement, which is fundamental to the functioning of the kidneys. Exactly how ion transport is linked to cyst formation is not fully understood. However, the fact that patients with PKD who receive kidney transplants do not develop cysts in these new organs indicates that the gene defects lead to abnormal development of the kidneys.

Although there is no cure for PKD, treatment is available for symptoms of pain, urinary tract infections, and high blood pressure that typically accompany this disease. As indicated earlier, kidney transplantation to replace failing kidneys has proven successful in restoring kidney function. Even more encouraging perhaps is the fact that with timely knowledge of the condition, kidney failure can be forestalled with proper diet and blood pressure control. Consequently, genetic testing is quite valuable, especially in cases where the family history of conditions is unknown. Testing and counseling are also necessary before undertaking a transplant.

Recessive PKD, also known as infantile PKD, is caused by mutations in the *PKHD1* gene on chromosome 6. As in the polycystins, the protein made by this gene seems to be a membrane protein that has the ability to interact with transcription factors. Even less is understood about the mechanism by which infantile PKD disease is caused. Because of its early onset, this version of PKD lends itself less readily to precautionary measures, although early diagnosis can still be useful to determine the need for transplantation or dialysis, which are the only courses to deal with renal failure. Prenatal ultrasounds can help detect the presence of renal cysts and liver scarring in cases where genetic tests indicate a familial susceptibility to the condition.

See also: Channel Proteins and Cross-Membrane Transport—Ion Channels, Transcription: RNA Polymerases and Transcription Factors (Ch. 4).

PORPHYRIA

Perhaps best known as the disease that afflicted King George III ("Mad King George," 1738–1820) of England and other members of the royal family, porphyria actually refers to a group of different disorders caused by abnormalities in the chemical steps leading to the production or breakdown of heme. Heme is the iron-containing portion of proteins such as hemoglobin, which as you may remember from our discussion of blood performs the essential function of binding oxygen (via the heme group) and transporting it to various tissues. Heme compounds may also be found in other tissues such as the liver, where heme is synthesized, or the brain and muscles, which require a lot of oxygen.

There are two major categories of porphyrias, depending on which part of the body is affected. *Cutaneous porphyrias* affect the skin. People with cutaneous porphyria develop blisters, itching, and swelling of their skin when exposed to sunlight. *Acute porphyrias* affect the nervous system. Symptoms include pain in the chest, abdomen, limbs, or back; muscle numbness, tingling, paralysis, or cramping; vomiting; constipation; and personality changes or mental disorders. These symptoms appear intermittently. Attacks can develop over hours or days and last for days or weeks. The symptoms may also be triggered by different drugs (barbiturates, tranquilizers, birth control pills, and sedatives), fasting, smoking, drinking alcohol, infections, emotional and physical stress, menstrual hormones, and exposure to the sun.

The name porphyria is derived from the fact that the common feature in these diseases is an accumulation of different *porphyrins*, which are the intermediary compounds formed during the biosynthesis and degradation of the heme groups, as well as other metal-containing proteins involved in the control of electron transport systems. There are many different porphyrins in the body, but the one with the highest concentration is the porphyrin incorporated into hemoglobin. Other examples of porphyrin-containing compounds include the chlorophyll of plants (in combination with magnesium) and the cytochrome P-450 group of enzymes, which are essential for many metabolic processes throughout the living kingdom.

Chemically, porphyrins are large ring-like organic compounds that can combine with metal ions. The large ring structure of heme has a specific spatial configuration that enables it to hold an activated iron molecule in its center in such a way that it can interact with reactive molecules such as oxygen. This ring structure is also responsible for the red color of most porphyrins. Some porphyrins are colorless but will turn red or purple when exposed to sunlight. In fact, the terms "porphyrin" and "porphyria" are derived from the Greek word "porphyrus," meaning purple.

The heme synthetic pathway is quite complex and involves a number of biochemical reactions, each catalyzed by an enzyme. A multistep pathway with several porphyrin intermediates leads to the synthesis of the heme group, which must then combine with iron and globin in order for hemoglobin to bind and transport oxygen. When red cells die and are broken down, the porphyrin ring structures are broken down, again in a series of steps, by enzymes to form a long-chain molecule called bilirubin, which gives bile its yellow-green color. Most of the metabolic processes involving the porphyrins occur in the liver and in the bone marrow, and the bilirubin is excreted via the gall bladder.

As you can imagine, the production, remodeling, and destruction of these large organic compounds is complicated and requires a number of enzymes to catalyze each of a sequence of biochemical reactions. If any one of these enzymes is missing or is abnormal, the process goes awry. When this happens, the intermediate products build up in the blood and are excreted in the urine and stool, which may appear deeply colored (reddish or bluish) or show signs of darkening after standing in the light.

The rate of each specific chemical reaction in the synthesis and degradation of porphyrins is controlled by the concentration and activity of the enzyme system. As a result, they influence the concentrations of both the precursor and end products of the specific reaction. For the most part, the specific enzymes involved in porphyrin synthesis seem to be encoded by genes in a single locus. If the gene is defective or abnormal, the metabolic functions that it controls probably will be defective as well. Several problems can then develop. If the enzyme process is slowed, there may be a buildup of potentially toxic precursors. If the chemical reaction is too fast, the end products may accumulate in too high a concentration. Sometimes the abnormal enzyme systems change the direction of the reaction and produce abnormal metabolites. These precursors and end products can build up to toxic levels. Other water-soluble compounds may be carried by the blood to other tissues such as the skin, where they can absorb abnormal amounts of radiant energy and affect the body in a different way. Most compounds are simply excreted in the stool and urine in abnormal amounts without any clinical problem. During pregnancy, sometimes the abnormal compounds will not allow the developing fetus to survive. In some cases, the abnormality may not become apparent until after puberty or even middle age. Frequently, the enzyme abnormalities are influenced or induced by other stimuli, such as certain drugs (e.g., barbiturates and sulfonamides) or viral infections.

The genes for all enzymes in the heme pathway have been identified, as have many of the disease-causing mutations of the individual genes involved in different porphyrias. The risk that individuals in an affected family will have the disease or transmit it to their children is quite different depending on the specific type of porphyria. Often the children of porphyric patients may be at risk of inheriting their parent's disease. The disease may also appear without any antecedent identifiable family involvement.

The various diseases that are classified under the collective name of porphyria can be differentiated from one another on the basis of a combination of clinical symptoms and blood, urine, and stool tests. Unfortunately, this classification is somewhat unsatisfactory. It would be preferable to classify the porphyrias on the basis of the specific genetic defects giving rise to abnormal clinical and biochemical results. However, many of the results of genetic studies of porphyria have yet to be adapted for common diagnostic clinical use.

As mentioned earlier, the porphyrias are most commonly classified according to the specific tissues in which the major toxic effects appear. The skin, nervous system, and liver are the most frequently affected. Accumulations of intermediates called aminolevulinic acid (ALA) and porphobilinogen (PBG) can affect nerve endings and can cause a variety of neurovisceral symptoms and specific neurologic syndromes. Frequently, the nerves to the gastrointestinal tract are affected, leading to severe abdominal pain that is often severe enough to be confused with acute appendicitis and require exploratory surgery. There can be emotional and psychiatric problems such as anxiety, insomnia, agitation, confusion, paranoia, depression, and hallucinations. Another very important group of symptoms is related to the ability of the porphyrin ring structure to absorb and store radiant energy (typically ultraviolet light with a wavelength of about 400 nm). For the most part, this radiant energy is derived from exposure to bright sunlight. This energy buildup within the cells can damage the skin. This is called photosensitivity or phototoxicity. In an acute illness, exposure to sunlight can cause tingling, stinging, or burning skin discomfort during or soon after exposure, followed by redness, rashes, and blistering.

Although there is no cure for porphyria, there are some treatments available for the various conditions. Each form of porphyria is treated differently. Treatment may involve treating with heme, giving medicines to relieve the symptoms, or drawing blood. Some major kinds of porphyrias follow.

Acute Intermittent Porphyria (AIP)

Acute intermittent porphyria is perhaps the most severe of all of the porphyric syndromes in terms of its symptoms. This is probably the form of porphyria that afflicted King George III. The disease is inherited in an autosomal dominant fashion but is slightly more common in females than in males.

Acute intermittent porphyria is caused by various mutations of a single gene on chromosome 11 that controls the activity of the enzyme porphobilinogen deaminase. This enzyme catalyzes the joining of four porphobilinogen molecules into a linear chain that is then converted into the ring structure characteristic of the porphyrin molecule. In patients with AIP, the activity of the enzyme is decreased both in red blood cells and in liver cells. Interestingly, this decreased activity is still high enough for normal body functions, so this deficiency does not manifest itself unless some other stimulus blocks the enzyme system. When an attack occurs, the activity of the enzyme becomes further impaired, there is a rapid accumulation of the precursor compounds PBG and ALA, and the patient becomes acutely ill. The mechanisms for the development of the neurologic symptoms have yet to be clearly defined, but they appear to be related to an abnormal accumulation of ALA at the nerve endings that acts either as a direct neurotoxin or interferes with neurotransmission.

Symptoms include abdominal pain and cramps, nausea and vomiting, diarrhea or constipation, urinary retention, and muscle pain and weakness. Hallucinations, confusion, acute psychiatric syndromes, and occasionally seizures can occur. Sometimes, during an acute attack, the urine color turns brownish-red after exposure to bright sunlight because of the condensation of high concentrations of PBG to red porphyrin complexes. This discoloration of the urine is often an important clue to help in the diagnosis of this disease. Diagnosis is based primarily on clinical signs and symptoms and positive urine-screening tests with increased levels of PBG and ALA.

Variegate Porphyria (VP)

Variegate porphyria is a type of porphyria that is associated with the symptoms of the neurovisceral crises similar to those of the patient with AIP as well as with a classic photosensitive skin disorder. The defective gene is located on chromosome 1 and expresses the enzyme protoporphyrinogen oxidase (PPO), which controls one of the final stages of heme synthesis. In patients with this disease, the activity of this enzyme is reduced by at least 50 percent. The disease rarely appears before puberty and is most common in young adults, but it may suddenly occur at any age, including in the elderly. If both parents carry the abnormal gene so that the patient is homozygous, the disease will present in early childhood and be rather severe. However, the outlook for the heterozygous individual is generally relatively good.

The neurovisceral crises give symptoms similar to those of patients suffering from AIP as well as skin sensitivity to solar radiation. These changes include skin fragility, erosions, and blisters during the acute attack and abnormal pigmentation, skin thickening, and excessive hair growth with chronic exposure. The precipitating factors are also similar to those of AIP.

Porphyria Cutanea Tarda (PCT)

Porphyria cutanea tarda is the most common of all the porphyrias. It is primarily a skin disease and is caused by decreased activity of the enzyme that catalyzes the final step in the heme biosynthetic pathway, called uroporphyrinogen decarboxylase (UROD). This enzyme is present primarily in the liver and, to a lesser extent, in the red blood cells. When its concentration is decreased or its activity inhibited, there is an overaccumulation of uroporphyrin and other highly carboxylated porphyrins that are concentrated in the skin. Due to their propensity for accumulating radiant energy, these compounds can inflame the tissues and cause symptoms in the skin.

There appear to be two distinct types of PCT. About 20 percent of cases are caused by an autosomal dominant trait associated with deficient activity of UROD both in the red blood cells and the liver. The onset of this inherited disease is usually delayed into adulthood, although cases can occur in children. About 80 percent of cases of PCT do not have a familial history and are called sporadic, toxic, or acquired PCT. There may, however, be a genetic defect (not necessarily inherited) in many of these cases that is related to liver disease in general. Excessive alcohol ingestion has long been recognized as an important cause, possibly due to the development of chronic liver disease. Similarly, viral infections, particularly the HIV and hepatitis C viruses, have also been implicated in the disease.

The predominant symptom is photosensitivity, with abnormalities on the areas of the skin exposed to light such as the face, the arms, and the backs of the hands and wrists. There is irritation and blistering followed by increased skin fragility, hair growth, scarring, and pigment deposition. It is often difficult to differentiate PCT from VP, but this differentiation can be important because the treatment of the two is different. The treatment of PCT is usually quite successful. The aggravating factors should be removed or controlled. The antimalarial drug chloroquine given in low doses has proven to be effective, and the sunscreen skin lotions with beta-carotene are also helpful.

Erythropoietic Protoporphyria (EPP)

Erythropoietic protoporphyria is caused by an enzyme defect in the last step of heme synthesis, which involves the insertion of the iron atom into the middle of the ring structure. The enzyme is called *ferrochelatase* and is encoded on chromosome 18. The disease is characterized by the accumulation of protoporphyrins in red blood cells, which then spill over into the plasma as the red blood cells mature. Because protoporphyrin is not very soluble in water, it is not excreted in the urine but in the feces. The disease can appear in childhood and is more common in males than in females.

The symptoms are exacerbated by sunlight, which causes burning, itching, swelling, and redness of the skin. Blistering and skin ulcers along with an increase in hair growth and pigmentation can follow chronic sun exposure. Occasionally liver disease may develop, and gall bladder disease is a common problem because the high concentration of protoporphyrin in the bile will lead to gallstone formation. Treatment with vitamin A, beta-carotene, and other sunscreens improves the tolerance to sunlight. The use of bile acid binding resins such as activated charcoal may help eliminate the protoporphyrins from the body.

PRIMARY IMMUNE DEFICIENCIES (PIDs)

As we saw in Chapter 4, the immune system is a diffuse and complex collection of soluble substances (molecules), cells, and organs that cooperate with one another to defend the body against various pathogens and other threats from the environment. If this system is compromised in any way, we lose the ability to fight off these pathogens. Immunodeficient individuals are very susceptible to infections, not just by the usual suspects (i.e., the bacteria and viruses commonly known to cause infections) but also a number of microbes (called opportunistic pathogens) whose presence often goes unnoticed by most of us. We are all familiar with the devastating effects of AIDS, an immunodeficiency characterized by the loss of T cells. However, AIDS, as its name indicates, is an *acquired* immunodeficiency caused by a virus we call the human immunodeficiency virus (HIV). The examples discussed here fall into a different category—the *primary* immunodeficiencies—so named because they are caused by some internal, inheritable defect in one of the components of the immune system.

Primary immunodeficiencies manifest a wide variety of symptoms across a broad spectrum of severities, depending on the specific cells and tissues that the mutations affect. Generally, T cell defects have the worst effects. This is because the different types of T cells both mediate the cellular immune responses *and* play a role in regulating humoral (antibody-mediated) immune responses, where B cells are the principal actors. The main signs of immunodeficiency include general susceptibility to infections, recurring and often prolonged episodes of colds, fevers, and diarrhea, and a higher than normal severity of these symptoms. According to the National Institute of Allergy and Infectious Diseases (NIAID) of the NIH, the number of people suffering from the most serious forms of primary immune deficiency ranges from 25,000 to 50,000 annually in the United States. Furthermore, these estimates likely overlook milder forms of primary immunodeficiency, which often go undiagnosed, and the real numbers are probably much higher. For the most part, these disorders are a result of loss-of-function mutations, namely mutations in which the normal function of the protein is disrupted. Primary immunodeficiencies are thus recessive conditions because the normal gene on the second chromosome compensates for a deficiency caused by its mutated partner. In instances where the affected gene is on the X chromosome, the disease is X-linked and occurs more frequently in boys.

Chronic Granulomatous Disease (CGD)

Chronic granulomatous disease is an inherited disorder of the phagocytic cells—macrophages and neutrophils—of the immune system. The normal function of these cells is to engulf and destroy particulate pathogens such as bacteria and fungi. After they ingest the pathogen, the phagocytes kill the organism through enzymatic pathways that convert oxygen to molecules such as hydrogen peroxide and superoxide, which are toxic.

Chronic granulomatous disease is caused by mutations in one of four different genes. All four genes encode enzymes (or parts of enzymes) that are involved in the oxygen-metabolizing pathway mentioned earlier. They include genes that code on the peptide chains of cytochrome b-245 (on chromosomes X and 16) and two proteins called the neutrophil cytosolic factors 1 and 2 (NCF1 and NCF2 on chromosomes 7 and 1, respectively) that are subunits of a large enzyme complex called NADPH oxidase. The function of normal NADH oxidase is to produce a large burst of superoxide to the cell, which kills the engulfed organisms. X-linked CGD is the most widely occurring form of the disease and accounts for about 70 percent of all diagnosed cases. The other mutations manifest less frequently because of their autosomal recessive pattern of inheritance.

Patients with CGD commonly suffer from infections of known pathogens such as *E. coli* and the staphylococci, as well as opportunistic bacteria such as *Serratia* and *Pseudomonas*. Fungal infections associated with this disease are *Candida* (particularly associated with pneumonias) and *Aspergillus*. Repeated infections often lead to the formation of characteristic tumor-like masses called granulomas at the sites where the phagocytes come into contact with the pathogen (e.g., skin, lungs, lymph nodes, gastrointestinal and urinary tracts, liver, and bones). The granulomas, which consist of dead white blood cells, tend to heal slowly and drain pus for a long time after the initial infection.

Doctors may suspect CGD in babies who show recurring and persistent signs of low resistance to infections associated with symptoms of fever, rashes, coughing, boils, gum disease, and swollen lymph nodes. They can confirm their diagnosis with certain lab tests to look for high white blood cell counts, low RBC counts (anemia), and high levels of immunoglobulins that apparently have no protective effect. Symptoms typically appear at a young age, although the disease may manifest itself as late as adolescence. The worst of the long-term effects of CGD (e.g., a blockage of the intestines due to the granulomas) can be avoided if the disease is diagnosed early enough and antibiotics are taken to control recurrent infections.

Hyper IgM Immunodeficiency Syndrome

The name of this immune deficiency might seem paradoxical at first glance because the word "hyper" indicates an excess—not a deficiency—of an antibody. However, the problem in this disease is that the excess IgM is produced at the cost of other immunoglobulins—specifically IgA and IgG—that protect certain tissues and mucosal surfaces more effectively than IgM. The result of this imbalance of antibodies is that hyper IgM individuals are especially susceptible to recurrent bacterial infections, especially by opportunistic pathogens such as *Cryptosporidium* and *Pneumocystis* species. Patients are also at higher risk for developing cancers such as non-Hodgkin's lymphoma, as well as autoimmune diseases, early in life.

As we saw in Chapter 4, in a normal immune response to a new antigen, B cells first produce IgM antibodies and only later switch to producing the more effective IgG, IgA, or IgE antibodies. The B cells do this by splicing out the portions of the gene coding for the IgM-constant regions in response to signals from T cells that instruct them to make the switch. Disruptions in either signaling or splicing can cause disease, so, like other genetic diseases we have discussed (e.g., CGD), hyper IgM is the outcome of mutations in more than a single gene. Mutated genes giving rise to this disease phenotype include:

- A gene for tumor necrosis factor, TNFSF5, that normally enables T cells to transmit signals to B cells by binding with receptors called CD40 on the B cell surfaces. This gene is on the X chromosome and is the most common cause of the visible disease, which affects more males than females.
- A gene for a receptor of the tumor necrosis factor on chromosome 20.
- A gene on chromosome 12 for a member of the family of cytidine deaminase enzymes. This enzyme is an RNA editing enzyme in B cells that responds to signals from the T cells to splice out the M-constant chain genes.

Treatment for hyper IgM immunodeficiencies is possible by regular intravenous (IV) injections of the missing IgG antibodies and antibiotics for tackling infections. In experiments with mice, a synthetic version of the molecule CD40 that mediates communication between T and B cells has proved useful in improving the ability of the immunodeficient mice to switch from IgM to IgG production. Currently, scientists are investigating the possibility of using a similar approach in human patients also.

Severe Combined Immunodeficiency (SCID)

Severe combined immunodeficiency is perhaps one of the most well-known primary immune disorders. Readers might better recognize SCID by its more common name "bubble boy disease," which was coined sometime in the 1970s or 1980s when a young boy named David Vetter managed to survive this disease by living in a plastic bubble. Born with this congenital defect that left his immune system unable to respond to anything, Vetter was placed in this bubble in an attempt to keep him in a completely germ-free environment. He lived in it for 12 years but tragically died in 1984 due to complications following a bone marrow transplant.

The defining characteristic of SCID is a massive deficiency of the lymphocytes—both B and T cells—which are the main actors in an immune response to a pathogen. Symptoms usually appear within the first 3 months of life, with the onset of serious, often life-threatening infections, especially pneumonia, meningitis, and blood infections. Chronic skin infections, yeast infections in the mouth and diaper areas, diarrhea, and chronic hepatitis also develop. Common and mostly nonfatal childhood diseases such as chicken pox and measles have far more drastic outcomes than usual.

Mutations in at least three different genes lead to the same spectrum of symptoms:

- The most common form of SCID (about half of all diagnosed cases and also the type from which David Vetter suffered) is an X-linked condition caused by defects in the gene for a portion of an interleukin-2 receptor. A defect in this gene interferes with the signal transduction pathways involved in the development of the lymphocytes. Lymphocytes are unable to respond to messages or signals delivered by the interleukin-2 and so do not mount proper immune responses.

- A second form of SCID is caused by mutations in a gene on chromosome 19 for the signaling molecule JAK3. The activation of JAK3 is actually the next step following receptor binding (described earlier) in the interleukin-2 mediated pathway for lymphocyte development. Its inactivation or deficiency has the same consequences as those of the IL-2 receptor.

- The third form of SCID is unrelated to signal transduction and follows the pattern of storage diseases (see the entry on inborn errors of metabolism) in its mechanism. Mutations occur in a gene on chromosome 20 for an enzyme called adenosine deaminase (ADA) involved in the metabolism of the nucleotide adenosine, and its deficiency or inactivation results in the accumulation of the enzyme substrates (adenosine). Immature lymphocytes are particularly sensitive to this substance and hence fail to develop properly, resulting in SCID. Interestingly, a mutation in the same gene, leading to excess ADA rather than its deficiency, also causes disease, specifically a hemolytic anemia caused by the destruction of red blood cells.

The overall occurrence of SCID is quite low—NIAID reports estimate that the frequency is about 1 in every million people—but its outcome is very dramatic. David Vetter's tragic death notwithstanding, the most successful treatment option available so far is bone marrow transplantation. Gene therapy for SCID is still in the preclinical research stages. Because bone marrow transplants have a better chance of success when performed early in infancy, newborn screening for SCID is recommended although not mandatory.

See also: Signal Transduction (Chs. 2 and 4); Inborn Errors of Metabolism.

X-Linked Agammaglobulinemia (XLA)

As its name indicates, X-linked agammaglobulinemia is an immune disorder characterized by the almost complete lack of gamma globulins (antibodies) in the bloodstream. Patients with this disease have very low numbers of mature and functional B cells and so cannot produce antibodies. In addition to lowering the person's resistance to bacterial and viral infections, X-linked agammaglobulinemia also renders most vaccinations ineffective in preventing diseases.

The affected gene, which is present on the X chromosome, encodes a tyrosine kinase enzyme that is involved in the signal transduction pathways regulating the maturation of antibody-producing B cells. It is also known as the Bruton type of tyrosine kinase (BTK), and the disease is sometimes referred to as Bruton agammaglobulinemia. Infants born with XLA are particularly prone to pus-producing bacterial infections of the inner ear, lungs, and sinuses, as well as serious septicemia and systemic infections. Although they are susceptible to chronic viral infections such as hepatitis, polio, and ECHO viruses, short-term viral infections (e.g., chicken pox and measles) seem to present less of a threat. Lymphoid glands such as the tonsils and adenoids are often small or absent altogether, and the patients may exhibit signs of poor growth (both in height and weight). The only viable therapeutic option available at present is the administration of regular injections of both specific and nonspecific gamma globulins.

RETINITIS PIGMENTOSA (RP)

Retinitis pigmentosa (RP) is the name for a group of related but genetically heterogeneous inherited diseases of the eye that are characterized by a loss of vision caused by the progressive destruction of the retina, primarily the photoreceptors (i.e., the rods and cones). This damage is often, but not always, accompanied by a dystrophy of the pigment-carrying epithelial cells of the retina (hence the descriptor of pigmentosa). Retinitis pigmentosa can manifest at any age, and its progression varies widely from patient to patient, but a majority of the affected population becomes legally blind by the age of 40. Individuals are considered legally blind if their best visual acuity, even with corrective lenses, is 20/200 or worse in their better eye (20/20 is considered to be the best), or if their field of vision, regardless of acuity, is restricted to a radius of 10 degrees (most of us have a field of nearly 180 degrees).

Different types of RP arise from mutations to different genes that are specifically expressed in the retina (see the discussion on vision in the entry on the senses in Chapter 4 for details about these genes). According to the nature and location of mutations, the disease may be dominant or recessive and preferentially damage one type of receptor or cell over the other. Consequently, the type of vision loss is quite variable. For instance, as we saw in Chapter 4, the rod cells, which function under conditions of dim light, are situated toward the rim or periphery of the retina. This means that when rod cells are the target of dystrophy, for example because of mutations in the *RHO* gene, the patients will experience a gradual diminishing of their night and peripheral vision capabilities. Treatment with vitamin A (retinol) has proven useful in slowing vision loss in this type of RP. A specific locus of chromosome 17, called *RP17*, has also been associated with an autosomal dominant form of RP caused by rod dystrophy, although no functional gene has been identified at this locus so far.

As might be expected, a dystrophy of the cone cells leads to defects in color vision. A particularly severe form of RP, usually ending in total blindness, is an X-linked disorder caused by mutations in the *RPGR* gene, which encodes a protein that regulates the activity of GTPase. Although the gene is expressed all over the body, the mutations appear to restrict their harmful effects to the retina. This and other dystrophies of the cone cells differ from cases of color blindness in that the latter are caused by defects in the light-absorbing pigments and are not typically associated with retinal degeneration.

SICKLE-CELL ANEMIA

One of the most common genetic disorders seen in the United States, sickle-cell anemia is a serious condition that affects red blood cells (RBCs) in the body. Normally the blood circulating in the body has flexible, doughnut-shaped RBCs that can squeeze through the narrow blood vessels and reach even the deepest tissues to provide them with oxygen. In the case of sickle-cell anemia, some of the red blood cells become hard and sticky, assuming a rigid crescent or sickle shape (hence the name) that makes them clog up the blood vessels. (See the photo of normal versus sickled RBCs.) This clogging causes an increased destruction of red blood cells (hemolysis) at the site, which in turn leads to a generalized anemia (i.e., a decrease in the overall blood supply). Consequently, tissues are deprived of oxygen and soon begin to get damaged, often irreversibly. Outward manifestations of the disease include weakness, severe pain, especially in regions such as the bones, and a gradual loss of functioning of organs at different sites in the body.

The cause of sickle-cell anemia is a single base substitution in the beta hemoglobin gene, which results in the replacement of a valine with a glutamic acid in the sixth position of the peptide chain that this gene encodes. When red blood cells containing this altered hemoglobin (hemoglobin S) release their oxygen to the tissues, they change shape from round to sickled and start to block normal blood flow.

Sickle-cell anemia is an autosomal recessive disease in that it only manifests itself when an individual is homozygous for the *HbS* gene (i.e., has inherited two copies of the mutant gene, one from each parent). People who are heterozygous (and thus have only one mutant gene) are said to carry the sickle-cell trait. They seem to be able to produce enough normal hemoglobin to avoid the symptoms of sickle-cell anemia. The disease can only be transmitted genetically (i.e., from parent to offspring). Although there is no cure for the disease, various symptomatic treatments are available, such as regular blood transfusions and folic acid to alleviate anemia and intravenous hydration, pain medication, antibiotics, and oxygen

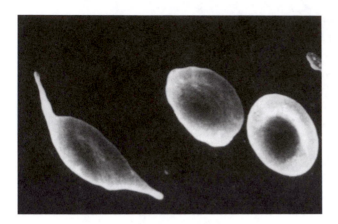

Sickled RBC (left) with normal RBCs shown for comparison. [© Roseman/Custom Medical Stock Photo]

for generalized symptoms. More recently, certain drugs have been developed to act as a lubricant, allowing sickled cells to flow more easily through small blood vessels. This makes early, even prenatal, diagnosis of the disease quite important.

Why has the hemoglobin S mutation persisted in human populations? After all, given that the mutation seems lethal, one might reasonably expect it to have been bred out over generations. Yet, it is quite widely found, not only in people of African origin (in whom it seems most common, at least in the United States) but also among Arabs, Greeks, Italians, Latin Americans, and people from the Indian subcontinent. The reason, most geneticists believe, is that the mutation may provide a defense against malaria. Studies on populations in tropical areas have shown that people with the sickle-cell trait are more resistant to malaria than those who have no mutation at all. Evidently, evolution has considered the incidence of sickle-cell disease in some proportion of the population (e.g., 1 in 400 African Americans have the disease) an acceptable risk in the battle against malaria, a deadly infectious disease.

See also: Inheritance, Mutation (Ch. 2); Blood—Hemoglobin (Ch. 4).

THALASSEMIA

The term thalassemia does not refer to a single condition but rather a group of genetic disorders of the blood that manifest similar clinical symptoms. The main symptom associated with these diseases is anemia—or a decrease in functional red blood cells in the body—and the underlying basis is a loss of function in the genes for hemoglobin. Because some thalassemias can be treated by transfusion, early or prenatal diagnosis is very helpful.

There are different types of thalassemias that may be distinguished from one another on the basis of the type and number of hemoglobin genes that are affected. As described in Chapter 4 (see the entry on blood), a hemoglobin molecule is made of two chains each of alpha- and beta-globins, which are encoded by four and two genes, respectively. A defect in one or more of these genes can lead to the development of different categories of thalassemia.

Alpha-thalassemias are caused by defects in one or more of the four alpha-chain genes on chromosome 16. There are four manifestations, in order of increasing severity, corresponding with the number of defective genes. Individuals with one alpha gene missing or deleted are silent or asymptomatic carriers, whereas those missing two genes have smaller than normal RBCs and show slight signs of anemia. When three alpha genes are deleted, the person develops *hemoglobin H disease*. This is characterized by the production of small, misshapen RBCs and fragments of these cells in the circulating blood, leading to severe anemia. In order to compensate for anemia, the blood-producing organs of the body, such as the spleen and bone marrow, begin to work overtime, resulting in an overproduction of immature RBC precursors. As these precursor cells accumulate in the site of production, they cause the organs to expand, as evidenced by the development of bony abnormalities, especially involving areas of the cheeks and forehead. The spleen may grow to 10 times its normal size. This hyperactivity is often accompanied by an appearance of malnourishment in the patients, which is a by-product of the bulk of the body's energy being funneled toward blood cell production. The condition associated with a loss of function of all four alpha-globin genes is called *hydrops fetalis*. Babies with this defect rarely make it to term and instead succumb to severe anemia and other complications before birth.

The *beta-thalassemias* arise from a reduction or absence of the beta-globin chains. Unlike the alpha-globin genes, the beta-globin genes are rarely deleted completely, and the thalassemias are most often the result of a suppression of gene expression (i.e., the gene product, the beta-hemoglobin protein chain, is decreased in amount). The degree of gene suppression is variable and does not show a one-to-one correspondence with symptoms as do the alpha-thalassemias. There are three main categories of this condition based on the clinical symptoms, although the actual spectrum of symptoms is broader. *Thalassemia minor*, which is nearly asymptomatic, arises from the reduced function in one of the two beta-globin genes. *Thalassemia intermedia* is a condition of intermediate severity, with symptoms of anemia and the acceleration of the RBC-producing activity, resulting in spleen enlargement and weakening, easily deformable bones (as described in hemoglobin H disease). *Thalassemia major* results when there is virtually no beta-globin produced, resulting in very severe anemias that are only treatable by repeated blood transfusions.

See also: Blood—Hemoglobin (Ch. 4).

TRIPLE X SYNDROME

Earlier in this chapter we discussed Klinefelter syndrome, a trisomy of the sex chromosomes that results in males with extra X chromosomes. The triple X, or XXX, syndrome is the equivalent condition in females (i.e., a trisomy of the sex chromosomes occurring as a result of the fertilization of an improperly segregated sperm cell (XX) with an egg, or an X-containing sperm with an abnormal XX egg). It is the most common X-chromosome disorder in females and occurs at a frequency of 1 in 1,000–3,000 births.

As in the case of Klinefelter syndrome, triple X is not apparent at birth and varies very widely with respect to the type and severity of symptoms. Characteristic physical features often include increased space between the eyes and vertical folds of skin covering the inner corner of the eye (which further enhances the spaced-apart look), unusually tall stature, and a disproportionately small head. Puberty is frequently delayed. Not much is known about infertility in XXX girls, but from most accounts they appear to have normal fertility levels and also give birth to normal (i.e., XX or XY) children. In other words, the tendency toward trisomy is not an inherited trait.

Development is largely normal in most cases. However, some XXX girls may exhibit learning disabilities, behavioral problems, and a delay in the development of certain motor skills. If not given adequate care, they tend to be quiet and do not assert themselves. Very rarely, symptoms might be severe enough to cause seizures and mental retardation. For the most part, the average intelligence among XXX children is within the range considered normal, although it may be slightly less than average. Women born with this syndrome have normal life spans, and indeed many XXX women do not even know they have the condition unless they are specifically tested for it.

Like other chromosomal abnormalities, XXX is not curable. But because there are supporting treatments to cope with different symptoms, it may be useful for certain risk groups, such as older women in their first pregnancy, to have karyotypic diagnoses during pregnancy to prepare themselves for the future should their daughters be born with the condition. The most common diagnostic test is amniocentesis, the same process used to identify Down syndrome and other trisomies.

See also: Chromosome, Genetic Screening, Meiosis (Ch. 2); Klinefelter Syndrome.

TURNER SYNDROME

Turner syndrome is a rare chromosomal disorder in which one of the two X chromosomes in the affected patient—always a female—is either missing (a condition known as X-monosomy) or incomplete. The exact cause is unknown, but the disorder is believed to arise from an error during meiosis—specifically a failure of sex chromosomes to separate properly. The result of this "nondisjunction" is that the daughter cells, either sperm or egg cells (depending on the sex of the parent in which meiosis is disrupted), have an abnormal haploid profile. In other words, the gamete does not have a full, normal complement of 23 chromosomes, and its X chromosome is either missing or is present only in part. When this type of abnormal cell participates in fertilization with a normal X-containing gamete donated by the second parent, the resulting embryo develops Turner syndrome.

In comparison with other conditions caused by defects in meiosis (e.g., Down syndrome and the sex chromosome trisomies), Turner syndrome patients may have variable genotypes. Patients exhibit a genetic mosaicism, whereby some of their cells have a normal 46 chromosome complement but others are missing one X.

In 1938, a physician named H. H. Turner described this condition for the first time after examining a young woman with markedly short stature, a webbed neck, and sexual infantilism (i.e., the lack of sexual development at puberty). The genetic link to X-monosomy, however, was not established until 1959, when karyotyping became possible. Different patients with Turner syndrome have varied symptoms with a wide range of severity. Heart defects and kidney abnormalities may be found in serious cases, whereas at the other end of the spectrum, patients may show virtually no symptoms. The severity of symptoms seems to be related to the extent of chromosomal damage as well as the mosaic patterns.

Turner syndrome occurs at a very low frequency across all ethnic populations. About 98 percent of pregnancies with Turner syndrome fetuses abort spontaneously, and only about 1 in every 3,000 live female babies born has the condition. Although there is no cure, hormone replacement therapy can make up for deficiencies in the growth hormones estrogen and progesterone that are seen in Turner syndrome patients.

WAARDENBURG SYNDROME

Waardenburg syndrome is a genetic disease first described in the 1960s by a Dutch eye doctor named Petrus Waardenburg, who noticed that people with two different colored eyes often had problems with hearing. Upon further investigation of over 1,000 families with histories of deafness, he found that there were certain other common physical characteristics among people who shared these two traits and that they tended to cluster within families. Other signs of Waardenburg syndrome, as the disease was named, include other types of unusual pigmentation, such as white eyelashes and forelocks, as well as a tendency for premature graying. The eyes, when not differently colored, are often a peculiarly brilliant shade of blue. Waardenburg syndrome also presents characteristic facial features such as an unusually wide-bridged nose, connected eyebrows, and low frontal hairlines. Deafness in these patients varies from moderate to profound. The degree and combination of features in different patients are quite different.

Based on genetics, researchers have identified at least four distinct types and three subtypes of Waardenburg syndrome, types 1, 2, 2A, 2B, 3, and 4. Of these, most cases of types 1, 2, and 3 are autosomal dominant, whereas type 4 is autosomal recessive.

WS1 and WS3 are linked to mutations in the *PAX3* gene on chromosome 2. This gene is a member of a family of transcription factors called paired-box transcription factors that play a critical role during the development of the fetus. *PAX3*, in particular, regulates aspects of the development of the face and ear. Type 1 is caused by mutations within the gene, and WS3 (also called Waardenburg–Klein syndrome) involves the deletion of part of chromosome 2, including *PAX3* and adjacent genes. The latter is usually more severe and is associated with limb abnormalities in addition to typical WS features.

The different forms of WS2 are also caused by mutations to a gene for a transcription factor, this one called *MITF* (for *Mi*crophthalmia-associated *T*ranscription *F*actor) and located on chromosome 3. The *MITF* gene product is specifically involved in regulating the differentiation and development of pigment cells (melanocytes) in the epithelium of the retina as well as the control of genes that code for enzymes involved in melanin biosynthesis.

The mechanics of WS4, also called Waardenburg–Hirschsprung disease or Waardenburg–Shah disease, are probably the least understood. This syndrome is associated with colon problems that originate from the time of birth. It has been traced to mutations in several genes: *Sox10* on chromosome 22, *EDN3* on chromosome 20, and *EDNRB* on chromosome 13. *Sox10* probably encodes a transcription factor involved in the regulation of embryonic development and in the determination of the cell fate. *EDN3* codes for a cellular growth factor called endothelin-3 that is essential for the development of neural crest cells (i.e., cells that give rise to the nervous system) in the embryo. *EDNRB* expresses the G-protein-coupled receptor for endothelin-3. Mutations in *EDN3* and *EDNRB* interfere with the signaling pathways initiated by the endothelin and hence block proper development.

See also: Deafness (Congenital/Hereditary), Hirschsprung's Disease.

WERNER SYNDROME

Werner syndrome is a rare genetic disease characterized by symptoms of premature aging beginning in adolescence or early adulthood. Physical characteristics include a variety of features that usually appear later in life, such as wrinkled skin and baldness, as well as the premature onset of age-related diseases such as atherosclerosis, muscular atrophy, cataracts, diabetes mellitus, and osteoporosis. One of the major causes of death, a common outcome by about the age of 40, is myocardial infarction. The disease is inherited in an autosomal recessive manner.

The exact genetic mechanisms that cause Werner syndrome are not yet properly understood, although the mutations have been mapped to the *RECQL2* gene on chromosome 8 for a helicase-like protein. The helicases are a family of DNA-unwinding proteins, and RECQL2 shows homology to this family (in particular, to a bacterial version of this enzyme), although its activity in unwinding DNA has not yet been ascertained either in humans or in model systems. Mutations in this gene appear to result in the formation of an altered gene product that leads to defective DNA metabolism, which may lie at the heart of the disease symptoms. More recently, homology studies have identified a yeast protein called SGS1 that is similar to the human RECQL2. Mutations in SGS1 have been seen to decrease the life span of yeast cells and give rise to other signs of aging. These yeast cells may be useful as a model for studying the process of aging in humans and for understanding the mechanism of Werner syndrome and related conditions. It should be noted also that cells from Werner syndrome patients have been shown to have a shorter life span in culture than do normal cells.

The idea that mutated helicases lead to defective DNA metabolism and hence human disease is reinforced by a second, even rarer genetic disorder, known as *Bloom syndrome*, that has been mapped to the *RECQL3* gene on chromosome 12. The normal RECQL3 gene product is another example of a helicase-like protein with as yet undetected unwinding activity in humans. Like Werner syndrome, Bloom syndrome is an autosomal recessive disorder. Rather than premature aging, it is characterized by proportionate pre- and postnatal growth deficiency, sun-sensitive skin that becomes mottled with hypo- and hyperpigmented areas, a predisposition to cancers, and chromosomal instability.

See also: DNA Synthesis/Replication and Repair—Helicases (Ch. 4).

XERODERMA PIGMENTOSUM (XP)

Readers familiar with the 2001 horror movie *The Others* may be surprised to learn that the deathly allergy to sunlight suffered by of the children of the protagonist (played by Nicole Kidman) is a real, though rare, disease and not a figment of the screenwriter's imagination. In fact, the disease was documented as early as the nineteenth century (in the 1880s) when a physician named Kaposi coined the term xeroderma pigmentosum (XP) to describe the characteristic dry, flaky, and red blistering of the skin within a few minutes of exposure to the sun.

Xeroderma pigmentosum is the result of defects in the genes for DNA repair mechanisms, specifically genes involved in nucleotide excision repair (see the entry on DNA synthesis/replication and repair in Chapter 4). The affected person is highly sensitive to ultraviolet (UV) light, and hence to sunlight, which is our most abundant source of UV light on Earth. In addition to the characteristic blistering, individuals with XP show signs of premature aging, especially of exposed regions of the skin, lips, eyes, mouth, and tongue, and are more prone to cancers of these tissues and to blindness, progressive deafness, and neurological degeneration. The DNA damage is irreversible and life-threatening and reduces the normal life expectancy by 30 years. If the disease is diagnosed early, then symptoms can be controlled by preventing exposure to UV radiation, for instance by staying indoors during the daytime and shielding the skin and eyes. Fortunately, XP has a very low frequency (e.g., 1 in 250,000 in the United States). It is recessive in nature, so there are more carriers than people with the disease.

Mutations causing XP have been found to be widely distributed throughout the genome. They are grouped into at least eight different complementation groups, labeled XPA through XPG, and one XP variant. A complementation group is a set of mutations, usually in a single region of a chromosome, that cause a particular genetic disease. In laboratory experiments, the DNA from the cells of one complementation group can reverse the effects of a different complementation group when introduced into cells of the latter but cannot cure the phenotype of the cells from its own group. Over 90 percent of all XP cases belong to the complementation groups A, C, D, and variant.

Further Reading and Resources

Readers may find the following resources useful for further information about the human genome. These books and Web sites point to still more detailed sources of information.

BOOKS

Bishop, Jerry E., and Michael Waldholz. *Genome: The Story of the Most Astonishing Scientific Adventure of Our Time—The Attempt to Map All the Genes in the Human Body*. Updated ed. Simon & Schuster, 1999.

Dawkins, Richard. *The Selfish Gene*. New ed. Oxford University Press, 1989.

DeSalle, Rob, and Michael Yudell. *Welcome to the Genome: A User's Guide to the Genetic Past, Present, and Future*. Wiley-Liss, 2005.

Gelehrter, Thomas D., Francis S. Collins, and David Ginsburg. *Principles of Medical Genetics*. 2nd ed. Williams and Wilkins, 1998.

Keller, Evelyn Fox. *Century of the Gene*. Harvard University Press, 2001.

Kevles, Daniel J., and Leroy Hood. *The Code of Codes: Scientific and Social Issues in the Human Genome Project*. Harvard University Press, 1992.

Lewontin, Richard. *The Triple Helix: Gene, Organism, and Environment*. Harvard University Press, 2000.

Ridley, Matt. *Genome: The Autobiography of a Species in 23 Chapters*. Harpercollins, 1999.

Shreev, James. *The Genome War: How Craig Venter Tried to Capture the Code of Life and Save the World*. Knopf, 2004.

Watson, James D., with Andrew Berry. *DNA: The Secret of Life*. Knopf, 2003.

WEB SITES

Biotechnology Industry Organization. http://science.bio.org/.

The GDB Human Genome Database—Worldwide database for the annotation of the human genome. http://www.gdb.org/.

GeneCards—A database of human genes, their products, and their involvement in diseases. http://bioinfo.weizmann.ac.il/cards/index.shtml.

The Human Genome on the Wellcome Trust site. http://www.wellcome.ac.uk/en/genome/.

The Human Genome Organization. http://www.gene.ucl.ac.uk/hugo/.

The Human Genome Project at the Sanger Institute. http://www.sanger.ac.uk/HGP/.

Human Genome Project Information. http://www.ornl.gov/sci/techresources/Human_Genome/home.shtml.

National Center for Biotechnology Information (NCBI). Genes and Disease. http://www.ncbi.nlm.nih.gov/disease/.

———. Human Genome Resources. http://www.ncbi.nlm.nih.gov/genome/guide/human/.

———. Online Mendelian Inheritance in Man (OMIM). http://www.ncbi.nlm.nih.gov/entrez/query.fcgi?db=OMIM.

National Human Genome Research Institute. http://www.genome.gov.

Nature Genome Gateway. http://www.nature.com/genomics/.

Science magazine's functional genomics Web site. http://www.sciencemag.org/feature/plus/sfg/.

World Health Organization—The Genomic Resource Centre. http://www.who.int/genomics/en/.

Index

About the Authors

TARA ACHARYA is a consultant with the Rockefeller Foundation working on international health issues. She was a research associate at the University of Toronto and has worked as a scientist in the biotechnology industry.

NEERAJA SANKARAN is a freelance writer with training in microbiology and the history of science. She is the author of *Microbes and People: An A-to-Z of Microorganisms in Our Lives* (Oryx, 2001).

√